Fundamentals of
Radiation Chemistry

Fundamentals of Radiation Chemistry

A. Mozumder
University of Notre Dame
Notre Dame, Indiana

Academic Press

San Diego   London   Boston   New York   Sydney   Tokyo   Toronto

This book is printed on acid-free paper. ∞

Academic Press
A Harcourt Science and Technology Company
525 B Street, Suite 1900, San Diego, California 92101-4495, USA
http://www.apnet.com

Academic Press
24-28 Oval Road, London NW1 7DX, UK
http://www.hbuk.co.uk/ap/

Library of Congress Catalog Card Number: 99-60404

International Standard Book Number: 0-12-509390-X

PRINTED IN THE UNITED STATES OF AMERICA
99 00 01 02 03 04 BB 9 8 7 6 5 4 3 2 1

*Dedicated to the memory
of my parents*

CONTENTS

7 Spur Theory of Radiation Chemical Yields: Diffusion and Stochastic Models

PREFACE

Chemical effects from the absorption of charged-particle irradiation were observed almost immediately following the discoveries of X-rays and the electron in the last decade of the nineteenth century. The field, though, remained unnamed until 1942, when Milton Burton christened it "radiation chemistry." At present, it has developed into a vigorous discipline embracing radiation physics on one hand and radiation biology on the other. The purpose of this book is to give a coherent account of the development of this field with stress on the fundamental aspects.

The writing has been carefully organized so that no more than an undergraduate background is necessary for the reader to follow the essential results. Some of the theoretical arguments and the ancillary mathematics can be omitted for the first reading without loss of continuity. Chiefly, this book is aimed at upper undergraduate or graduate students as a supplement to general reading in physical chemistry. I hope that it also contains material that will be useful for both experimental and theoretical research. This book gives considerable importance to the electron, both as irradiation and as a reactant species. The electron occupies a central position in radiation chemistry, because a large number of secondary electrons are always produced in the absorption of high-energy radiations, regardless of the nature of the primary irradiation.

The notation used in this book is that normally encountered in the day-to-day work of a radiation chemist—eV, G value, and the like. For the hydrated electron, though, e_h is used instead of the frequently employed notation e_{aq}^-, an equivalence the reader is urged to keep in mind.

Although the book is concerned with the fundamental aspects of radiation chemistry, a chapter on applications has been added at the end. David Packer, executive editor for chemistry at Academic Press, is thanked for this as well as for various helpful suggestions and cooperation.

It is a pleasure to acknowledge appreciation for John L. Magee and Milton Burton, who initiated the author in radiation chemistry. I am indebted to my colleagues in research, especially Jay A. LaVerne and Simon M. Pimblott, for allowing me to incorporate certain aspects of their work. Mike Pecina and Mark Cozzi have been very helpful and cooperative with computer programs and data processing. The figures in the book have been made possible with active help from Gordon Hug. G. Ferraudi is thanked for certain technical aspects of book writing. Considerable secretarial help and computer assistance have been received from J. Mickelson and C. Trok. Finally, my heartfelt thanks are given to my wife Runa—without her constant encouragement, it is doubtful that this book would have seen the light of day.

A. Mozumder

CHAPTER 1

Introduction

1.1 HISTORICAL AND CLASSICAL RADIATION STUDIES

Lind (1961) defines radiation chemistry as *the science of the chemical effects brought about by the absorption of ionizing radiation in matter*. It can be said that in 1895, along with X-rays, Roentgen also discovered the chemical action of ionizing radiation. He drew attention to the similarity of the chemical effects induced by visible light and X-rays on the silver salt of the photographic plate. This was quickly followed by the discovery of radioactivity of uranium by Becquerel in 1896. In 1898, the Curies discovered two more radioactive elements—polonium and radium.

At the core of Becquerel's discovery also lies the similarity of actions of radioactive radiation and visible light on the photographic plate (Becquerel, 1901). There is some indirect evidence of chemical effects of ionization going back to the first quarter of the nineteenth century. In 1815, Berzelius studied light emission from materials containing (what is now known as) radioactive elements. In 1860, Andrews and Tait studied the effect of electrical discharges in gases, and in 1874, Thenard actually condensed gaseous acetylene under silent discharge. Later, Berthelot's extension of this work proved a remarkable point, namely, that both formation and decomposition of compounds can be brought about by electrical discharge.

In 1899, the Curies first reported the coloration of glass and porcelain and the formation of ozone from oxygen by radioactive radiation. Giesel (1900) noted that the coloration of alkali halides under these radiations was similar to the effect of cathode rays; he also observed the decomposition of water. P. Curie and Debierne (1901) observed continuous evolution of hydrogen and oxygen from solutions of radium salts. Ramsay and his associates studied reactions induced by radon, and W. H. Bragg (1907), analyzing the data of Ramsay and Soddy (1903), came to the conclusion that the number of water molecules decomposed was almost equal to the number of ionizations produced by radon in air. Although this appeared to him as a numerical conspiracy, the observation had far-reaching implications indeed.

During the period 1907–1919, a number of investigations were carried out with radon, including those by Cameron and Ramsay, Usher, Wourtzel, and by Lind. In 1917, Kailan started the study of the chemical effects of penetrating (β-and γ-) radiations on organic molecules. Lind, who had already started a systematic study of gases (Lind, 1912), continued his work with α-radiation from radon in the twenties. In the mid-twenties, Mund and his associates investigated simple gases, and probably they were the first to study polymers. In 1925, Coolidge, using high-voltage cathode rays, demonstrated that similar chemical effects could be brought about by sources not involving nuclear transformation. In the 1930s, H. Fricke made systematic studies of the effect of X-rays on aqueous solutions. About the same time, it was realized that polymerization of simple monomers could be effected by X-, γ-, or neutron radiations. It was also noticed that electron bombardment changes the properties of polymers, especially thermal properties. Much of the early work in radiation chemistry has been summarized by Lind (1961) and by Mund (1935).

During the Second World War, there was greatly increased activity in the study of the effects of radiation on water and aqueous solutions on one hand, and on the materials of the nuclear reactor on the other. The first arose naturally from the necessity of assessing biological effects of radiation. The latter was related to performance of reactors under operating conditions. A good deal of research was done under the Manhattan Project in the United States. Similar work was also carried out elsewhere—for example, at the Evergreen Laboratories at Montreal, Canada, and by Weiss, in the open, at Newcastle-on-Tyne. In 1942, M. Burton christened the field as radiation chemistry, realizing that it had remained nameless for 47 years after Roentgen's discovery (Burton, 1947, 1969). Modern radiation chemistry started vigorously after the Second World War, with the fundamental understanding of radiation's interaction with matter as an objective.

About 1910, M. Curie suggested that ions were responsible for the chemical effects of radioactive radiations. Soon thereafter, mainly due to the pioneering work of Lind on gases, the notation M/N was introduced for a quantitative measure of the radiation effect, where N is the number of ion pairs formed and M is the number of molecules transformed—either created or destroyed. This notation, referred to as the ion pair yield, was most conveniently employed in gases where N is a measurable quantity. However, for some time the same usage was extended to condensed systems assuming that ionization did not depend on the phase. This, however, is not necessarily correct. The notation G was introduced by Burton (1947) and others to denote the number of species produced or destroyed per 100 eV absorption of ionizing radiation. In this sense, it is *defined* purely as an experimental quantity independent of implied mechanism or theory. Burton is also credited with the invention of the symbol $\rightsquigarrow$ to be read as "under the action of ionizing radiation," but without implying any specific mechanism.

A theory of radiation-chemical effects based on local heating as a result of energy absorption was first proposed by Dessauer (1923). Although this theory was later discarded on the basis that the local rise of temperature is too small and its duration too short, it is plausible that in some cases, especially involving high-LET (linear energy transfer) radiations, local heating may have some influence. During the thirties and the forties, Lea (1947) sought for a workable theoretical model with application to radiobiology in view. He made use of Bethe's stopping power theory (see Sect. 2.3.4) for estimating energy absorption but regarded the ionized electron in liquid water as essentially free. This idea was further cultivated by Gray and by Platzman through the fifties, and together it is sometimes called the Lea–Gray–Platzman model. On the other hand, Magee and his associates, working since the fifties, stressed the recapture of the ionized electron, followed by the formation of reactive radicals through dissociation processes, and final formation of products by the diffusion and reaction of these intermediates. In a certain sense, both models coexist at the present time. The theories of radiation chemistry seek to explain radiation-chemical effects on the basis of well-established principles of physical chemistry and chemical physics.

1.2   RELEVANCE OF RADIATION CHEMISTRY TO BASIC AND APPLIED SCIENCES

The study of radiation chemistry may be viewed from three directions—namely, life, industry, and basic knowledge. Human as well as animal and plant lives are constantly under the effect of environmental radiation. According to one estimate (Vereshchinskii and Pikaev, 1964), the environmental radiation dose at sea level is ~1 mr/day, where 1 r (roentgen) is that dose of X- or γ-radiation that produces 1 electrostatic unit of charge of either sign in 1 cm^3 of air at STP. While this may ordinarily be considered as insignificant, its importance over geological periods must be recognized. Additionally, doses are received from explosions, both accidental and deliberate. Furthermore, there is a widespread belief that aging and indeed the genetic evolution of man may be significantly influenced by ambient radiation. Understanding of these effects belongs to radiation biology, which depends on radiation chemistry for basic information.

In industry, radiation is applied both as an initiator and as a control mechanism on one hand, and as a sustainer of reactions on the other. Among the many industrial uses of radiation, one may mention food preservation, curing of paints, manufacture of wood–plastic combinations, syntheses of ethyl bromide, of ion exchange materials, of various graft copolymers, and of materials for textile finishing. In addition, there are important uses of tracers in various process industries and in mining and metallurgy.

From the viewpoint of basic knowledge, experiments in radiation chemistry are invaluable. Charged particles impart to a molecule energies that are largely inaccessible by any other means. The chemistry of ions, excited states, and radicals brings the investigator closer to the basic understanding of matter and its interactions than is possible by other branches of conventional chemistry. Radiation chemistry is a proving ground for chemical and, indeed, physical theories, and it is not unreasonable to expect that in turn it will continue to enrich the basic disciplines. This field, where the primary interaction is electronic in nature, should be distinguished from radiochemistry, which is the chemistry of radioactive elements. Although a sharp distinction does not exist between radiation chemistry and photochemistry, in practice a photon energy in the range ~30–50 eV may be taken as the upper limit of photochemistry. Photochemical changes follow from excitation to well-defined quantum states, whereas in radiation chemistry a broad spectrum of states is generally excited. Currently, radiation chemistry has acquired an interdisciplinary character, overlapping with physics, chemistry, biology, and industrial applications.

1.3   SCOPE AND LIMITATION

This book lays emphasis on the fundamental aspects of the chemical consequences of charged particle interactions with matter, particularly in the condensed phase. No details will be given about experimental apparatus or procedure, but results of experiments are discussed in relation to theoretical models. The role of the electron both as a radiation (primary and secondary) and as a reactant has been fully treated. Wherever necessary, physical theories have been discussed in detail with understanding of radiation-chemical experiments in view.

REFERENCES

Becquerel, H. (1901), Compt. Rend. *133*, 709.
Bragg, W. H. (1907), Phil. Mag. *13*, 333.
Burton, M. (1947), J. Phys. Colloid Chem. *51*, 611 (1947).
Burton, M. (1969), Chem. Engg. and News *Feb. 10*, 86.
Curie, P., and Debierne, A. (1901), Compt. Rend. *132*, 770.
Dessauer, F. (1923), Z. Physik. *20*, 288.
Giesel, F. (1900), Verhandl. deut. physik. Ges. *2*, 9.
Lea, D. E. (1947), *Actions of Radiations on Living Cells*, Macmillan, New York.
Lind, S. C. (1912), J. Phys. Chem. *16*, 564.
Lind, S. C. (1961), *Radiation Chemistry of Gases*, Reinhold Publishing Corporation, New York.
Mund, W. (1935), *L'action chimique des rayons alpha en phase gazeuse*, Hermann, Paris.
Ramsay, W., and Soddy, F. (1903), Proc. Roy. Soc. A *72*, 204.
Vereshchinskii, I. V., and Pikaev, A. K. (1964), *Introduction to Radiation Chemistry* (translated from Russian), Daniel Davey, New York.

CHAPTER 2

Interaction of Radiation with Matter: Energy Transfer from Fast Charged Particles

2.1 PARTICLES AND RADIATIONS: LIGHT AND HEAVY PARTICLES

The terms particle and radiation are used interchangeably in quantum mechanics (respectively first and second quantization). Electron-positron, muons, and X- and γ-rays constitute the light group. Other radiations (e.g., protons, α-particles, fission fragments) are considered to be in the heavy group, although in some radiological cases higher-energy protons behave like light particles. Photons with sufficiently high energy can cause ionization of a medium in three different ways. These are

1. The photoelectric effect, in which the photon is absorbed and an electron is produced with kinetic energy equal to the difference between the photon energy and the binding energy of the electron (Einstein equation).

2. The Compton effect, in which the photon is scattered with significantly lower energy by a medium electron, which is then ejected with the energy differential.

3. Pair production, in which the photon is annihilated in a nuclear interaction giving rise to an electron-positron pair, which carries the energy
of the photon less twice the rest energy of the electron.

As the photon energy increases, its dominant interaction changes gradually
from photoelectric to Compton to pair production. For example, in the case of
^{60}Co-γ rays in water, having mean energy of about 1.2 MeV, almost the entire
interaction is via the Compton effect. Since the scattered γ-ray has a high probability of penetrating the sample without further interaction, the resultant radiation effect is attributable to recoil electrons. Compton electrons have a fairly
uniform distribution in energy except for a sharp peak at about three-fourths of
the photon energy (Klein and Nishina, 1929); thus, the average energy of the
recoil electrons is about half the γ-ray energy.

The electron itself is frequently used as a primary source of radiation, various kinds of accelerators being available for that purpose. Particularly important are pulsed electron sources, such as the nanosecond and picosecond pulse
radiolysis machines, which allow very fast radiation-induced reactions to be
studied (Tabata *et al.*, 1991). Note that secondary electron radiation always constitutes a significant part of energy transferred by heavy charged particles. For
these reasons, the electron occupies a central role in radiation chemistry.

X-rays, often used in radiation chemistry, differ from γ-rays only operationally; namely, X-rays are produced in machines, whereas γ-rays originate in
nuclear transitions. In their interaction with matter, they behave similarly—that
is, as a photon of appropriate energy. Other radiations used in radiation-chemical studies include protons, deuterons, various accelerated stripped nuclei, fission fragments, and radioactive radiations (α, β, or γ).

It is useful to group the radiations as low-, intermediate-, and high-LET (linear energy transfer) according to their specific rate of energy loss in water (*vide
infra*). Electrons together with β-, X-, and γ-rays constitute low-LET radiation.
Alpha-particles, stripped nuclei, and fission fragments belong to high-LET; protons, deuterons, and so forth, represent intermediate-LET cases.

2.2   PRINCIPAL CONSIDERATIONS
RELATED TO ENERGY TRANSFER
FROM CHARGED PARTICLES

2.2.1   MECHANISM OF ENERGY TRANSFER

As a rule, energy is transferred from a fast charged particle by electrostatic
interaction with the electrons of the molecule. Exceptions are found at very
low or very high speeds (*vide infra*). At ultrarelativistic speeds, the electromagnetic interaction compares significantly with the electrostatic interaction.

However, this additional interaction is largely masked by interaction with the nuclei, generating bremsstrahlung, which is the major carrier of the energy transfer at highly relativistic speeds.

The hamiltonian for the electrostatic interaction has, in the momentum representation, an exponential form in the product of momentum transfer and electron coordinate. For small momentum transfers, the linear term dominates because of the orthogonality of the stationary states of the molecule. Thus the electronic states that are excited under fast charged particle impact are the same as those reached by light absorption, and the same selection rules apply. (Of course, this does not mean that the *cross section* for charged particle impact is the same as for photoabsorption.) The validity of this so-called *optical rule* depends on the smallness of the momentum transfer. It is generally valid for high-speed incident heavy charged particles and electrons, except for cases in which a relatively high energy secondary electron is produced. At low speeds, the optical rule breaks down.

On energy absorption from the incident particle, a molecule is raised to one of its excited electronic states. If the energy transfer is sufficiently large, ionization may occur in competition with neutral dissociation. With large energy transfer the secondary electron acquires high kinetic energy, creating its own track and transferring energy to other molecules much like the primary. Thus electrons of higher generations will constantly be produced as long as the supply of energy lasts. To the molecules of the medium, however, the entire process is very quick, and all they experience is the total impact of a shower of charged particles with ubiquitous electrons of a given spectral character.

2.2.2 BEHAVIOR OF DEPOSITED ENERGY WITH RESPECT TO LOCALIZATION

A fast charged particle deposits energy in a geometrically well-correlated manner, forming a more or less straight track. A fundamental question related to the observation of α-tracks in cloud chambers may be stated as follows. According to Gamow's theory, an α-particle leaks out of the nucleus as a spherical wave. Why, then, do the ionizations lie on a straight track instead of being random in three dimensions? The answer, provided by Mott (1930) using the second order perturbation theory, is that the first ionization is indeed random; after that, all succeeding events are well correlated as if produced by a classically moving particle. After the first interaction, the incident particle is no longer represented by a spherical wave. Tracks of other charged particles including the electron can be similarly described.

Track theory starts with localized energy loss. On the other hand, attention has been frequently drawn to the role of delocalized energy loss in radiation chemistry. Fano (1960) estimated from the uncertainty relation that an energy radiobiological systems, the time of appearance of a damage may be subjective depending on the cell cycle being studied. This is partly due to the generations of biochemical and biological transformations that must take place before the damage is rendered "visible." It is recognized that these processes occur both in the directions of amplification and repair. Certain biochemical reactions, such as that of OH with sugars, can occur in the nanosecond time scale, whereas the biochemical O_2 effect may take microseconds. While DNA strand breakage may be considered an early biological effect, other damage of a more permanent nature may take ~1 day to ~40 years if genetic effects are considered. These, then, are the early times for such specific effects, which should be compared with ~1 µs, the time scale for track dissolution in liquid water (Turner *et al.*, 1983).

Figure 2.1 shows schematically the distribution of deposited energy among various degrees of freedom as a function of time. Initially, all the energy is in electronic form at a time consistent with the uncertainty principle. In the characteristic time of molecular vibrations, most of the energy appears in that form, with a small part appearing as low-grade heat and products of molecular dissociation. After a local temperature is established, a fraction of the energy has been transformed into heat. Ultimately, on a much longer time scale, most of the energy appears in the forms of heat, separated charges, and dissociation products. In many cases, especially for low-LET radiations, a significant rise in local temperature is not expected. The treatment given in Figure 2.1 is only approximately valid; many details are unknown.

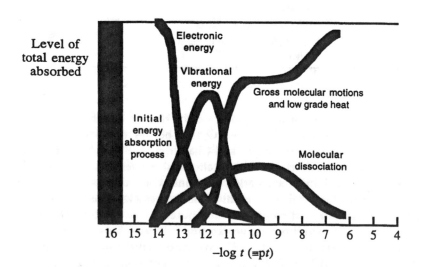

FIGURE 2.1   Distribution of deposited energy among different degrees of freedom as a function of time, represented by $pt = -\log t$ (sec). Note that, for exothermic reactions, low-grade heat can terminate above the level of absorbed energy. Luminescence can only alter the picture in a minor way. Reproduced from Mozumder (1969a), by permission of John Wiley & Sons, Inc.©

2.3   THEORY OF STOPPING POWER OF FAST CHARGED PARTICLES

2.3.1   GENERAL DESCRIPTION OF DEGRADATION PROCESSES

A number of processes are responsible for degrading the energy of a charged particle, depending on the charge and the velocity of the incident particle and the nuclear charges of the medium atoms. In most cases, the penetrating particle is thermalized, or significantly slowed down, before it undergoes a chemical reaction. It is remarkable that very few mechanisms of energy loss compete at a given speed. There are specific velocity regions in which individual mechanisms of energy loss predominate. For ultrarelativistic electrons, the dominant stopping mechanism is the production of bremsstrahlung. The ratio of energy losses due to radiative (bremsstrahlung) and collisional (electronic) mechanisms at an energy of E MeV is given approximately by $EZ/800$ (Bethe and Heitler, 1934), where Z is the nuclear charge. Thus, for electrons in water, bremsstrahlung becomes significant above 100 MeV. By contrast, electronic stopping dominates the velocity range from $\sim 0.99c$ down to about the speed of a least bound electron around the incident heavy particle, or to about the first electronic excitation energy for incident electron. For positive ions of still lower energy, charge exchange dominates the electronic stopping power, but the latter is superseded by nuclear collisions at low speeds; finally, the particle is thermalized by vibrational and elastic collisions.

For subexcitation electrons (Platzman, 1955, 1967), the dominant stopping mechanism is the excitation of molecular vibrations, often through the intermediary of temporary negative ions. In some cases, the process is very effective and several quanta in a given mode of vibration may be excited. At very low electron energies, the effectiveness of this process decreases because of low cross section and for energetic reasons. The demarcation is not sharp but may be around a few tenths of an electron-volt for molecular media (Mozumder and Magee, 1967). Thermalization of electrons of still lower energy depends on the state of aggregation. In a gas, it may require a large number of elastic collisions. In a nonpolar condensed medium, fewer collisions involving intermolecular vibrational excitation may be sufficient. In polar gases, rotational excitation is an important mechanism, and in hydrogen-bonded systems, excitation of bond vibration is a plausible mechanism. In any case, a very low energy electron may get trapped, react chemically, or become attached to a molecule either before or after it is thermalized.

2.3.2   STOPPING POWER AND LET

Energy transfer from a charged particle can be considered from two points of view: (1) the particle—that is, its charge, range, penetration, etc.—which constitutes radiation physics; or (2) the matter—receiving the energy to produce chemical changes, charge separation, luminescence, etc.—which is essentially radiation chemistry.

The stopping power of a medium toward a penetrating charged particle is the *energy loss suffered by that particle per unit path length*, whereas the linear energy transfer (LET) refers to the energy *received by the medium* in the immediate vicinity of the particle track. Their difference arises because, even at the shortest time, significant amounts of energy may be removed from the track vicinity (see Sect. 2.2.2). Zirkle *et al.* (1952) introduced the concept of LET, revising an earlier term (linear energy absorption). The International Commission on Radiation Units and Measurements advocates the use of the symbol L_A for the LET or restricted linear collision stopping power for energy transfers below a specified value, A (ICRU, 1970). Thus, L_{100} indicates LET for energy losses less than 100 eV. The symbol L_∞ denotes all possible energy loss; it is numerically equal to the stopping power. Note that stopping power refers to the medium and LET to the penetrating particle. Sometimes, the term *radiation quality* is used to denote a specific kind of radiation belonging to LET in a restricted interval. Table 2.2 gives approximate LETs of some common charged particle radiations in water.

2.3.3   BOHR'S CLASSICAL THEORY

Theories of stopping power are best described for incident heavy charged particles that are fast but not ultrarelativistic. Under these conditions: (1) the interaction is basically electrostatic; (2) the interaction can be treated as a perturbation; and (3) energy losses occur through quasi-continuous inelastic collisions, so that the average stopping power is a good representation of the statistical process of energy loss. All stopping power theories recognize these simplifications.

The first useful theory of stopping power was developed by N. Bohr[2] (1913, 1948). Bohr's theory describes the collision in terms of the classical impact parameter b, defined as the distance of closest approach between the incident particle and a medium electron were there no interaction (see Figure 2.2). The incident particle[3] of charge ze and velocity v is considered as interacting only over the segment AOB of the path with peak force $-ze^2/b^2$ and for a duration $2b/v$. With this consideration, the momentum transferred to the electron is $(-ze^2/b^2)(2b/v) = -2ze^2/bv$. Applying Gauss's theorem over the entire path, Fermi (1940) showed that this expression for momentum transfer is exact for a free electron. The energy transfer is now given by

$$Q = \frac{-(2ze^2/bv)^2}{2m} = \frac{2z^2e^4}{mb^2v^2}. \qquad (2.1)$$

---

[2] Theories prior to Bohr, such as that of J. J. Thompson, are only of historical interest.

[3] e and m refer respectively to the magnitude of the charge and the mass of the electron.

TABLE 2.2   Approximate LETs of Various Qualities of
Radiation in Water

Radiation quality	LET (eV/Å)
^{60}Co-γ (500-KeV electron)	~0.02
10-KeV δ-ray	0.3
Tritium β (6 KeV average)	0.5
1-MeV proton	3
Po-α (5.3 MeV)	9
Stripped carbon nucleus (10 MeV/amu)	17
Fission fragments[a]	100–1000

[a]LET of a fission fragment starts at a very high value but is
quickly reduced because of electron capture, resulting in an average of ~200–300 eV/Å.

The differential cross-section of this process for the range of impact parameters between b and $b - db$ is given geometrically from Eq. (2.1) as follows:

$$d\sigma = -\pi d(b^2)$$

$$= \frac{2\pi z^2 e^4}{mv^2} \frac{dQ}{Q^2}. \tag{2.2}$$

Equation (2.2) is just the Rutherford cross section for scattering of, strictly
speaking, free charges. To apply this to atomic electrons that are not free but can
be excited[4] with energy E_n, Bohr surmised the sum rule

$$\sum_n f_n E_n = ZQ \tag{2.3}$$

for the same momentum transfer as in the case of the imaginary free electron.
In Eq. (2.3), f_n is the oscillator strength for the transition, Z is the atomic number, and the summation also includes integration over the continuum states. By
definition, the stopping power $-dE/dx$ is the energy lost to all the electrons over
unit path length and over all permissible energy transfers. That is, from Eqs.
(2.2) and (2.3),

$$\frac{dE}{dx} = N \int d\sigma \sum f_n E_n = \frac{2\pi z^2 e^4 NZ}{mv^2} \ln \frac{Q_{max}}{Q_{min}}. \tag{2.4}$$

---

[4] Excitation includes ionization.

Further, Bethe defines I via the relation

$$Z \ln I = \int f'(\varepsilon) \ln \varepsilon \, d\varepsilon. \tag{2.7}$$

Using these relations in (2.6b), one gets

$$\left(\frac{dE}{dx}\right)_g = \kappa NZ \ln \frac{2mv^2}{I}. \tag{2.8}$$

For knock-on collisions, one uses the Rutherford cross section for free electrons, and the number of free electrons is taken equal to the integral of the oscillator strength up to the energy loss $\in$ (dispersion approximation). Thus,

$$d\sigma_{ko} = \frac{\kappa n(\varepsilon) \, d\varepsilon}{\varepsilon^2}; \qquad n(\varepsilon) = \int^\varepsilon f'(\rho) \, d\rho.$$

The contribution of knock-on collisions to the stopping power is now given by

$$\left(\frac{dE}{dx}\right)_{ko} = N \int \varepsilon \, d\sigma_{k.o.}$$

$$= \kappa N \left[\left| \ln \varepsilon \int^\varepsilon f'(\rho) \, d\rho \right|^{\varepsilon = 2mv^2} - \int^{2mv^2} f'(\varepsilon) \ln \varepsilon \, d\varepsilon \right].$$

In this equation, the small lower limit is of no consequence. Also, $2mv^2$ is sufficiently large to include essentially all oscillator strengths. Therefore, this equation simplifies to

$$\left(\frac{dE}{dx}\right)_{k.o.} = \kappa NZ \ln \frac{2mv^2}{I}, \tag{2.9}$$

which is equal to the glancing contribution as given in Eq. (2.8). This equality is sometimes called the *equipartition principle*, which is not exact but reflects the approximations in the procedure. Combining (2.8) and (2.9) and comparing with (2.5), we see that the stopping number in the Bethe equation is given by $B = Z \ln(2mv^2/I)$.

The Bethe equation has a wide range of validity except for slow, highly charged heavy particles such as fission fragments. On the other hand, the various approximations used in this theory need justification or correction when applied to a real system. For example, the velocity criterion is actually more restrictive than for just the validity of the Born approximation. Dalgarno and Griffing (1955) argued that the Bethe equation consistently overestimates the stopping power by extending the sum rules to energetically inaccessible states

TABLE 2.2   Approximate LETs of Various Qualities of
Radiation in Water

Radiation quality	LET (eV/Å)
^{60}Co-γ (500-KeV electron)	~0.02
10-KeV δ-ray	0.3
Tritium β (6 KeV average)	0.5
1-MeV proton	3
Po-α (5.3 MeV)	9
Stripped carbon nucleus (10 MeV/amu)	17
Fission fragments[a]	100–1000

[a]LET of a fission fragment starts at a very high value but is
quickly reduced because of electron capture, resulting in an average of ~200–300 eV/Å.

The differential cross-section of this process for the range of impact parameters between b and $b - db$ is given geometrically from Eq. (2.1) as follows:

$$d\sigma = -\pi d(b^2)$$

$$= \frac{2\pi z^2 e^4}{mv^2} \frac{dQ}{Q^2}. \tag{2.2}$$

Equation (2.2) is just the Rutherford cross section for scattering of, strictly
speaking, free charges. To apply this to atomic electrons that are not free but can
be excited[4] with energy E_n , Bohr surmised the sum rule

$$\sum_n f_n E_n = ZQ \tag{2.3}$$

for the same momentum transfer as in the case of the imaginary free electron.
In Eq. (2.3), f_n is the oscillator strength for the transition, Z is the atomic number, and the summation also includes integration over the continuum states. By
definition, the stopping power $-dE/dx$ is the energy lost to all the electrons over
unit path length and over all permissible energy transfers. That is, from Eqs.
(2.2) and (2.3),

$$\frac{dE}{dx} = N\int d\sigma \sum f_n E_n = \frac{2\pi z^2 e^4 NZ}{mv^2} \ln \frac{Q_{max}}{Q_{min}}. \tag{2.4}$$

---

[4] Excitation includes ionization.

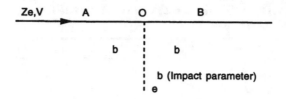

FIGURE 2.2   Bohr's semiclassical description of stopping power in terms of the impact parameter. The coulomb interaction is taken at its peak value over the track segment AOB and zero outside.

In Eq. (2.4) the maximum energy transfer is given by the kinematic relation, ignoring the small atomic binding, as $Q_{max} = 2mv^2$. Bohr argued that the collision must be sudden in order for the energy transfer may take place; that is, the collision time $2b/v$ must be $\leq g\hbar/E_1$, the reaction time of the atom, where $\hbar$ is Planck's constant divided by 2π and E_1 is a typical atomic transition energy. This gives the maximum impact parameter as $\hbar v/2E_1$ and Q_{min} as $8z^2e^4E_1/[(mv^2)(\hbar^2v^2)]$ from (2.1). With these values for Q_{max} and Q_{min} one gets, from (2.4),

$$\frac{dE}{dx} = \frac{4\pi e^4 z^2 N}{mv^2} B, \tag{2.5}$$

where B, the stopping number, is given in Bohr's theory by

$$Z \ln \frac{2mv^2}{E_1} \cdot \frac{\hbar v}{4ze^2}.$$

The impact parameter is not an observable. Despite this caveat, Bohr's stopping power formula has a wide region of validity, and if the *typical atomic transition energy E_1* is appropriately defined, its appearance is formally similar to the quantum mechanical formula of Bethe (*vide infra*). However, the conditions of validity of the classical and quantum expressions are complementary rather than identical. The quantum treatment depends on the validity of the Born approximation, which requires that $v \gg v_0$ where $v_0 = ze^2/h$ is the velocity of an electron in a 1s-orbital around the incident particle. For the validity of classical mechanics one requires, on the other hand, $v \ll v_0$ (Bohr, 1948; Bloch, 1933a; Williams, 1945). For heavy, slow particles, Bohr's formula therefore has an inherent advantage. Also, note that when $v \leq v_0$, the incident (positively charged) particle starts to capture electrons. For these reasons, stopping of fission fragments, especially in light media, is better described by the Bohr formula.

Frequently, one can display a stopping power formula as shown in Eq. (2.5), in which case the quantity within the parenthesis is called the kinematic factor since it contains no property of the target atom. The latter appears, in some average fashion, in the stopping number B. In the case of Bohr's theory, the appearance takes the trivial form of E_1, the typical transition energy. However, since the stopping power depends logarithmically on E_1, its exact evaluation is not critical. Often a reasonable estimation suffices.

2.3.4   THE BETHE THEORY

Bethe's (1930, 1932, 1933) quantum mechanical theory of stopping power considers energy and momentum transfers as basic variables and applies Born's (first) approximation to the problem of atomic collision. This approximation requires that the particle velocity be larger than the speed of an electron in an 1s- orbital around itself. However, a more restrictive condition is usually applied for the ease of mathematical handling—namely, it is assumed that the particle velocity is large compared with the speeds of the atomic electrons. Under these conditions, charge exchange is negligible (i.e., the incident positive ion travels as a fully stripped ion) and it is possible to obtain a closed formula for the rate of energy loss. In the following discussion, we will present a simplified treatment of the Bethe theory with its applications and limitations. Detailed derivation and ramifications will be found elsewhere (Livingstone and Bethe, 1937; Bethe and Ashkin, 1953; Fano, 1963; Inokuti, 1971).

It is expedient to classify the inelastic collisions as (1) *glancing*, involving larger impact parameters with a small energy loss per encounter; and (2) *knock-on*, involving smaller impact parameters with a large energy loss per encounter (Magee, 1953; Mozumder, 1969). The differential cross section for glancing collisions resulting in an energy transfer between ε and $\varepsilon + d\varepsilon$ is given in the nonrelativistic Born approximation as

$$d\sigma_g = \kappa \frac{f'(\varepsilon)}{\varepsilon} \ln \frac{2mv^2}{\varepsilon} d\varepsilon, \qquad (2.6a)$$

where $\kappa = 2\pi z^2 e^4/mv^2$ and $f'(\varepsilon)$ is the oscillator strength per unit energy interval at ε. In the following, excitation includes ionization and integration includes summation over the discrete states. The part of the stopping power due to glancing collisions is now obtained from (2.6a) as follows:

$$-\left(\frac{dE}{dx}\right)_g = N \int \varepsilon \, d\sigma_g = \kappa N \left[(\ln 2mv^2) \int f'(\varepsilon) \, d\varepsilon - \int f'(\varepsilon) \ln \varepsilon \, d\varepsilon \right]. \qquad (2.6b)$$

By the Thomas-Kuhn sum rule,

$$\int f'(\varepsilon) \, d\varepsilon = Z.$$

Further, Bethe defines I via the relation

$$Z \ln I = \int f'(\varepsilon) \ln \varepsilon \, d\varepsilon. \tag{2.7}$$

Using these relations in (2.6b), one gets

$$\left(\frac{dE}{dx}\right)_g = \kappa NZ \ln \frac{2mv^2}{I}. \tag{2.8}$$

For knock-on collisions, one uses the Rutherford cross section for free electrons, and the number of free electrons is taken equal to the integral of the oscillator strength up to the energy loss ε (dispersion approximation). Thus,

$$d\sigma_{ko} = \frac{\kappa n(\varepsilon) \, d\varepsilon}{\varepsilon^2}; \qquad n(\varepsilon) = \int^{\varepsilon} f'(\rho) \, d\rho.$$

The contribution of knock-on collisions to the stopping power is now given by

$$\left(\frac{dE}{dx}\right)_{ko} = N \int \varepsilon \, d\sigma_{k.o.}$$

$$= \kappa N \left[\left. \ln \varepsilon \int^{\varepsilon} f'(\rho) \, d\rho \right|^{\varepsilon=2mv^2} - \int^{2mv^2} f'(\varepsilon) \ln \varepsilon \, d\varepsilon \right].$$

In this equation, the small lower limit is of no consequence. Also, $2mv^2$ is sufficiently large to include essentially all oscillator strengths. Therefore, this equation simplifies to

$$\left(\frac{dE}{dx}\right)_{k.o.} = \kappa NZ \ln \frac{2mv^2}{I}, \tag{2.9}$$

which is equal to the glancing contribution as given in Eq. (2.8). This equality is sometimes called the *equipartition principle*, which is not exact but reflects the approximations in the procedure. Combining (2.8) and (2.9) and comparing with (2.5), we see that the stopping number in the Bethe equation is given by $B = Z \ln(2mv^2/I)$.

The Bethe equation has a wide range of validity except for slow, highly charged heavy particles such as fission fragments. On the other hand, the various approximations used in this theory need justification or correction when applied to a real system. For example, the velocity criterion is actually more restrictive than for just the validity of the Born approximation. Dalgarno and Griffing (1955) argued that the Bethe equation consistently overestimates the stopping power by extending the sum rules to energetically inaccessible states (i.e., to states of energy greater than $2mv^2$). They illustrated the point by numerical calculation of stopping power of protons in a gas of H-atoms with energy below 100 KeV. However, at low energies where this effect should be important, it is masked by drastic reduction of stopping power due to electron capture.

In Bohr's theory, only estimates of maximum and minimum impact parameters are necessary. Better computations are required for determining the transverse distribution of lost energy or the effect of secondary electrons. The minimum impact parameter according to classical mechanics is ze^2/mv^2; from angular momentum consideration in quantum mechanics, it is h/mv. In practice, the larger of these two is taken. Also, the impulse approximation used by Bohr for the maximum impact parameter is not an absolute rule; energy transfer beyond b_{max} falls off exponentially (Orear et al., 1956; Mozumder, 1974).

2.3.5 CORRECTIONS TO THE BASIC STOPPING POWER FORMULA

Bethe's formula requires that the velocity of the incident particle be much larger than that of the atomic electrons, a condition not easily fulfilled by the K-electrons except in the lightest elements. The required correction, called the shell correction, is denoted by subtracting a quantity C from the stopping number. In the penetration of high-Z material, even L-shell correction may be required. In that case, C denotes the sum total of all shell corrections. The subject of shell correction has been extensively treated by several authors, and various graphs and formulas are available for its evaluation (see, e.g., Bethe and Ashkin, 1953).

When the incident particle has relativistic speeds, the maximum energy transfer increases from $2mv^2$ to $2mv^2(1 - \beta^2)$, where $\beta = v/c$. Also, at these speeds, electromagnetic interaction needs to be considered along with the electrostatic interaction if accuracy is desired (Fano, 1963). Incorporation of these effects results in the addition of a term $-z[\beta^2 + \ln(1 - \beta^2)]$ to the stopping number. For small β, this correction is proportional to β^4, its importance being <0.1% for velocities less than 5×10^9 cm/s.

Fermi (1940) pointed out that as $\beta \to 1$ the stopping power would power would approach ∞ were it not for the fact that polarization screening of one medium electron by another reduced the interaction slightly. This effect is important for the condensed phase and is therefore called the density correction; it is denoted by adding $-Z\delta/2$ to the stopping number. Fano's (1963) expression for δ reduces at high velocities to

$$\delta = \ln \frac{\hbar^2 \omega_p^2}{I^2(1 - \beta^2)} - 1,$$

where $\omega_p = (4\pi N e^2 Z / m)^{1/2}$ is the so-called plasma frequency. At extreme relativistic velocities, the divergence of the stopping power due to kinematic reasons is reduced to half by the density correction. This half is the restricted energy loss, which smoothly rides to a constant called the *Fermi plateau*. The unrestricted half diverges anyway, but its effect is overshadowed by energy loss attributable to nuclear encounters. Including the shell, density, and relativistic corrections, the stopping number of a heavy charged particle may be written as

$$B = Z\left[\frac{\ln 2mv^2}{I} - \beta^2 - \ln(1 - \beta^2) - \frac{C}{Z} - \frac{\delta}{2}\right], \qquad (2.10)$$

In the ionization produced by an incident electron, there is no way to distinguish between the two outgoing electrons except on the basis of energy. Thus, by definition, the maximum energy transfer, ignoring atomic binding, for an electron is $(1/4)mv^2$ instead of $2mv^2$. On this basis, Bethe gives the electron stopping number in the nonrelativistic case as[5] $B = Z \ln[mv^2(e/2)^{1/2}/2I]$. At the same speed, the stopping number of an electron is comparable to that of a singly charged heavy particle but is always less by the quantity $\ln[4(2/e)^{1/2}]$, or 1.233.

At moderate energies, the electron can acquire relativistic speeds. Including this effect as well as corrections due to shell and density effects, the electron stopping number may be written as

$$B = \frac{Z}{2}\left[\ln\frac{mc^2\beta^2 E}{2I^2(1 - \beta^2)} - \left(2\sqrt{1 - \beta^2} - 1 + \beta^2\right)\ln 2\right.$$

$$\left. + (1 - \beta^2) + \frac{1}{8}\left(1 - \sqrt{1 - \beta^2}\right)^2 - 2\frac{C}{Z} - \delta\right], \qquad (2.11)$$

where E is the electron energy at velocity βc. The shell correction for the electron is not exactly the same as for a heavy particle, but the difference is not significant.

2.3.6 STOPPING POWERS OF COMPOUNDS AND MIXTURES

The theory of stopping power is developed strictly for an atom. To extend it to molecules, one uses the Bragg rule, which simply equates the stopping number of a molecule to the sum of the stopping numbers of its constituent atoms. This means that I for a molecule is given by the geometrical averaging of I of its atoms over their electron numbers. For most compounds and mixtures, Bragg's rule applies very well, to within 2-3%. However, the contribution of an atom to I of a molecule is not necessarily the same as its free atom value. Still,

---

[5] Here and in the following, *e* denotes the base of natural logarithm.

the contribution is about the same in various molecules. The H atom is a good example in this respect. The success of the Bragg rule has been traced to two reasons: (1) the nature of chemical binding is similar for the same atom; and (2) the transitions that have most oscillator strengths involve energies far in excess of chemical binding. H_2 and NO are important exceptions to the Bragg rule. There is evidence that the Bragg rule gradually breaks down at low energies.

2.4 DISCUSSION OF STOPPING POWER THEORIES

2.4.1 MEAN EXCITATION POTENTIAL

The most important nontrivial quantity in the stopping power formula of Bethe is the mean excitation potential I defined in Eq. (2.7). It summarizes the properties of the target in terms of energies and oscillator strengths of transitions and thereby allows a neat separation of the stopping power formula in terms of a kinematic factor and a stopping number. Its direct evaluation, however, requires knowledge of ground and excited state atomic wavefunctions. Such direct calculation is feasible only for a few atoms of low Z. In the case of hydrogen, exact calculation gives $I - 15$ eV. In other cases, it is usual to consider I as an experimental parameter to be adjusted by comparing calculated and measured ranges (*vide infra*) in some suitable cases—for example, that of a proton of high energy.

Bloch (1933a,b) first pointed out that in the Thomas-Fermi-Dirac statistical model the spectral distribution of atomic oscillator strength has the same shape for all atoms if the transition energy is scaled by Z. Therefore, in this model, $I \propto Z$; Bloch estimated the constant of proportionality approximately as 10-15 eV. Another calculation using the Thomas-Fermi-Dirac model gives $I/Z = a + bZ^{-2/3}$ with $a = 9.2$ and $b = 4.5$ as best adjusted values (Turner, 1964). This expression agrees rather well with experiments. Figure 2.3 shows the variation of I/Z vs. Z.

From this, however, it should not be concluded that the statistical model of the atom is a very good one. As Fano (1963) has pointed out, I appears only as a logarithm and an error δI in the computation of I shows up as a relative error in the stopping power as $\sim(1/5)\delta I/I$. Besides, it is an average quantity and can be approximated reasonably well without knowing the details of the distribution.

There is a relation between I and the complex dielectric constant $\varepsilon(\omega)$ at an angular frequency ω, which can be written as follows (Fano, 1963):

$$\ln I = -\frac{2}{\pi \omega_p^2} \int_0^\infty \omega \ \mathrm{Im}[\varepsilon^{-1}(\omega)] \ln \hbar\omega \ d\omega.$$

approximately twice the ionization potential. Notice that the final range distribution has been convoluted from the intermediate distributions with a demarcation energy of 110 eV.

Stopping power refers to the energy loss of a penetrating particle per unit absorber thickness. Although this kind of experiment is common to heavy ions, it is rare for fast electrons. The stopping power actually is an average over many discrete inelastic collisions and, as such, for electron penetration the absorber thickness should be sufficiently thick, over which the elastic collisions are generally not negligible. This makes the path length considerably greater than the absorber thickness, and the measured stopping power is then an upper limit to the true value. At very low energy, when the electron undergoes nearly isotropic scattering, stopping power obtained in this manner may be meaningless.

These problems and the approximate derivation of stopping power from range measurements have been discussed by LaVerne and Mozumder (1984). Range is usually measured in a transmission experiment, although other methods, such as substrate fluorescence, ionization, or even enzyme inactivation, have been used. Typically, monoenergetic electrons are allowed to pass through an absorber, and the fraction of transmitted (or otherwise detected) electrons is determined. As the absorber thickness is increased, the fraction of transmitted electrons decreases. The mean range is determined from the point of inflection on a graph of transmitted

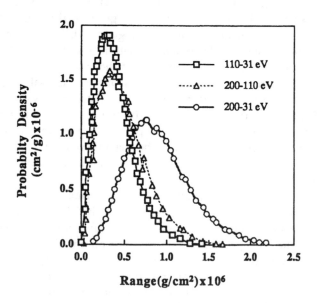

FIGURE 2.4   Range distribution of 200-eV electrons in N_2. See text for details. Reproduced from Mozumder and LaVerne (1985), with permission from Am. Chem. Soc.©

fraction as a function of energy, and the practical range is found by extrapolating that linear portion to zero transmission. Stopping power can then be computed from the reciprocal of the derivative of range with respect to energy.

Whiddington (1912) seems to be the first to give a range-energy relationship for low-energy electrons in the form $R = aE^m$, with a constant a specific to the material penetrated. An equation of this form is called Whiddington's rule. While Whiddington found $m = 2$ for various foils, later experiments gave m in the range of ~1.3 to ~1.8 depending on the situation. These experiments cover a wide range of energies (~0.1– ~50 KeV) in a variety of media including air, N_2, O_2, other gases, and various metallic foils, but usually not water because of clustering. These experiments have been reviewed by LaVerne and Mozumder (1984, see Table 1), which shows that good agreement may be obtained if both experimental and theoretical ranges are interpreted properly.

In some experiments, such as that by Grosswendt and Waibel (1982) using N_2 or that by Cole (1969) using collodion, there is evidence of excessive scattering. To compare theory and experiment in such cases, it is necessary to introduce an energy-dependent scaling factor for the transmission fraction. Mozumder and LaVerne (1985) have calculated this scaling factor through the computation of the mean cosine of the angle between the radial range and the penetration in the original direction of electron motion. Figure 8 of this reference gives a fair comparison between so-scaled experiment and theory.

If the sample penetrated is sufficiently thin, the energy distribution of the emergent particles is not Gaussian, but has a long tail in the high-energy region (Landau, 1944; Vavilov, 1957), a fact that is attributable to insufficient number of knock-on collisions produced in the sample. The Landau-Vavilov distribution has been extensively tabulated (NAS-NRC, 1964), and it has also been experimentally verified (Maccabee et al., 1968). This distribution is important in radiation biology. Radiation effects in a cell depend on the energy lost within the cell, which shows a large, asymmetrical fluctuation because of the small size of the cell.

2.5   PHENOMENA AT LOW VELOCITIES

2.5.1   HEAVY PARTICLES: CHARGE EXCHANGE AND NUCLEAR STOPPING

Positive ions start to capture electrons from the medium when their velocity is comparable to that of an electron in an 1s-orbital around itself. On further slowing, at first the captured electron is soon lost, and then another electron is captured. Thus cycles of capture and loss continue until it is energetically impossible to lose the captured electron. If the incident particle is multiply charged, another charge exchange cycle will soon be set up, and so on until the particle is reduced to a neutral atom.

Charge exchange cross sections depend on a high inverse power of velocity (Bohr, 1948); thus, at high speeds, they are insignificant. At low speeds, however, they dominate the stopping power, since the latter depends on the square of the charge. Basically, the charge exchange cycles are composed of a single capture followed by a single loss. Occasionally, two similar events may follow in succession. At a given velocity, an effective charge may be defined by $z_{eff}^2 = \langle z^2 \rangle$, where the latter is the equilibrium average of the square of the ionic charge at that velocity. For particles of high nuclear charge, such as fission fragments, the charge distribution may approach a gaussian form, but for smaller nuclear charge, a wide variation is expected. Some values of z_{eff} are (1) protons, 0.70 and 0.88 at 30 and 100 KeV, respectively; (2) α- particles, 1.6 at 1 MeV; and (3) carbon ions, 4.9 and 5.6 at 0.9 and 2 MeV/amu, respectively (Mozumder, 1969a). It is remarkable that the charge distribution is relatively independent of the nature of the medium penetrated.

At a low incident speed of the heavy ion, the electronic stopping power rapidly falls and is superseded by energy loss due to nuclear collisions (Ziegler, 1980–1985). The stopping power due to the latter process has a characteristic maximum of its own, and on further slowing down, elastic collision between completely screened charges offers the remaining mechanism of energy loss. It has been estimated (Mozumder, 1969a) that the maximum stopping power of water for protons due to nuclear collisions is ~1 eV/Å at an energy of ~100 eV.

2.5.2 ELECTRONS: CONJECTURES REGARDING ENERGY LOSS OF SLOW ELECTRONS

Although the motion of slow electrons is uncomplicated by charge exchange, its stopping power is still difficult to evaluate because of the gradual breakdown of the Born–Bethe approximation. On the other hand, this subject is of utmost importance in radiation chemistry. Mozumder and Magee (1966) constructed the ranges of low-energy electrons in water from stopping numbers of protons at *the same velocity* obtained from an earlier work of Hirschfelder and Magee (1948). With this oversimplified picture, Mozumder and Magee obtained ranges that are intermediate between those computed by Lea (1947) with a straightforward application of the Bethe formula using $I = 45$ eV and by Seitz (1958) using a quadratic range energy relationship. Later, a somewhat improved calculation was performed by Mozumder (1972) using a crude oscillator distribution of water owing to Platzman (1967) with a truncation at the maximum transferable energy. The root-mean-square penetration of low-energy electrons was also obtained in the same treatment by using a modified Thomas-Fermi potential for scattering and by extending a diffusion treatment owing to Bethe *et al.* (1938). The importance of scattering is indicated by Bethe *et al.* through the *Umwegfaktor*, which is defined to be the ratio of mean range to rms penetration, Its value for a typical low-energy electron is between 2 and 3.

The Born-Bethe approximation for low-energy electrons requires correction for two reasons. First, the integrals defining the total oscillator strength and the mean excitation potential have to be truncated at the maximum transferable energy because of kinematic inaccessibility of the very highly excited states. This is easily done, resulting in the replacement of Z and I by Z_{eff} and I_{eff}, respectively (LaVerne and Mozumder, 1983).

The second correction stems from the nonvanishing of the momentum transfer for inelastic collision of a low-energy electron. Using a quadratic extension of the *generalized oscillator strength* in the energy-momentum plane, this correction has been given by Ashley (1988; Ashley and Williams, 1983) still using only the DOSD. With Ashley's correction, the stopping power $S(E)$ of a medium at low electron energy E is given by

$$S(E) = \frac{4\pi\rho e^4}{mv^2} \int_0^{E/2} d\gamma \, f(\gamma) G\left(\frac{\gamma}{E}\right),$$

where $f(\gamma)$ is the DOSD at excitation energy γ and $G(\gamma/E)$ is given in terms of a tabulated elliptic integral of the first kind (Pimblott and LaVerne, 1991; Pimblott et al., 1996). In most cases, unless the electron energy is very low when the stopping power is uncertain because of excitation of dipole forbidden states and other complications, the stopping power can be simplified to give (LaVerne and Mozumder, 1983)

$$S(E) = 2\chi N Z_{eff}\left[\ln\left(\frac{mv^2}{I_{eff}}\sqrt{\frac{\bar{e}}{8}}\right) - \frac{\bar{\varepsilon}}{2E}\ln\frac{4\bar{e}E}{\bar{I}}\right],$$

where

$$\chi = 2\pi e^4/mv^2,$$

$$Z_{eff} = \int_0^{\varepsilon_{max}} f(\varepsilon) \, d\varepsilon,$$

$$Z_{eff} \ln I_{eff} = \int_0^{\varepsilon_{max}} f(\varepsilon) \ln \varepsilon \, d\varepsilon,$$

and $\bar{\varepsilon}$ and $\bar{I}$ are defined as follows:

$$Z_{eff}\bar{\varepsilon} = \int_0^{\varepsilon_{max}} \varepsilon f(\varepsilon) \, d\varepsilon; \qquad Z_{eff}\bar{\varepsilon} \ln \bar{I} = \int_0^{\varepsilon_{max}} \varepsilon \ln \varepsilon f(\varepsilon) \, d\varepsilon.$$

In these equations, e and m are respectively the electron charge and mass, v is the electron velocity at energy E, $\bar{e}$ is the base of natural logarithm, and ε_{max} is the maximum transferable energy.

The stopping power of gaseous and liquid water have been calculated using the preceding formalism (Pimblott *et al.*, 1996) and compared with other calculations by Paretzke *et al.* (1986) and by Kaplan and Sukhonosov (1991), using more approximate and indirect procedures. The results are shown in Figs. 2.5a and b. In Pimblott *et al.*'s calculation, the DOSD of gaseous water has been taken from Zeiss *et al.* (1975) and that for liquid water has been extracted from the UV reflectance data of Heller *et al.* (1974) with appropriate extension above 26 eV, the upper limit of that experiment (LaVerne and Mozumder, 1986).

Integration of the inverse stopping power between initial and final energies gives the CSDA range (*vide supra*). The penetration is quite different due mainly to elastic scattering. Ignoring straggling, LaVerne and Mozumder (1983) calculated the penetration distribution based on the theories of Lewis (1950) and of Goudsmit and Saunderson (1940), both originating in the Boltzmann transport equation. These authors have shown that the most important quantity in the penetration problem is the angular distribution of the unit velocity vector $\vec{v}$ when the electron has lost a specified amount of energy.

Consider track segments within each of which the scattering cross section remains sensibly constant. Taking the mean number of scatterings in two adjacent segments to be v_1 and v_2, and denoting the angular distribution of $\vec{v}$ relative to immediate prior scattering by $f(\theta)$ and $f(\theta^1)$, respectively, Goudsmit and Saunderson give

$$f(\theta) = (4\pi)^{-1} \sum_0^\infty (2l + 1)G_l P_l(\cos \theta)$$

and similarly for $f(\theta^1)$. Here P_l is the lth Legendre polynomial, $G_l = \exp(-v_1 Q_l^1)$, and $Q_l^1 = \langle P_l(\cos \theta_1) \rangle$; similarly for G_l^1, where θ_1 represents the angle at a single local scattering. Using cylindrical symmetry, statistical independence of sectional scatterings, and the persistence property of Legendre polynomials, one gets $\langle P_l(\cos \Theta) = \langle P_l(\cos \theta) \rangle \langle P_l(\cos \theta^1) \rangle$, where Θ is the final direction of $\vec{v}$ relative to the initial. The distribution of $\vec{v}$ at the end of the two segments is given by

$$f(\theta) = (4\pi)^{-1} \sum_0^\infty (2l + 1)\Gamma_l P_l(\cos \Theta),$$

where $\Gamma_l = G_l G_l^1 \exp(-v_1 Q_l^1 - v_2 Q_l^2)$. Generalizing this procedure to a continuous track with initial and final energies E_i and E_f, and denoting the final polar angle of $\vec{v}$ by ξ, one obtains its distribution as follows:

$$f(\xi) = (4\pi)^{-1} \sum_0^\infty (2l + 1)k_l(E)P_l(\cos \xi),$$

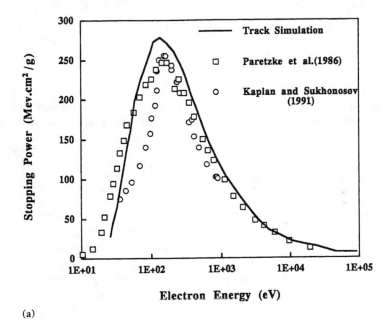

(a)

(b)

FIGURE 2.5   Density-normalized stopping power of water (MeV.cm²/g) as a function of energy, according to Pimblott *et al.* (1996) [track simulation, full curve], Kaplan and Sukhonosov (1991) [circles], and Paretzke *et al.* (1986) [squares and diamonds]. (a) Gas phase; (b) liquid phase. Reproduced from Pimblott *et al.* (1996), with permission from Am. Chem. Soc.©

where

$$\kappa_l = \int Q_l \, dv = \int_{E_f}^{E_i} N\sigma_e (1 - \cos\theta)\,[dE/S(E)].$$

Here dv is the mean number of scatterings in path length $dE/S(E)$, $S(E)$ being the stopping power at energy E, and σ_e is the scattering cross section. The last equation may be recast in a form more suitable for computation, using the differential scattering cross section $d\sigma_e/d\Omega(\theta)$, as follows:

$$\kappa_l = 2\pi N \int_{E_f}^{E_i} \frac{dE}{S(E)} \int_0^\pi \frac{d\sigma_e}{d\Omega(\theta)(\sin\theta)[1 - P_l(\cos\theta)]\,d\theta}.$$

Although the detailed spatial distribution is an arduous task, Lewis (1950) has shown, by a moment analysis, that some important averages such as $\langle Z \rangle$, $\langle Z \cos\theta \rangle$, and $\langle X^2 + Y^2 \rangle$ can be calculated in a straightforward manner. LaVerne and Mozumder (1983) generalized the procedure to give $\langle Z^2 \rangle$, $\langle X^2 \rangle = \langle Y^2 \rangle$, and the rms penetration $\langle r^2 \rangle^{1/2}$, by virtue of cylindrical symmetry. Finally, utilizing a result of random walk theory for a large number of scatterings (Chandrasekhar, 1943), the position distribution is given by

$$W(r) = (2\pi A)^{-1}(2\pi B)^{-1/2} \exp\left\{-\left[\frac{X^2 + Y^2}{2A} + \frac{(Z - \langle Z \rangle)^2}{2B}\right]\right\},$$

where $A = \langle X^2 \rangle$ and $B = \langle Z^2 \rangle - \langle Z \rangle^2$. The distribution is spheroidal with a center displaced by $\langle Z \rangle$ on the Z axis from the point of origin. The mean radial range is given from this distribution by

$$\langle r \rangle = 2\pi \int_0^\infty r^3 \, dr \int_0^\pi W(r) \sin\theta \, d\theta,$$

where now θ is the angle of the position vector relative to the Z axis—that is, the initial direction of motion. LaVerne and Mozumder used this procedure to calculate the mean path length and rms penetration of low-energy electrons in the gaseous phase of water. The stopping power was obtained from the DOSD by the Ashley procedure as outlined earlier in this section. The differential scattering cross section was fitted to available swarm and beam data (Itikawa, 1974) by adopting Molière's (1947, 1948) modification of Rutherford scattering with a prescribed screening function. The result, reproduced in Figure 2.6, shows a great difference between rms range and mean path length at low energies. Some difference persists even at higher energies. Detailed analysis shows the following.

1. The angular distribution of the velocity remains spheroidal—that is, the memory of the initial direction persists—until the electron loses

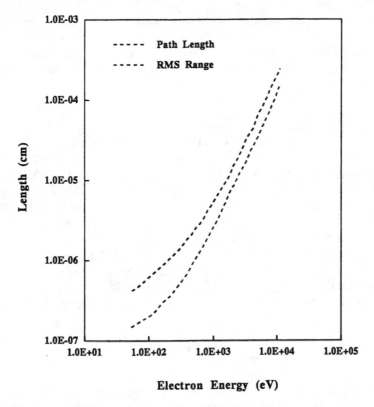

Electron Energy (eV)

FIGURE 2.6   Root-mean-square range and mean path length of low-energy electrons in gaseous water. From LaVerne and Mozumder (1983).

~70–80% of its initial energy. For example, the directional distribution of a 1-KeV electron does not approach to be spherical until the energy has been degraded to ~200 eV.

2. The electron penetrates a certain distance with relatively few scatterings and energy loss, after which considerable energy loss and excessive scattering set in, giving a nearly diffusive character. The mean penetration no longer increases significantly, although the rms penetration continues to increase until the particle stops.

3. The number of elastic collisions needed to give a nearly isotropic distribution increases with energy. For example, about 15 collisions are needed for an initial 1-KeV electron in water, whereas that number is about 74 for an initial 10-KeV electron.

We have described in this section and in Sect. 2.4.2 the electron penetration problem when either straggling or scattering was ignored. There is as yet no analytical framework that incorporates both scattering and straggling. Recently, however, Monte Carlo simulation has been developed for liquid water (Pimblott *et al.*, 1996) and for that purpose that takes into account the statistical fluctuations in energy loss and scattering angle consistent with all available experiments (*vide infra*).

2.5.3   PHASE EFFECT ON ELECTRON STOPPING

Range and straggling of electrons are determined mainly by the DOSD (see Sects. 2.4.2 and 2.5.2). Phase effects enter naturally through DOSD. In the gas phase, the DOSD is obtained from absorption and/or inelastic scattering experiments, and a fairly complete determination has been made for water by Zeiss *et al.* (1975). In the condensed phases, direct absorption measurement in the far UV and beyond is very difficult and uncertain. DOSD [$f'(\varepsilon)$] in such cases is indirectly obtained from reflectance measurement at the vacuum–liquid (or vacuum–solid) interface, using electromagnetic relationships. If the reflectance R is known at all energies E, then the phase angle can be obtained from the Kramers–Kronig relationship:

$$\phi(E) = \frac{E}{\pi} \, P\!\int_0^\infty \frac{\ln R(E')}{E'^2 - E^2} \, dE',$$

Where P indicates Cauchy's principal value for the integral. From R and ϕ, the real and imaginary parts of the refractive index are given respectively by $n = (1 - R)/(1 + R - 2R^{1/2}\cos\phi)$ and $k = (-2R^{1/2}\sin\phi)/(1 + R - 2R^{1/2}\cos\phi)$. Next, the real and imaginary parts of the dielectric function and the energy loss function are obtained respectively as $\varepsilon_1 = n^2 - k^2$, $\varepsilon_2 = 2nk$ and $\mathrm{Im}(-1/\varepsilon) = \varepsilon_2/(\varepsilon_1^2 + \varepsilon_2^2)$. Finally, the DOSD $f'(\varepsilon)$ is calculated from the relation $\mathrm{Im}[-1/\varepsilon(\omega)] = (h^2 e^2 NZ/2m)f'(\varepsilon)/\varepsilon$, where apart from the usual universal constants Z is the number of electrons in the molecule (=10 for water).

The crucial step is the calculation of the phase angle using the Kramers–Kronig relationship, which requires the reflectance at *all* energies. Since there is an upper limit for experimental reflectance measurement (usually ~28 eV), this calls for a fairly long extrapolation. The UV reflectance data of liquid water is obtainable from Heller *et al.* (1974) up to 26 eV, and those for hexagonal and amorphous ice at 80 K from Seki *et al.* (1981) and Kobayashi (1983), respectively, all up to 28 eV. Heller *et al.* (1974) suggest two analytical continuations beyond the experimental limit: one exponential and the other a power law function of the type $R(E) = R_0(E_0/E)^\beta$, where R_0 is the reflectance at the highest experimental energy E_0. Of these, the exponential function has been found inconvenient by LaVerne and Mozumder (1986), as it produces undesirable divergence in the integral for phase angle unless prevented by an arbitrary cut off at a high energy. These authors use the power law function by demanding a value of β that gives near transparency in the visible region (below 6 or 7 eV) and also gives the correct number of valence electrons (8.2 for water) on integration of the DOSD. For both ices, the so-determined value of $\beta = 3.8$, which is close to the theoretical limit of 4.0 for the excitation of core electrons (Phillip and Ehrenreich, 1964). This should be compared with $\beta = 5.2$ obtained by Heller *et al.* (1974) for liquid water.

The so-determined DOSD $f'(\varepsilon)$ for the gaseous and liquid phases of water is reproduced in Figure 2.7. To this, the contribution due to K excitation may be added in all phases by assuming free atomic excitation. There are two noticeable effects of condensation on DOSD. The first is a general loss of structure in going from the gas to the condensed phase. The second is an upward shift in the maximum excitation energy, from ~18 eV in the gas to ~21 eV in the condensed phases of water. At almost all energies, the main difference is between the gas and the condensed phases; there is little difference among the condensed phases except perhaps at the dominant peak. The phase effects tend to diminish with excitation energy, and these are barely noticeable beyond ~100 eV.

Figure 2.8 shows the range distribution of 100-eV incident electrons in water for different phases obtained by using the DOSD of Figure 2.7 and the procedure of Sect. 2.4.2. Notice that the density normalized ranges are smaller in the gaseous phase. While all range distributions are wide, those in the condensed phases are even wider. The distribution in the gas phase is clearly different from those in the condensed phases, but there is no significant difference among the condensed phases. Calculated mean ranges of a 100-eV electron in gaseous and condensed water are, according to LaVerne and Mozumder (1986), respectively 0.62 and 1.15 µg/cm². The corresponding stopping powers are respectively ~270.0 and ~225.0 MeV cm²/g.

It should be pointed out that there are confusing and inconsistent reports about phase effects on the stopping of low-energy particles. A detailed discussion of this has been given by LaVerne and Mozumder (1986). Earlier experiments found no phase effect on the stopping of α-particles. Later experiments with α-particles and protons showed that the density normalized stopping power was greater in the gas than in the condensed phases, and that this trend diminishes with increasing energy. These findings are consistent with the view presented here. Paretzke and Berger (1978) initially stated that the electron transport properties would relatively phase independent. Later, however, Turner *et al.* (1982) found that the density normalized stopping power to be greater in the liquid for electron energy >50 eV, which is exactly opposite to the views of LaVerne and Mozumder (1986).

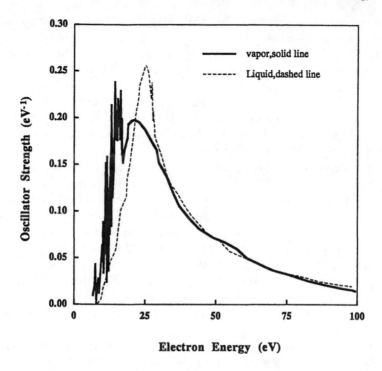

FIGURE 2.7   Differential oscillator strength distribution (DOSD) in the vapor and liquid phases of water. Reproduced from LaVerne and Mozumder (1986), with permission of Am. Chem. Soc.©

2.6   MISCELLANEOUS CONSIDERATIONS

2.6.1   THIN ABSORBERS

Reasonably thin absorbers may be interposed on the path of a high-energy heavy charged particle to reduce its energy. Provided that the incident particle is sufficiently energetic and that the absorber is not too thin, this will not create an excessive spread in the energy distribution of the emergent particles. A continuous slowing down approximation will still remain valid, and the emergent beam is still pretty well defined. Thus foils of Al, Au, and the like can be used to produce beams of suitable lower energy.

   If the medium is on the order of a few microns in thickness, the average energy loss of the penetrating charged particles will be somewhat less than what is calculated on the basis of the stopping power. Also, a substantial dispersion will be observed in the energy of the emergent particle if the incident energy is not high

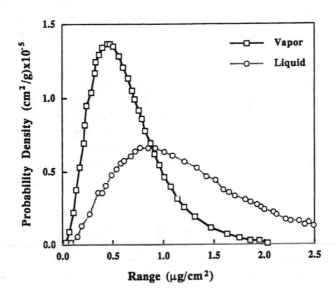

FIGURE 2.8   Range distribution of 100-eV electrons in the vapor and liquid phases of water. Reproduced from LaVerne and Mozumder (1986), with permission of Am. Chem. Soc.©

(see Sect. 2.4.2). For extremely thin samples (on the order of a few hundred angstroms thickness), the energy loss may be of collective plasma character (see Sect. 2.6.4). Usually in such cases, the lost energy is radiated or thermalized without affecting chemical transformation. The changeover from collective to atomic (or molecular) excitations occurs at around 0.1 μm thickness.

2.6.2   GENERIC EFFECTS

Knock-on collisions often generate secondary electrons with a limiting (energy)$^{-2}$ distribution. The process may continue in higher generations until energetically forbidden. Often, later generation electrons have higher LET than former because of lower energy. On the other hand, for high-LET primaries, the secondary electrons represent low-LET cases, Cases of protons, deuterons, and α- particles are intermediate. Thus it is seen that generic effects are important at both high- and low-LET, since radiation-chemical yields are, generally speaking, LET dependent.

2.6.3   CERENKOV RADIATION

When the speed of an electron exceeds the group velocity of light in a transparent medium, a faint bluish white light is emitted. This phenomenon, called the Cerenkov radiation, was discovered by Cerenkov (1934) and interpreted theoretically by Frank and Tamm (1937) and by Ginsberg (1940). The radiation is continuous from infrared through ultraviolet with an intensity roughly proportional to the inverse square of the wavelength (Jelly, 1958). The angle θ between the direction of Cerenkov radiation and the electron trajectory is given by $\cos^{-1}(1/\beta n)$, where n is the medium refractive index for the emitted light and $\beta = v/c$, v being the electron velocity. Therefore: (1) for a given refractive index, an electron must have a minimum velocity $v_{min} = c/n$ for emitting Cerenkov radiation; and (2) for ultrarelativistic particles, the maximum angle of emission is given by $\theta_{max} = \cos^{-1}(1/n)$.

Cerenkov radiation accounts for a very minor part of the energy loss of fast electrons. Its main importance is for monitoring purposes and establishment of a reference time, since it is produced almost instantaneously with the passage of the particle. Katsumura et al. (1985) have observed a very fast rise of solute fluorescence attributable to the Cerenkov effect; the G value for this process is estimated to be ~0.02.

2.6.4   PLASMON EXCITATION

Plasmons are collective electron oscillations in condensed matter brought about by longitudinal coulomb interaction (see Sect. 2.6.1). Other collective excitations include excitons, which are due to transverse interaction. From time to time, plasmons have been conjectured as a possible mechanism for radiation interaction in a condensed medium (Fano, 1960). It was invoked by Heller et al. (1974) for interpreting UV reflectance from the surface of liquid water. The reflectance data were converted to optical constants (n, k) from which the real and imaginary parts of the complex dielectric constant were computed: $\varepsilon_1 = n^2 - k^2$ and $\varepsilon_2 = 2nk$. With these, the energy loss function was calculated: $Im(-1/\varepsilon) = \varepsilon_2/(\varepsilon_1^2 + \varepsilon_2^2)$, which showed a broad peak around 21 eV. This was interpreted to be due to plasmon excitation, since the calculated plasmonic quantum in liquid water including all the electrons matched with this value. However, an earlier experiment by Daniels (1971) on the energy loss of fast electrons in ice showed no evidence of plasmons as the criterion of collective excitation was not met. On the other hand, it is generally believed that the 7-eV excitation in condensed benzene is due to an exciton (Killat, 1974), or other collective motion (Williams et al., 1969).

LaVerne and Mozumder (1993) carefully analyzed the necessary conditions for the occurrence of plasma excitation in water and found no convincing evidence for it. Originally, Fano based the criterion for collective excitation on an index U defined by $U(E) = (\hbar\omega_p)^2 f(E)/2E$, where $f(E)$ is the dipole oscillator strength density (DOSD) at excitation energy E. The condition $U \ll 1$ indicated charged particle excitation resembling that of a collection of independent molecules, whereas $U \gg 1$ called for collective excitation. Using rather accurate DOSDs, LaVerne and Mozumder (1993) found that in all phases and at unit density, U is neither $\ll 1$ nor, $\gg 1$, but that U generally lay in the range of 0.1–0.2 between 5 and 50 eV, which is the most important part of excitation spectrum. Thus the existence of collective excitation in water remains inconclusive on this basis.

More recent analysis by Fano (1992) gives a criterion for plasma excitation, which may be written as

$$(E_s - E_r)^{-1} \int_{E_r}^{E_s} U(E)\, dE > 1,$$

where the spectral range is indicated by $E_r < E < E_s$. Since this quantity can be interpreted as the mean value of U over the spectral interval, this criterion too is not fulfilled for liquid water, because $U(E)$ nowhere exceeds unity.

Various criteria for plasma resonance can be obtained by setting the dielectric function to be zero in the long-wavelength limit. Thus, in terms of ε_1 and ε_2, Ehrenreich and Philipps (1962) give the requirement of plasmon excitation as follows: (1) ε_1, $\varepsilon_2 \ll 1$; (2) $d\varepsilon_1/d\omega > 0$, $d\varepsilon_2/d\omega < 0$; and (3) $(\omega/\varepsilon_1)(d\varepsilon/d\omega) \gg 1$, $(\omega/\varepsilon)|d\varepsilon_2/d\omega| \gg 1$. There is evidence that such conditions are met in some condensed media including group III–V semiconductors (Ehrenreich and Philipps, 1962) but not in water (LaVerne and Mozumder, 1993).

Finally, the integral of the oscillator strength up to $E = 30$ eV only amounts to ~3.0 in both gaseous and liquid water, which falls much shorter than the value 10 if all the electrons were to participate in plasma excitation, giving an excitation energy ~21 eV.

REFERENCES

Ashley, J. C. (1988), J. Electron. Spectrosc. Relat. Phenom. 46, 199.

Ashley, J. C., and Williams, M. W. (1983), Report of the Rome Air Force Development Center, Griffiss Air Force Base, Rome, N.Y.

Bethe, H. A. (1930), Ann. Physik 5, 325.

Bethe, H. A. (1932), Z. Physik 76, 293.

Bethe, H. A. (1933), Ann. Physik 24, 273.

Bethe, H. A., and Ashkin, J. (1953), in Experimental Nuclear Physics, v. 1 (Segre, E., ed.), pp. 166–357, Wiley, New York.

Bethe, H. A., and Heitler, W. (1934), Proc. Roy. Soc. A146, 83 .

Bethe, H. A., Rose, M. E., and Smith, L. P. (1938), Proc. Am. Phil. Soc. 78, 573.

Bloch, F. (1933a), Ann. Physik 16, 285.

Bloch, F. (1933b), Z. Physik 81, 363.

Bohr, N. (1913), Phil. Mag. 25, 10.

Bohr, N. (1948), Kgl. Danske Vıed. Selskab. Mat-Fys. Medd. 18, 9.

Cole, A. (1969), Radiat. Res. 38, 7.

Cerenkov, P. A. (1934), Dokl. Akad. Nauk, 2, 451.

Chandrasekhar, S. (1943), Rev. Mod. Phys. 15, 1.

Dalgarno, A., and Griffing, G. W. (1955), Proc. Roy. Soc. A232, 423.

Daniels, J. (1971), Optics Commun. 3, 240.

Ehrenreich, H., and Philipps, R. H. (1962), in Report of the International Conference of the
    Physics of Semiconductors (Exeter), pp. 367–374, The Institute of Physics and The Physical
    Society, London.

Fano, U. (1960), in Comparative Effects of Radiations (Burton, M., Kirby-Smith, J. S., and
    Magee, J. L., eds.), pp. 14–21, Wiley, New York.

Fano, U. (1963), Ann. Rev. Nucl. Sci. 13, 1.

Fano, U. (1992), Revs. Mod. Phys. 64, 313.

Fermi, E. (1940), Phys. Rev. 57, 485.

Frank, I. M., and Tamm, I. (1937), Dokl. Akad. Nauk, 14, 109.

Ginsberg, V. L.(1940), ZETP 10, 589.

Goudsmit, S., and Saunderson, J. L. (1940), Phys. Rev. 57, 24.

Grosswendt, B., and Waibel, E. (1982), Nucl. Instrum. Method, 197, 401.

Hart, E. J., and Platzman, R. L. (1961), in Mechanisms in Radiobiology (Frossberg, A., and
    Errera, M., eds.), v. 1, ch. 2, Academic Press, New York.

Heller, J. M., Hamm, R. N., Birkhoff, R. D., and Painter, L. R. (1974), J. Chem. Phys. 60, 3483.

Hirschfelder, J. O., and Magee, J. L. (1948), Phys. Rev. 73, 207.

ICRU (1970), Report 16 published by the International Commission on Radiation Units and
    Measurements, Washington, D.C.

Inokuti, M. (1971), Revs. Mod. Phys. 43, 297.

Itikawa (1974), At. Data Nnel. Data Tables 14, 1.

Jelly, J.V. (1958), Cerenkov Radiation and Its Application, Pergamon, Elmsford, NY.

Kalarickal, S. (1959), Ph.D. Thesis, University of Notre Dame (Unpublished).

Kaplan, I. G., and Sukhonosov, V. Y. (1991), Radiation Res. 127, 1.

Katsumura, Y., Tabata, Y., and Tagana, S. (1985), Radiat. Phys. Chem. 24, 489.

Killat, U. (1974), J. Phys. C7, 2396.

Klein, O., and Nishina, Y. (1929), Z. Physik 52, 853.

Kobayashi, K. J. (1983), J. Phys. Chem. 87, 4317.

Landau, L. (1944), J. Phys. (USSR) 8, 201.

LaVerne, J. A., and Mozumder, A. (1983), Radiat. Res. 96, 219.

LaVerne, J. A., and Mozumder, A. (1985), J. Phys. Chem. 89, 4219.

LaVerne, J. A., and Mozumder, A. (1984), Radiat. Phys. Chem. 23, 637.

LaVerne, J. A., and Mozumder, A. (1986), J. Phys. Chem.90, 3242.

LaVerne, J. A., and Mozumder, A. (1993), Radiat. Res. 133, 282.

Lea, D. E. (1947), Actions of Radiations on Living Cells, p. 24, Macmillan, New York.

Lentle, B., and Singh, H. (1984), Radiat. Phys. Chem. 24, 267.

Lewis, H. W. (1950), Phys. Rev. 78, 526.

Livingstone, M., and Bethe, H. A. (1937), Revs. Mod. Phys. 91, 245.

Maccabee, H. D., Raju, M. R., and Tobias, C. A. (1968), Phys. Rev. 165, 469.

Magee, J. L. (1953), Ann. Rev. Nucl. Sci. 3, 171.

Magee, J. L., and Chatterjee, A.(1987), in Radiation Chemistry: Principles and Applications
    (Farhataziz and Rodgers, M. A. J., eds.), ch. 5, VCH Publishers, New York, NY.

Molière, G. (1947), Z. Naturforsch. Teil A2, 133.

Molière, G. (1948), Z. Naturforsch. *Teil A3*, 78.

Morrison, P. (1952), in *Symposium on Radiobiology: The Basic Aspects of Radiation Effects on Living Systems* (Nickson, J. J., ed.), pp. 1–12, Wiley, New York.

Mott, N. F. (1930), Proc. Roy. Soc. *A126*, 79.

Mozumder, A. (1969a), Advances Radiat. Chem. *1*, 1.

Mozumder, A. (1969b), J. Chem. Phys. *50*, 3153.

Mozumder, A. (1972), in *Proc. 3rd Tihany Symposium on Radiation Chemistry* (Dobo, J., and Hedvig, P., eds.), v. 2, pp. 1123–1132, Akademiai Kiado, Budapest.

Mozumder, A. (1974), J. Chem. Phys. *60*, 1145.

Mozumder, A. (1985), Radiation Res. *104*, S-33.

Mozumder, A., and LaVerne, J. A.(1984), J. Phys.Chem. *88*, 3926.

Mozumder, A., and LaVerne, J. A. (1985), J. Phys. Chem. *89*, 930.

Mozumder, A., and Magee, J. L. (1966), Radiat. Res. *28*, 203.

Mozumder, A., and Magee, J. L. (1967), J. Chem. Phys. *47*, 939. National Academy of Sciences–National Research Council (NAS–NRC) (1964), Publication 1133, Washington, D.C.

Orear, J., Rosenfeld, H., and Schluter, R. A. (1956), *Nuclear Physics*, p. 29, University of Chicago Press, Chicago.

Paretzke, H. G., and Berger, M.J. (1978), *Proc. 6th Symp. Microdosimerty*, pp. 749–758, Harwood, London.

Paretzke, H. G., Turner, J. E., Hamm, R. N., Wright, H. A., and Ritchie, R. H. (1986), J. Chem. Phys. *84*, 3182.

Phillip, H. R., and Ehrenreich, H. J. (1964), J. Appl. Phys. *35*, 1416.

Pimblott, S. M., and LaVerne, J. A. (1991), J. Phys. Chem. *95*, 3907.

Pimblott, S. M., LaVerne, J. A., and Mozumder, A. (1996), J. Phys. Chem. *100*, 8595.

Platzman, R. L. (1955), Radiat. Res. *2*, 1.

Platzman, R. L. (1967), in *Radiation Research* (Silini, G., ed.), pp. 35–36, North Holland, Amsterdam.

Seitz, F. (1958), Phys. Fluids *1*, 2.

Seki, M., Kobayashi, K., and Nakahara, J. (1981), J. Phys. Soc. Jpn. *50*, 2643.

Tabata, Y., Ito, Y., and Tagawa, S. (1991), *CRC Handbook of Radiation Chemistry*, chs. 2 and 3, CRC Press, Boca Raton, Fla.

Turner, J. E. (1964), in *Studies in Penetration of Charged Particles in Matter*, Publication 1133, National Research Council, pp. 99–101, Washington, D.C., *5*, 749.

Turner, J. E., Paretzke, H. G., Hamm, R. N., Wright, H. A., and Ritchie, R. H. (1982), Radiat. Res. *92*, 47.

Turner, J. E., Magee, J. L., Wright, H. A., Chatterjee, A., Hamm, R. N., and Ritchie, R. H. (1983), Radiat. Res. *96*, 437.

Vavilov, P. V. (1957), JETP (English translation) *5*, 749.

Warman, J. M., Kunst, M., and Jonah, C. D. (1983), J. Phys. Chem. *87*, 4292.

Whiddington, R. (1912), Proc. Roy. Soc. *A86*, 360.

Wiesenfeld, J. M., and Ippen, E. D. (1980), Chem. Phys. Lett. *73*, 47.

Williams, E. J. (1945), Revs. Mod. Phys. *17*, 217.

Williams, M. W., MacRae, R. A., Hamm, R. N.,and Arakawa, E. T. (1969), Phys. Rev. Lett. *22*, 1088.

Zeiss, G. D., Meath, W. J., Macdonald, J. C. F., and Dawson, D.J. (1975), Radiat. Res. *63*, 64.

Ziegler, J. F., ed. (1980–1985), *The Stopping and Ranges of Ions in Matter*, v. 1–6, Pergamon Press, Elmsford, NY.

Zirkle, R. E., Marchbank, D. F., and Kuck, K. D. (1952), J. Cell. Comp. Physiol. *39*, suppl. 1, 75.

CHAPTER 3

Structure of Charged Particle Tracks in Condensed Media

3.1 STOPPING POWER IN WATER FOR VARIOUS CHARGED PARTICLES

In Chapter 2, we presented stopping power theories. In this section, as a prelude to the structure of particle tracks, we will discuss some actual values in water, which is the most important medium both chemically and biologically. A wide span of energy (1–10^{10} eV) is considered, to emphasize various interaction mechanisms that dominate specific energy intervals. It is clear that at very high energies the nature of the energy loss process is relatively independent of the impinging particle, whereas at very low energies both the nature of the penetrating particle and that of the medium are important. Figure 3.1 shows the stopping power of water toward various charged particles plotted on a log-log scale. The present discussion is approximate and qualitative. Less approximate values may be found elsewhere.

3.1.1 THE ELECTRON

Energy loss of electrons above 100 MeV is dominated by nuclear encounters producing bremsstrahlung. This process, characterized by the radiation length R within which most of the energy is lost, is independent of particle energy.

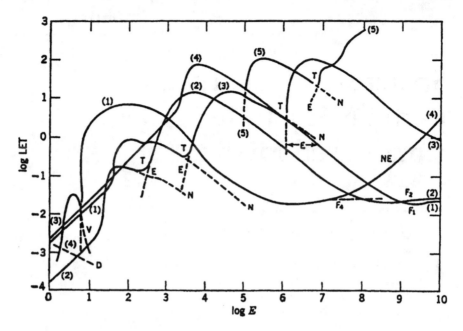

FIGURE 3.1   Stopping power of water for various charged particles over a wide span of energy;
1: electron, 2: (positive) muon, 3: proton, 4: carbon nucleus, and 5: fission (light) fragment. See
text for details. Reproduced from Mozumder (1969), by permission of John Wiley & Sons, Inc.©

Thus, the *average* stopping power is proportional to the initial energy except
for corrections due to atomic collisions (electronic excitation) near 10^8 eV. For
a medium of nuclear charge Ze and mass number A, the radiation length is given
by (Bethe and Ashkin, 1953)

$$R_l^{-1} = \frac{4e^6}{\hbar m^2 c^5} \frac{N}{A} Z(Z + 1) \ln(193\, Z^{-1/3}).$$

Here N is Avogadro's number. For low-Z media, R varies slowly with Z, having
the values 37.7, 37.1, and 24.4 cm² for air, water, and Al, respectively.

In the region 10^4–10^9 eV, where the energy loss is via electronic excitation
and ionization, Bethe's formula with corrections (Eq. 2.11) describes the stopping power quite accurately. In the interval 10^4–10^6 eV, the decrease of stopping
power with energy is attributable to the v^{-2} term. It reaches a minimum of ~0.02
eV/A at ~1.5 MeV; then it shows a relativistic rise before the restricted part rides
to the Fermi plateau at ~40 MeV.

In the interval 25–10^4 eV, stopping power has been evaluated according to
the procedure of Sect. 2.5.2. At lower energies, the computation is neither accurate nor certain. Extrapolation of electron range from 5 to 10 KeV using a power

law obtained as a best fit in a log–log plot shows good agreement with well-known ranges at higher energies. However, this fact does *not* guarantee that stopping power computation is good at lower energies. Smooth extrapolation has been made from 25 eV to the subexcitation energy, taken as 6 eV.

In the subexcitation region (<6 eV), the energy loss is mainly via vibrational excitation of the water molecule. It is assumed that this process has a broad maximum at ~4 eV and is reduced to insignificance at ~2 eV and at ~10 eV. Taking, on average, a cross section ~1 Å² per degree of freedom with a quantum ~0.3 eV, the maximum stopping power due to vibrational excitation is estimated as ~0.03 eV/Å. In Figure 3.1, we have used this maximum stopping power and a gaussian curve that reduces to insignificance at 2 and 10 eV.

At still lower energies, the loss mechanism is the interaction of the electron with the permanent dipoles of the water molecule. Fröhlich and Platzman (1953) estimated a constant *time rate* of energy loss due to this effect at ~10^{13} eV/s. The stopping power in eV/Å is then approximately given by $(1.7 \times 10^{-3})E^{-1/2}$, where the energy E is in eV.

3.1.2   THE MUON

For stopping considerations, the muon may be taken as a stable heavy particle, since the stopping time is much smaller than the natural lifetime. In the entire energy region 500–10^{10} eV, the stopping power for muons is given by that for protons at the same velocity, since both the cross sections for energy loss and charge exchange (which dominates below 2 KeV) depend only on the velocity, not on the mass of the incident particle. Below 500 eV, Rutherford collision with the nuclei is the dominant energy loss mechanism. The stopping power s_n due to this effect depends on the mass, as shown in the following formula due to Bohr (1948):

$$s_n = \frac{m_1}{m_2} \frac{2\pi N Z_1^2 Z_2^2 e^4}{E} \ln \gamma E . \tag{3.1}$$

Here the subscripts 1 and 2 refer to the incident particle and nuclei of the stopping medium, respectively; m and Ze denote mass and charge, and $\gamma = [m_2/(m_1 + m_2)]/Z_1 Z_2 s[(1/2)mv_0^2]$ with the screening constant s given by $s^2 = Z_1^{2/3} + Z_2^{2/3}$. It should be noted that (1) the full nuclear charge of the incident particle appears in Eq. (3.1), not the effective charge used for electronic stopping power calculation; and (2) in principle, s_n goes through a maximum separately for the protons and oxygen nuclei; however, for muons the proton contribution obliterates that from the oxygen nucleus, resulting in a single peak ~0.17 eV/Å at ~60 eV. Below about 50 eV, the final stopping is due to elastic collisions of the billiard ball type because of almost complete screening. The latter, estimated with a geometrical cross section, is shown by the straight-line part of curve 2 in Figure 3.1.

3.1.3   THE PROTON

Proton stopping power in the interval 10^5–10^{10} eV, given rather accurately by the Bethe formula (Sects. 2.3.4–2.3.6), is shown as curve 1 in Figure 3.1. This includes the relativistic rise and the Fermi plateau in the 10^9–10^{10} eV region. Below about 1 MeV, the logarithmic term in the stopping power formula induces a maximum at somewhat below 100 KeV. Gradual lowering of effective charge due to electron capture accentuates this maximum; however, this effect is unimportant above about 200 KeV. Another effect at low energies is that the maximum energy transfer ($2mv^2$) is not sufficiently large to contain all the oscillator strength of the molecule. This effect tends to reduce the stopping power by lowering Z and, at the same time, tends to increase it by reducing I. Here we have used a procedure analogous to that for low-energy electrons (Sect. 2.5.2) to correct for these effects.

   Charge-changing cross sections are denoted by σ_{ij} , where i and j respectively indicate initial and final charge states. Thus, $j < i$ means electron capture and $j > i$ means electron loss. Capture or loss of more than one electron at a time is rare and, therefore, ignored. Further, these cross sections depend more on the velocity and the ionic charge of the impinging particle and very mildly on the nature of the medium penetrated. Mozumder (1969) uses the experimental charge-changing cross sections in neon to arrive at the effective charge of protons in water. If the equilibrium probabilities of charge states at a given velocity are indicated by ϕ_j ($j = z, z - 1, \ldots, 1, 0, -1$), then one must have the following set of homogeneous equations:

$$\sum_i \sigma_{ij} \phi_i - \phi_j \sum_j \sigma_{ji} = 0. \tag{3.2}$$

   At a given velocity, these probabilities are uniquely determined from this equation subject to the normalization

$$\sum_j \phi_j = 1.$$

Thus, one can determine z_{eff}^2 through

$$\sum_j j^2 \phi_j$$

and use this value in the Bethe equation. For protons, $j = +1, 0,$ or -1 and the charge state probabilities are given by

$$\phi_{+1} = \frac{A}{A + B + D}, \quad \phi_0 = \frac{D}{A + B + D}, \quad \text{and} \quad \phi_{-1} = \frac{B}{A + B + D}, \tag{3.3}$$

where $A = \sigma_{\bar{1}1}(\sigma_{0\bar{1}} + \sigma_{0\bar{1}}) + \sigma_{01}\sigma_{\bar{1}0}$, $B = \sigma_{0\bar{1}}(\sigma_{1\bar{1}} + \sigma_{10}) + \sigma_{01}\sigma_{1\bar{1}}$, and $D = \sigma_{\bar{1}0}(\sigma_{10}$ $+ \sigma_{1\bar{1}}) + \sigma_{\bar{1}1}\sigma_{10}$. Above 40 KeV, H$^-$ is virtually absent and σ_{+1} and σ_0 are simply given by $\sigma_{01}/(\sigma_{10} + \sigma_{01})$ and $\sigma_{10}/(\sigma_{10} + \sigma_{01})$, respectively. Charge state probabilities for a proton in Ne have been computed (Mozumder, 1969) using experimental charge-changing cross sections (Allison, 1958). It is seen that these probabilities are sensitive to energy below about 200 KeV. Thusly calculated values of z_{eff}^2 are approximately 0.27, 0.50, and 0.93 at $E = 4$, 30, and 200 KeV, respectively. In Figure 3.1, Bethe's stopping power formula with the appropriate value of z_{eff}^2 and with corrected values of Z_{eff} and I_{eff} has been used down to 4 KeV.

Rutherford collisions dominate proton stopping in water in the interval 40–4000 eV, and the stopping power due to this effect has been computed using Eq. (3.1) with separate contributions coming from the protons and the oxygen nucleus of the water molecule. Proton-proton collision is a very efficient slowing down process, especially at low energies. This fact results in the low-energy peak of curve 1 in Figure 3.1. The peak at somewhat higher energy is due to proton-oxygen collision. Finally, in the region of 1–40 eV, stopping involves only billiard ball type collision, which is computed in the same way as for muons.

3.1.4   THE CARBON NUCLEUS

In the past, radiation-chemical experiments have been made with nuclei of relatively low atomic number. Variable-energy cyclotrons (VECs), heavy-ion linear accelerators (HILACs), tandem Van de Graaff accelerators, and other machines have been used for that purpose. Yields of molecular products and of ferric ion in the Fricke dosimeter have been measured for radiations of carbon through neon nuclei with energies up to about 10 MeV/amu (Schuler, 1967; Imamura et al., 1970; Schuler and Barr, 1956; Burns and Reed, 1970). At these energies, these particles have high LET in water. Subsequently, more experiments have been carried out with the Bevalac accelerator at Berkeley and with other accelerators elsewhere. These experiments have been reviewed by LaVerne and Schuler (1987b) and by LaVerne (1988, 1996). At the high-energy limit of a few hundred MeV/amu to ~GeV/amu obtainable with some machines, the accelerated particles may be of low, intermediate, or high LET depending on the nuclear charge, if all nuclei in the periodic table are considered. The present discussion is, however, limited to $Z \leq 10$ and incident energy ≤ 10 MeV/amu. The carbon nucleus is taken as a typical example.

The stopping power of water toward the carbon nucleus is shown as curve 3 in Figure 3.1. If the incident velocity of an ion is much in excess of zv_0 (i.e., its energy much greater than 0.025 z^2 MeV/amu), then the ion retains its full nuclear charge. The effective charge decreases with progressive slowing. If the charge-changing cross sections are known, the effective charge can be calculated by a procedure similar to that for protons (see Sect. 3.1.3). However, such is the case only at high and low velocities; rough approximations are required at intermediate velocities. In Sect. 3.4.4c, we will describe such an approximation, by which we calculate the effective charge and use it in Bethe's equation to obtain the stopping power in the region 10^6–10^{10} eV. We find that (1) the carbon nucleus retains its full charge above 10^8 eV, and (2) its electronic stopping power goes through a maximum at ~4 MeV, being accentuated by the electron capture process.

Stopping power in the region 2×10^3 to 10^6 eV is determined mostly by Rutherford collision between the nuclei, which goes through a maximum at about 6 KeV. At still lower energies, elastic billiard ball type collision is the operative mechanism, giving a linear plot down to 1 eV. Compared with the proton, this region (1 to 2×10^3 eV) is greatly extended for the carbon nucleus.

3.1.5   FISSION FRAGMENTS

So far, the highest available LET is by the fission of ^{235}U, which, however, produces a heterogeneous radiation consisting of fragments over a range of charge and mass and is also accompanied by γ-radiation. In the present discussion, we follow Bohr (1948) and divide the fragments into light and heavy groups with nuclear charge, mass number, and initial velocity as (38e, 95, 6v_0) and (54e, 140, 4v_0), respectively. Because of the low velocity and high nuclear charge of fission fragments, their stopping is dominated by charge exchange right from the beginning. Since they always carry a large number of electrons, their ionic charge is given rather well by the Thomas-Fermi model as $z_{eff} = z^{1/3}v/v_0$ (Bohr, 1948). Also, as a result of low velocity, the electronic stopping power for fission fragments is better computed via Bohr's formula (Eq. 2.5). This has been done, taking E_1 as the mean excitation potential; the result is shown as curve 5 of Figure 3.1 for the light fragment.

The maximum of *electronic* stopping power for fission fragments is usually not observable, since that occurs at energies higher than that provided by the fission process. Electronic stopping dominates until the fission fragments slow down to ~10–20 MeV in water, and thereafter Rutherford collision takes over. The maximum stopping power due to the latter process is seen to occur for the light fragment at about 300 KeV. *At a given energy,* the electronic stopping power for the lighter fragment is higher than that for the heavier fragment, but the reverse it true for the nuclear stopping power. The behavior of the electronic stopping power in this respect is apparently paradoxical, but it is understood in terms of excessive electron capture by the heavier fragment. The total stopping power of water is therefore nearly equal (~42 eV/Å) for both the light and heavy fragments at ~9 MeV energy.

Figure 3.1 shows the stopping power of water toward some typical charged particle radiations. For others, similar curves can be drawn using the procedure discussed in the previous sections; the task will be greatly simplified remembering that the electronic stopping power, including charge exchange, depends only on the velocity and not on the mass of the incident particle. At extremely high energies, elementary particles, such as protons, pions, and so forth, will have strong nuclear interactions producing high local energy loss. This effect produces a large amount of straggling; however, the stopping along the track between such interactions is given quite well by the ordinary stopping power theory.

3.2   FATE OF DEPOSITED ENERGY: IONIZATION, DISSOCIATION, TRANSFER, AND LUMINESCENCE

The stopping power gives the energy loss rate of the incident particle; this is essentially in the domain of radiation physics. The deposited energy appears in the molecules of the medium, at first in electronic form. The transformation and utilization of this energy in forming new products and destroying old ones are the concerns of radiation chemistry. In this section, only a brief qualitative description will be given. Some details will be presented later (see Chapter 4).

On energy absorption from the impinging particle, a molecule is raised to one of its spin-allowed (singlet) excited states. If the excitation energy exceeds the ionization potential, the minimum energy required for electron ejection, ionization may occur. In principle, spin-forbidden (triplet) states may be excited by the impact of slow electrons. Although a shower of slow electrons is expected for any incident high-energy radiation, Stein (1967, 1968) concluded that the participation of triplet states would be ineffective in liquid state radiolysis. This does not necessarily mean that such states are not produced; it implies that, if produced, these eventually decay to the ground state in a nonradiative process that is also chemically not significant. This approximation is recognized in almost all theories of aqueous radiation chemistry—namely, the primary species are assumed to be formed from singlet (excited or ionized) states.

Even if the energy absorbed by a molecule is greater than its ionization potential, prompt ionization may not occur; instead, the molecule is raised to a neutral superexcited state due to the excitation of a more strongly bound (inner) electron. The ultimate fate of a superexcited state is competitive between delayed ionization and neutral dissociation; in the latter case, the excess energy is invested as the kinetic energy of the dissociating fragments. Platzman (1962) has stressed the importance of superexcited states for isolated molecules. In view of the relatively large ionization yield in liquid water, it seems that superexcited states are not produced in any abundance in that medium. However, the situation may be different in organic liquids.

The nature and the energy of the positive ion produced in an ionization are also of some consequence. In the case of H_2O, mass spectroscopic evidence indicates that the main positive ions are H_2O^+ (~80%) and OH^+ (~20%), with minor contributions from H^+ and O^+. There are reasons to believe that in the liquid phase all these ions end up as H_3O^+ via different ion-molecule reactions. This, then, is the positive ion species to be considered in aqueous radiolysis. Based on fragmentary oscillator strength distribution in the water molecule, Platzman (1967) concluded that the positive ion produced in the interaction of high-energy charged particles would have an average excitation energy of ~8 eV. Later, the less approximate analysis of Pimblott and Mozumder (1991) placed that value to be ~4 eV.

Molecular excited states are either dissociative or may undergo various nondissociative transformations including radiative, nonradiative, and energy transfer processes. The excited states of water are highly dissociative. The well-known opacity of water in all phases starting at wavelengths shorter than ~180 nm is due to dissociation into the ground states of H and OH. Dissociation into other modes, requiring somewhat higher energy, may result in H + OH* (excited state) or H + H + OH, and so forth, whereas dissociation into H_2 and O can occur throughout the excitation spectrum. In most theoretical models of liquid water radiolysis, only neutral dissociation into H + OH and H_2 + O are considered, roughly in the ratio of 4 : 1. In the latter case, the thusly produced singlet oxygen almost invariably undergoes reaction with the nearest water molecule, yielding two OH radicals: $O + H_2O \rightarrow 2OH$.

Excited states of hydrocarbon molecules often undergo nondissociative transformation, although dissociative transformation is not unknown. In the liquid phase, these excited states are either formed directly or, more often, indirectly by electron-ion or ion-ion recombination. In the latter case, the ultimate fate (e.g., light emission) will be delayed, which offers an experimental window for discrimination. A similar situation exists in liquid argon (and probably other liquefied rare gases), where it has been estimated that ~20% of the excitons obtained under high-energy irradiation are formed directly and the rest by recombination (Kubota et al., 1976).

Excited states of a typical organic molecule can undergo various radiative and nonradiative transformations as follows:

1. Internal conversion—that is, without change of multiplicity—to the lowest singlet, or to the ground state ($S_n \rightarrow S_1$ or S_0)
2. Fluorescence, including delayed fluorescence ($S_1 \rightarrow S_0 + h\nu$)
3. Intersystem crossing—that is, with a change in multiplicity ($S_1 \rightarrow T_1$ or $T_1 \rightarrow S_0$)
4. Phosphorescence ($T_1 \rightarrow S_0 + h\nu$)

For the fate of the excited states in condensed media, we must add to this list energy transfer processes. These are broadly classified as radiative (or "trivial"), coulombic (mainly dipole-dipole interaction), or electron-exchange processes.

In the radiative process, a real photon is emitted by the donor molecule and the same is absorbed by the acceptor molecule. The effectiveness of this process depends on, among other factors, the degree of overlap between the emission spectrum of the donor and the absorption spectrum of the acceptor.

For the coulombic interaction, the dipole-dipole term usually dominates over dipole-quadrupole and higher-order terms. Forster's (1959, 1960) theory of energy transfer mediated by dipole-dipole interaction gives a rate varying inversely as the sixth power of donor (D)-acceptor (A) distance. This transfer extends over relatively long distances and is in competition with deactivation of D^* (internal conversion and/or intersystem crossing). The distance at which the energy transfer rate equals the deactivation rate is called the critical transfer distance (R_0). For various aromatic pairs of (D^*-A) in benzene solution as examples, R_0 is found to lie in the range ~3.0–4.5 nm. Of course, the transfer processes allowed by this mechanism are those in which the individual spins are conserved, namely, $^1D^* + {}^1A \longrightarrow {}^1D + {}^1A^*$ and $^1D^* + {}^3A \longrightarrow {}^1D + {}^3A^*(T_n)$.

In contrast to the dipole–dipole interaction, the electron-exchange interaction is short ranged; its rate decreases exponentially with the donor-acceptor distance (Dexter, 1953). This is expected since, for the electron exchange between D^* and A, respective orbital overlap would be needed. If the energy transfer is envisaged via an intermediate collision complex or an exciplex, $D^* + A \longrightarrow (D - -A) \longrightarrow D + A^*$, then Wigner's rule applies: there must be a *spin common factor* between the various combinations of spins formed before and after the transfer, which then is the spin of the intermediate complex. Thus, both the triplet–triplet transfer, $^3D^* + {}^1A \longrightarrow {}^1D + {}^3A^*$, and the singlet-singlet transfer, $^1D^* + {}^1A \longrightarrow {}^1D + {}^1A^*$, are collisionally allowed. While the triplet-triplet transfer is allowed under exchange, it is (doubly) forbidden under dipole-dipole interaction; however the singlet-singlet transfer is allowed by both interactions.

Energy transfer effects are important in radiation chemistry and in photochemistry. Early experiments by Franck and Cario (1922) and by West and Paul (1932) showed respectively the energy transfer from an excited state of Hg to a hydrogen molecule and from an excited state of benzene to an alkyl iodide molecule in solution, resulting in chemical change in the acceptor in both cases. Later, the extensive researches of Manion and Burton (1952) demonstrated the protective action of benzene against cyclohexane decomposition in benzene solutions. Apparently, the decomposition yield decreased in benzene solution as a result of energy transfer, the transferred energy being mostly degraded to heat. The discussion in this section has been intended to give some general idea of the various processes that are involved with track effects in radiation chemistry.

3.3   DISTRIBUTION OF ENERGY ON A MOLECULAR TIME SCALE

The molecular time scale may be taken to start at ~10^{-14} s following energy absorption (see Sect. 2.2.3). At this time, H atoms begin to vibrate and most OH in water radiolysis is formed through the ion–molecule reaction $H_2O^+ + H_2O \rightarrow H_3O^+ + OH$. Dissociation of excited and superexcited states, including delayed ionization, also should occur in this time scale. The subexcitation electron has not yet thermalized, but it should have established a quasi-stationary spectrum; its mean energy is expected to be around a few tenths of an eV.

This time scale extends to ~10^{-13} s or a little longer, during which the intermediate species formed in the radiolysis of water are e^- (epithermal), H, OH, H_3O, and perhaps some undissociated excited and superexcited states of H_2O. None of these species are expected to be fully thermalized. A significant amount of the transferred energy is expected to be found in molecular motions. Vibrations are excited not by momentum transfer, but partly via the Franck-Condon effect because most electronic excitations change the vibrational state of the molecule. Vibrations may also be excited via temporary negative-ion formation by the impact of slow electrons (see, e.g., Herzenberg and Mandl, 1962). According to one estimate (Magee and Mozumder, 1973) the deposited energy at this time scale is distributed, for gas phase water radiolysis, among the various components as follows: (1) positive ions, including ionization energy and excitation energy of the positive ions (71.7%); (2) directly excited states (4.9%); (3) superexcited states, excluding those that autoionize (13.8%); (4) vibrationally excited states (8.9%); and (5) kinetic energy of epithermal electrons (0.7%).

The foregoing analysis was for gas-phase water radiolysis. Similar estimates may be made for the condensed phases and for other media when the relevant yields and energetics become progressively available. On the whole, these species will gradually thermalize and become available for track reactions. Such reactions are greatly influenced by track structure, which is taken up in the following section.

3.4   CHARGED PARTICLE TRACKS IN LIQUIDS

3.4.1   GENERAL DESCRIPTION

In a formal sense, a track is generated by correlated energy loss events along the direction of the momentum of the penetrating particle. Figure 3.2 shows the formation of a track according to Mott in the second-order perturbation theory

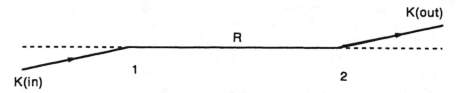

FIGURE 3.2 Track formation according to Mott (1930). Simultaneous excitation of atoms at 1 and 2 has negligible probability in second-order perturbation theory unless the interatomic separation vector $\bar{R}$ is well aligned with the incoming and outgoing momentum vectors of the incident particle. Reproduced from Mozumder (1969), by permission of John Wiley & Sons, Inc.©

(Darwin, 1929; Mott, 1930). The incident particle has momentum $\hbar K_0$ before any interaction; its momentum after exciting atoms 1 and 2 respectively into the nth and mth states is represented by $\hbar K_{nm}$. Mott showed that the entire process has negligible cross section unless the angular divergences are comparable to or less than $(K_0 a)^{-1}$, where a denotes the atomic size. As Darwin (1929) correctly conjectured, the wavefunction of the system before any interaction is the uncoupled product of the wavefunctions of the atom and of the incident particle. After the first interaction, "*these wavefunctions get inextricably mixed and each subsequent interaction makes it worse.*" Also, according to the Ehrenfest principle, the wavefunction of the incident particle is localized to atomic dimensions after the *first interaction*; therefore, the subsequent process is adequately described in the particle picture.

Charged particle tracks in liquids are *formally* similar to cloud chamber or bubble chamber tracks. In detail, there are great differences in track lifetime and observability. Tracks in the radiation chemistry of condensed media are extremely short-lived and are not amenable to direct observation. Also, it must be remembered that in the cloud or bubble chamber, the track is actually seen at a time that is many orders of magnitude longer than the formation time of the track. The manifestation occurs through processes extraneous to track formation, such as condensation, formation of bubbles, and so forth. In a real sense, therefore, charged particle tracks in radiation chemistry are metaphysical constructs.

3.4.2 PHENOMENA REQUIRING A TRACK MODEL

There is a large body of known radiation-physical and radiation-chemical phenomena, the existence or explanation of which requires a track model. With the exception of the consideration of electron escape (free-ion yield), these phenomena have been summarized by Fano (1963) in an excellent review. Mozumder (1969) has also reviewed them from the viewpoint of radiation chemistry.

The areas where the use of the track model has been found particularly expedient are (1) LET variation of product yields in the radiation chemistry of liquids; (2) the yield of escaped ions and its variation with particle LET; (3) energy loss in primary excitations and ionizations; (4) radiation-induced luminescence; and (5) particle identification.

3.4.3   TRACK EFFECTS IN RADIATION CHEMISTRY

In radiation chemistry, the track effect is synonymous with LET variation of product yield. Usually, the product measured is a new molecule or a quasi-stable radical, but it can also be an electron that has escaped recombination or a photon emitted in a luminescent process. Here LET implies, by convention, the initial LET, although the actual LET varies along the particle track; also, the secondary electrons frequently represent regions of heterogeneous LET against the background of the main particle.

In theory, one assumes the formation of radicals before the chemical stage begins (see Sect. 2.2.3). These radicals interact with each other to give molecular products, or they may diffuse away to be picked up by a scavenger in a homogeneous reaction to give radical yields. The overlap of the reactive radicals is more on the track of a high-LET particle. Therefore, the molecular yields should increase and the radical yields should decrease with LET. This trend is often observed, and it lends support to the diffusion-kinetic model of radiation-chemical reactions.

The quantitative aspects of track reactions are involved; some details will be presented in Chapter 7. The LET effect is known for H_2 and H_2O_2 yields in aqueous radiation chemistry. The yields of secondary reactions that depend on either the molecular or the radical yield are affected similarly. Thus, the yield of Fe^{3+} ion in the Fricke dosimeter system and the initiation yield of radiation-induced polymerization decrease with LET. Numerous examples of LET effects are known in radiation chemistry (Allen, 1961; Falconer and Burton, 1963; Burns and Barker, 1965) and in radiation biology (Lamerton, 1963).

3.4.4   STRUCTURE OF TRACKS

By track structure is meant the distribution of energy loss events and their geometrical dispositions. Naturally, track structure becomes rather important for second-order reactions in the condensed phase. Track structure, coupled with a reaction scheme and yields of primary species, forms the basis of radiation-chemical theory.

Mozumder (1969) arbitrarily classified ionizing radiations as of low, intermediate, or high LET depending on the LET value falling below 10^{-1}, lying between 10^{-1} and 10, or exceeding 10 eV/Å, respectively. Electrons of energy greater than ~30 KeV constitute the majority of low-LET radiations in water. Since secondary electrons always carry a significant fraction of energy for any ionizing radiation, these constitute regions of LET usually different from that generated by the main particle. Thus, against a background of overall low-LET radiation, there may be regions of higher LET, and vice versa.

3.4.4a Low LET

There is virtually no oscillator strength for electronic transition in the water molecule below about 6.5 eV (see Figure 2.7), while the mean excitation potential is about 65 eV. On the other hand, the LET of fast (~1 MeV) electrons in water is ~0.02 eV/Å. This means that the energy loss events are, on the average, spaced by ~3000 Å or more. This idea gives rise to the spur theory for low-LET track structure (Samuel and Magee, 1953), according to which the track is to be viewed as a random succession of *spurs,* or localized energy loss events. If the spurs are taken as spherical beads, then the track would look like a *string of beads.* Even the adjacent spurs are so far apart that there is practically no overlap of the reactants between them. Thus, the yields are calculated simply on the basis of isolated spurs. Samuel and Magee calculated the ratio of forward (or molecular) to radical yields in water radiolysis using a gaussian distribution of reactants and a distribution of spur size in energy derived from cloud chamber data (Wilson, 1923). Being a one-radical model, it could not distinguish among the different molecular products formed, and Samuel and Magee were obliged to multiply their forward yield by an assumed factor to eliminate the back formation of water.

Ganguly and Magee (1956) retained the idea of a string of beads and extended it to include (1) LET variation along the track, (2) overlap between spurs, and (3) a scavenger reaction in competition with recombination. However, they ignored the spur size distribution in energy. Both these models use one-radical, prescribed diffusion treatment—that is, the radical distribution is assumed to be *always* gaussian. The importance of these models lies not so much in predicting quantitative yields, but in establishing the spur picture in water radiolysis. Later, Schwarz (1969) made the Ganguly–Magee model quantitative by introducing multiradical treatment and using realistic rate and diffusion coefficients.

Nearly all computations of radiation-chemical yields use either diffusion kinetics (see, e.g., Schwarz, 1969) or stochastic kinetics (Zaider *et al.,* 1983; Clifford *et al.,* 1987; Pimblott, 1988; Paretzke *et al.,* 1991; Pimblott *et al.,* 1991). Diffusion kinetics uses deterministic rate laws and considers the reactions to be (partially) diffusion controlled while the reactants are also diffusing away. Stochastic kinetics considers random flights of reactants, and the reactions (often taken pairwise) occur probabilistically subject to overall stationary rates. Clifford *et al.* (1982, 1986) have shown that significant differences exist between the results of stochastic and diffusion kinetics, these differences being greatest with the smallest number of reactants. Track structure is a necessary ingredient in the calculation of radiation-chemical yields by either diffusion or stochastic kinetics. In this section, we will concern ourselves only with geometric and energetic aspects of track structure; other topics will be taken up in Chapter 7.

Mozumder and Magee (1966a, b) classified energy deposition in water by fast electrons, including secondary electrons of all generations. They also addressed the problem of constructing an energy-loss cross section from the oscillator distribution. According to them, there are three categories of energy loss with little overlap at low or intermediate LET. These are (1) spurs (up to 100 eV), (2) blobs (100–500 eV), and (3) short tracks (500–5000 eV). In principle, spurs originate from glancing collisions in which any secondary electron has little kinetic energy and is quickly slowed to subexcitation energies. Blobs and short tracks are produced when the secondary electron has sufficient energy to ionize, but these events are close enough that they cannot be considered isolated. Short tracks are considered approximately cylindrical in view of the relatively high energy of the progenitor electron, whereas blobs are approximately spheroidal or spherical like spurs. A secondary electron with energy above 5 KeV (a branch track) produces ionizations that are sufficiently far apart not to overlap; it creates spurs, blobs, and short tracks mimicking the primary particle. The demarcations between spurs, blobs, and short tracks are a matter of convenience; in reality, there is only a continuous distribution of energy loss events. However, the classification is useful for applying spatial attributes to the energy loss processes that greatly facilitate diffusion-kinetic or stochastic-kinetic calculations.

Figures 3.3 and 3.4 show sections of low-LET tracks chosen to show a blob and a short track, respectively. Numerically, spurs dominate over blobs and short tracks. On the other hand, the fraction of energy held up in the extra-spur entities is significant, and in a real sense these represent LET effect in electron tracks.

Mozumder and Magee (1966a, b) obtained the fractions of energy deposited in water by constructing the life history of electron tracks via a Monte Carlo procedure using a combination of synthesized cross sections and a random number generator to give the locations and amounts of energy losses. The energy partition itself can be obtained through integral equations if such cross sections are simple analytic functions. Additionally, the Monte Carlo method gives information regarding the distribution of energy among the spurs. Mozumder and Magee found the division of energy among spurs, blobs, and

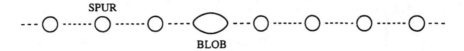

FIGURE 3.3   Section of a low-LET track selected to show a blob. On high-energy electron tracks, spurs outnumber blobs by about 50 : 1.

short tracks for a 1-MeV electron in water to be 0.67 : 0.11 : 0.22 and that for
^{60}Co-γ radiation to be 0.64 : 0.12 : 0.24. Their procedure was rudimentary:
Monte Carlo simulation was limited; the cross sections were derived from an
unrealistic oscillator distribution, required only to obey a selected well-known
sum rule; and elastic collisions were entirely neglected.

Later, these shortcomings were remedied following the procedure of Sects.
2.5.2 and 2.5.3 (Pimblott *et al.*, 1990). Finally, elastic collisions were also
included (Pimblott *et al.*, 1996), which showed that (1) the trajectory of even
an energetic electron is not a straight line; (2) lower-energy electrons suffer significant deviation from the initial trajectory such that the path length can exceed
axial or radial penetration by as much as a factor ~3.0; and (3) the CSDA (continuous slowing down approximation) range can be taken as the path length for
electron energies above 2 KeV. According to Pimblott *et al.*(1990), the energy
division among spurs, blobs, and short tracks for a 1-MeV electron in liquid
water is 0.75 : 0.12 : 0.13, and those in gaseous water and ice are respectively
0.76 : 0.10 : 0.14 and 0.69 : 0.16 : 0.15. Figure 3.5 shows Pimblott *et al.*'s (1990)
calculation of the energy partition among the track entities as a function of the
primary electron energy. Figures 3.6 and 3.7, also from Pimblott *et al.* (1990),
show respectively the spur histograms in number and energy arising out of the
total absorption of a 1-MeV electron in water.

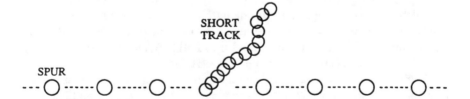

FIGURE 3.4   Section of a low-LET track selected to show a short track. On high-energy electron tracks, spurs outnumber short tracks by about 500 : 1.

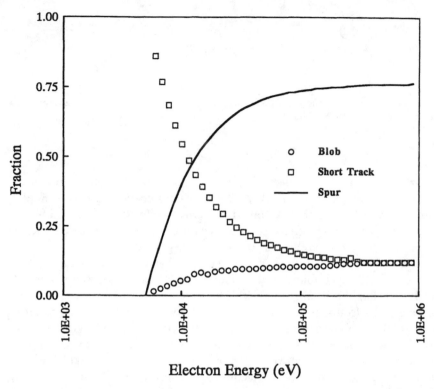

FIGURE 3.5   Energy partition among the track entities (spurs, blobs, and short tracks) as a function of electron energy in liquid water. Data from Pimblott *et al.* (1990), with permission of Am. Chem. Soc.©

3.4.4b   Intermediate LET

Both light and heavy particles can have intermediate LET in water. Electrons of energy less than about 30 KeV belong to this group—for example, β-decay processes in ^{3}H and ^{37}Ar. Of the heavy particles in this group, the main contributors are protons of energy less than about 70 MeV and most α-particles, although some alphas may have high LET. Stripped nuclei of Li through C present marginal cases. They have intermediate LET if their energy is greater than about 10 MeV/amu. For lower energies, they progressively run into the high-LET regime. Highly accelerated (several hundred MeV/amu) heavy ions of high nuclear charge also can have intermediate LET in water.

Tritiated water in solution has been used in radiolysis; scavenger studies are rare, but a few are known (Appleby and Gagnon, 1971; Lemaire and Ferradini, 1972). Electrons from the β-decay of tritium have a broad spectrum between 0 and 18 KeV, with a peak at 5.5 KeV. Over this distribution, the energy is partitioned between spurs, blobs, and short tracks as 0.2 5 : 0.08 : 0.67, which

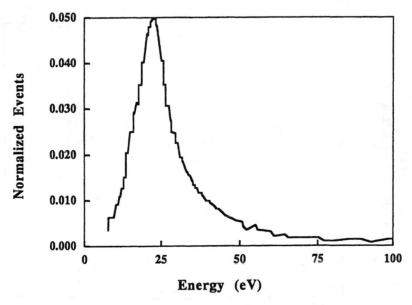

FIGURE 3.6 Spur histogram in normalized number of events for liquid water. The histograms are relatively insensitive to incident energy over the interval 10 KeV to 1 MeV. From Pimblott *et al.* (1990), with permission of Am. Chem. Soc.©

corresponds to a mean electron energy of 7.5 KeV (Pimblott *et al.*, 1990). Thus, for most purposes, the tritium-β radiation should have the attributes of a short track (but see Samuel and Magee, 1953).

Results of the scavenging studies with tritium are somewhat confusing. Appleby and Gagnon (1971) found H_2 to be less scavengable under tritium than under ^{60}Co-γ radiation using $CuSO_4$ as a scavenger. The opposite was found to be true by Lemaire and Ferradini (1972) for the scavenging of H_2O_2 using halides. Both groups of workers attributed their observations to the structure of tritium tracks! Tritium tracks are very short, being only 8.9 and 0.9×10^{-3} g/cm^2, respectively, at maximum and median energies (see Table 2 in Pimblott *et al.*, 1996). Thus, end effects may not be negligible. There is a tendency among practicing radiation chemists to treat tritium radiation in terms of the spur theory, which is not justified.

Hummel *et al.* (1966) have used radiations from ^{37}Ar to determine the free-ion yield in *n*-hexane (see Sect. 9.3.1), but no molecular product has yet been measured with this radiation, which is highly desirable in view of its mono-energetic (2400 eV) character. Mozumder (1971) has developed a diffusion theory for ion recombination for (initially) multiple ion-pair cases, which can be applied to ^{3}H and ^{37}Ar radiations. According to this theory, the track is cylindrically symmetric to start with. As neutralization proceeds, the track

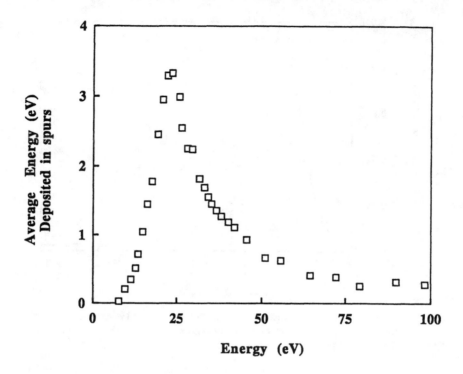

FIGURE 3.7   Spur energy spectrum for liquid water. Ordinate represents (from the data of Figure 3.6) average energy deposited in spurs of energy between adjacent steps, where the number of spurs is normalized to unity. Average energy of all spurs is the sum of the ordinates (in this case, 43.4 eV). From Pimblott *et al.* (1990), with permission of Am. Chem. Soc.©

degenerates at a later stage into a collection of spurs, out of which electrons can escape neutralization. In this connection, it should be remembered that no charge can escape from a truly cylindrical field; therefore, such a dichotomy of track structure is indicated.

Tracks of α-particles and MeV protons are long and cylindrical. Samuel and Magee (1953) found, however, that no unequivocal answer could be obtained for the probability of molecular yield formation since, in a truly cylindrical geometry, no radical can escape recombination in the limit $t \rightarrow \infty$. So they carried their calculation only up to such times as is required for adjacent tracks to overlap by diffusion. According to this procedure, the track develops in two stages; the first stage is short but accounts for most of the recombination. Such a procedure is dose-rate dependent, although weakly and this is meaningful only when no scavenger is present. In the presence of a scavenger, there is no ambiguity in the yield calculations, as shown by Ganguly and Magee (1956) and by Kuppermann and Belford (1962).

No serious attention is generally paid to the secondary electrons produced on proton and α-particle tracks. It happens that in most such cases the mean LET of the secondary electrons is about the same as that of the main particle (Mozumder, 1969). Therefore, if track reactions depend on LET only (which is an oversimplification), then there is a built-in safety feature for these models. On the other hand, this is not true for low- or high-LET tracks, where there is a real difference between LETs of secondary electrons and the main particle. Figure 3.8 shows the picture of an MeV-proton track in water together with the secondary electrons. The energies, angles of ejection and ranges of secondary electrons, shown parenthetically, are obtained via a Monte Carlo procedure using approximate cross sections. The track diameter is obtained from an estimated range of 100-eV electrons in water (*vide infra*). The scattering of the secondary electrons has not been shown.

It is clear from the present discussion that LET alone does not completely determine track structure. Track length (or particle velocity) is also a contributing factor. For example, at the same intermediate LET, high-energy heavy-ion tracks would be cylindrical while those for the KeV electrons would be better described as either spheroidal or as partially overlapped spurs (Samuel and Magee, 1953).

3.4.4c High LET

High-LET radiations in water are represented by fission fragments and atomic nuclei of relatively low Z. The subject is of interest to radiation chemistry, but even more to radiation biology. Earlier high-LET work was limited to (except for fission fragments) $Z = 10$ and incident energies up to ~10 MeV/amu. Fission fragments were employed for radiolysis by Boyle *et al.* (1955), Sowden (1959), and Boyd (1963). Accelerated or otherwise energetic atomic nuclei were used by (1) Schuler and Barr (1956), for ferric ion yield measurement with $^{10}B(n, \alpha)^7Li$ radiation;

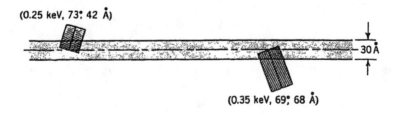

(0.25 keV, 73°, 42 Å)

30 Å

(0.35 keV, 69°, 68 Å)

FIGURE 3.8 A typical MeV proton track in water. See text for details. Reproduced from Mozumder (1969), by permission of John Wiley & Sons, Inc.©

(2) Schuler (1967), for the same with C ions; (3) Burns and Reed (1970), for H_2 and other product yields in cyclohexane with C ions, Ne ions, and other radiations; and (4) Imamura *et al.* (1970) using C and N ions. Subsequently, high-LET radiolysis has been extended by various groups using a variety of heavy ions with energies up to several hundred MeV/amu (see, e.g., LaVerne and Schuler, 1983, 1984, 1986, 1987a; Appleby *et al.*, 1985). LaVerne (1988) has given an extended bibliography of studies with heavy-particle radiolysis including theoretical research. The experimental work is concerned with different yield measurements—for example, H_2 yield in benzene, H_2O decomposition yield, and the yield of Fe^{3+} in the Fricke dosimeter system. The subject has also been briefly reviewed by LaVerne and Schuler (1987b) and by Appleby (1987).

Charge exchange is important all along the high-LET tracks. The effective ionic charge is determined by cross sections of electron capture and loss, which depend predominantly on the ionic velocity. Electron loss may be simply described by an ionization of the incident ion in its own reference frame due to the impact of medium electrons and nuclei. Following Bohr (1948), Mozumder *et al.* (1968) wrote the cross section for this process as[1]

$$\sigma_l = \frac{2\pi e^4 (Z^2 + Z)}{mv^2} \left(\frac{1}{I_S} - \frac{1}{2mv^2} \right), \tag{3.4}$$

where I_s is the binding energy of the (least bound) electron in the ion and $2mv^2$ is the maximum energy transfer.

In Eq. (3.4), the contributions of both the nuclei and the electrons are considered. The maximum energy transfer for electrons is $(1/4)mv^2$, but since their contribution to the loss cross section is small, no great error is committed in taking the maximum energy transfer as $2mv^2$ for all ionizations represented in Eq. (3.4). Electron capture is a three-body process best visualized as ionization of a molecule of the medium with the ejected electron having a speed v at least equal to that of the incident ion, followed by capture of that electron in an orbit around the impinging ion. Mozumder *et al.* (1968) modified an earlier formula of Bohr (1948) and wrote the capture cross section as

$$\sigma_c = \frac{3\pi Z a_0^2 z^5 (v_0/v)^7}{n^3}, \tag{3.5}$$

where n is the principal quantum number of the orbital in which the capture takes place.

From the high inverse power dependence of σ_c on v as seen from Eq. (3.5), it is clear that capture probability increases very rapidly with slowing down. The equilibrium ionic charge can be estimated at a given velocity from Eqs. (3.4) and (3.5). Since charge exchange is a nonequilibrium phenomenon,

---

[1]In this chapter, ze denotes ionic charge and Z the number of electrons in a molecule of the medium.

Mozumder *et al.* (1968) instead chose to develop the tracks to the end by a Monte Carlo procedure to obtain the range and range extension (due to electron capture, etc.). Numerical differentiation of the range gave them stopping powers, from which they obtained the effective ionic charge by comparison with the Bethe formula. In any case, it turns out that ions retain their full charge to a high degree when their energy is ≥ 5 MeV/amu and excessive charge exchange takes place for energies below 1 MeV/amu. This statement is not absolute but depends somewhat on the nuclear charge. At the same energy per amu, charge transfer is more important for higher-z particles. The phenomenon of charge exchange itself contributes to energy loss; in most cases, however, this is negligible. The main effect of charge exchanges is to alter the stopping power through the ionic charge.

Note that the charge-exchange cross sections given here are rather approximate. In some special cases, better approximations may be available; however, modifications of Bohr's formulas were considered adequate for use over a large range of particle charge and energy.

The high-LET track model of Mozumder *et al.* (1968) consists of a cylindrical *core* surrounded by a *penumbra*.[2] The core is generated by continuous excitations and ionizations by the primary particle, mainly in glancing collisions; some of the secondary electrons will deposit some or all of their energy in the core. This is the basic high-LET region. Some of the secondary electrons, having sufficient energy, penetrate the core and deposit their energy outside. The LET of these electrons, roughly speaking, is about an order of magnitude less than that of the main particle. They constitute the penumbra, which is then the region of lower LET in an overall high-LET situation.

Figures 3.9 and 3.10 show the tracks of an oxygen ion and a fission fragment in water, respectively. On the oxygen ion track, the production of secondary electrons is seen to be sporadic. Their energy and point of origin are obtained from a Monte Carlo procedure using estimated cross sections. If the initial energy of the secondary electron is ε, then the ejection angle is given by $\cos^{-1}[(\varepsilon/2mv^2)^{1/2}]$. The electron range is obtained by integrating the stopping power as given in Sects. 2.5.2 and 2.5.3. The penumbra has an overall cylindrical symmetry, although the appearance of the secondary electrons is stochastic in nature. In the case of fission fragments, the secondary electrons are so dense that the penumbra also appears continuous and cylindrically symmetric. The radius of the envelope is estimated from the mean range of the emergent electrons.

The earlier evaluation of the core radius was in terms of Bohr's impulse condition (see Sect. 2.3.3) at (relatively) high energies. This gives the core radius as 30 Å at a particle energy of 10 MeV/amu. For much lower energies, this relation is unrealistic, since electrons ejected in glancing collisions penetrate

---

[2]Originally this was called an envelope; the term was later changed to penumbra in conformity with later work of others.

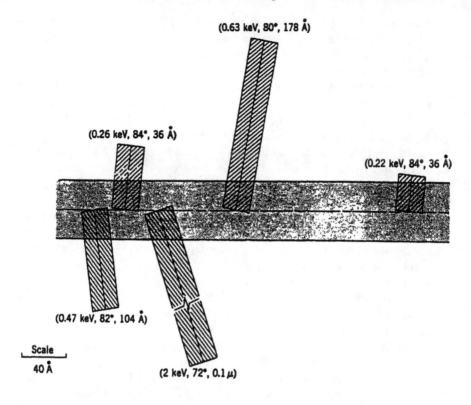

FIGURE 3.9 A typical O^{8+} ion in water. Secondary electron tracks (without scattering) are shaded; the core region is dotted. Figures in parentheses denote ejected electron energy, classical ejection angle, and estimated range (qualitative). Reproduced from Mozumder (1969), by permission of John Wiley & Sons, Inc.©

further. A minimum core radius was then taken to be equal to the range of a 100-eV electron in water, estimated as 15 Å (Mozumder *et al.*, 1968).

Later, a more satisfactory definition was given by Mozumder and LaVerne (1987). This is explained for a 1-MeV C ion track in water as shown in Figure 3.11. *A* is the earliest radial distribution of deposited energy in electronic form, consistent with particle LET, obtained from the impact parameter analysis of Bohr (1913, 1915), which, however, uses the realistic oscillator distribution of liquid water (Heller *et al.*, 1974). Secondary electrons generated from ionization events redistribute their kinetic energy; this redistribution is considered uniform within a volume determined by the maximum radial penetration of the electron starting at the classical ejection angle. Scattering and inhomogeneous energy distribution are ignored in this simplified treatment.

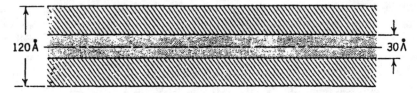

FIGURE 3.10   A typical fission fragment track in water, showing dense overlap of secondary electrons in the penumbra region (shaded). Core is shown dotted. Reproduced from Mozumder (1969), by permission of John Wiley & Sons, Inc.©

The redistributed secondary electron energy is given by C, and $B = A - C$ represents losses to nonionizing events. B is further redistributed over the diameter of a water molecule to give D. The resultant distribution at the end of the physical stage (see Sect. 2.2.3) is $C + D$, which compares with the Rutherford distribution over the important distance scale ~10 nm. *Notice that here the energy deposition is given by eV/molecule, rather than per unit volume, to emphasize energy deposition in molecules.*

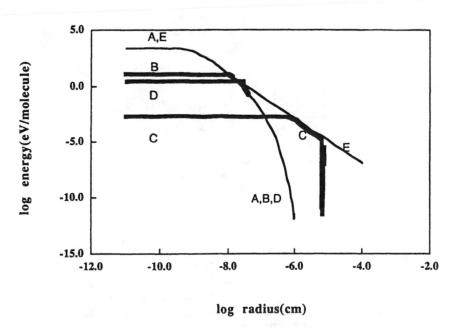

log radius(cm)

FIGURE 3.11   Core definition. After redistribution of energy at the molecular scale, core size is suggested by the distance at which energy transported by secondary electrons just exceeds that due to every other channel (see text for details). From Mozumder and LaVerne (1987).

The *physical core size* now suggests itself as the distance at which the transported energy deposition by secondary electrons just exceeds those due to every other channel. In the present example, this occurs at 1.4 nm. The chemical stage is signaled by reactive species in the overlapping spurs. Clearly, the minimum core radius is given by that of the electron spur, taken to be 2.5 nm (Schwarz, 1969). Bohr's adiabatic criterion (Sect. 2.3.3) gives an impact parameter $b = 2.5$ nm at an ion energy of 5 MeV/amu. At lower ion energies, the core radius remains fixed at this value, provided there is sufficient spur overlap to make a continuous core (see Figure 3.12). Considering an average energy loss ~60 eV (Green *et al.*, 1988), the minimum required LET would be 12 eV/nm. In water, this occurs for p, α, and C ions at 3.7, 17, and 420 MeV/amu, respectively. Heavy ions of higher energy will not form a continuous core.

Magee and Chatterjee (1980) give a different criterion for the size of a chemical core depending on a scavenger reaction in competition with radical recombination. Its nature, however, is extraneous to the physical track structure.

To calculate the energy partition between the core and the envelope, Mozumder *et al.* (1968) considered the equipartition of deposited energy between glancing and knock-on collisions (Sect. 2.3.4). Of the ejected electrons

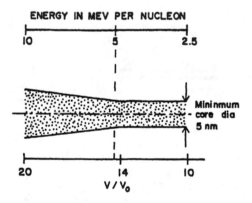

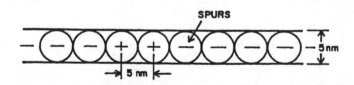

FIGURE 3.12   Core size as a function of energy, and the criterion for a continuous core. See text for explanation. From Mozumder and LaVerne (1987).

produced in knock-on encounters, some are totally stopped inside the core, whereas others penetrate it after depositing some energy in the core. Electrons of the first group are ejected almost normal to the track axis, because their energy is small compared with $2mv^2$.

With this knowledge and using the Rutherford cross section, the energy deposited in the core by the electrons of the first group is given by

$$\xi = \frac{1}{2} \int_{E_u}^{E_i} dE \, \frac{\ln\left[\rho(Y)/\varepsilon_0\right]}{\ln(4E/\mu\varepsilon_0)}, \tag{3.6}$$

where E_i and E_u denote the energy per amu at the start and at velocity u, respectively. Here ε_0 is the minimum energy of secondary electrons, taken as 100 eV, μ is the proton to electron mass ratio, and $\rho(Y)$ is the energy of the electron that has range equal to the local core radius, Y. Of course, $\xi = 0$ for $E_i \leq E_u$.

Electrons of the second group of energy ε are ejected at an angle $\theta = \cos^{-1}[(\varepsilon/2mv^2)^{1/2}]$, and their energy deposition in the core over the entire track is given by

$$\eta = \frac{1}{2} \int_{E_f}^{E_i} dE \left[\frac{Y}{\ln(4E/\mu\varepsilon_0)} \right] \int_{\rho(Y)}^{4E/\mu} S(\varepsilon) \csc\theta \, \frac{d\varepsilon}{\varepsilon^2}, \tag{3.7}$$

where E_f is the final energy. Its actual value is of no great consequence, but for computational purposes it is taken as 0.25 MeV/amu. The stopping power of the electron of initial energy ε, averaged over its path inside the core, is represented by $S(\varepsilon)$. This quantity and $\rho(Y)$ are obtained from low-energy electron stopping power (Sect. 2.5.2).

Total energy deposition in the core is now obtained by adding together (1) half the initial energy (equipartition principle); (2) ξ and η from Eqs. (3.6) and (3.7); and (3) energy losses in charge-exchange processes, if desired. The result is shown in Figure 3.13 as a percentage of energy deposited in the core versus primary energy. A nearly universal curve is obtained. The difference between the various particles at the same energy/amu is due to charge exchange.

Scattering of secondary electrons has been ignored in this calculation. Later, it was included by Chatterjee et al. (1973) on an approximate diffusional basis for electron energies below about 1600 eV. However, the qualitative features of Figure 3.13 were retained.

At high energies the core expands, resulting in large energy deposition inside the core. At low energies the core contracts, but the contraction is limited (see Figure 3.12); on the other hand, the secondary electrons cannot penetrate the core effectively because of their low energy. Again, therefore, a large percentage of energy is deposited in the core. Thus, one would expect that at an intermediate energy the percentage in the core will be a minimum. Figure 3.13 shows this minimum at an incident energy ~2.5 MeV/amu. The computations of

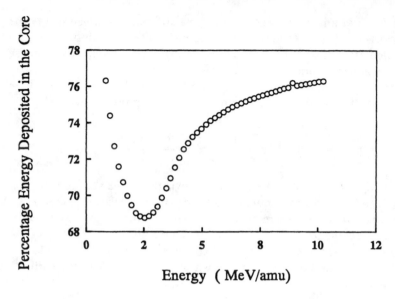

FIGURE 3.13   Percentage of energy deposited in the core of heavy ions as a function of incident energy. Variation with ionic charge is relatively minor for $z = 6–10$. From Mozumder *et al.* (1968), with permission of Am. Chem. Soc.©

Chatterjee *et al.* (1973) indicate a uniform radial energy density in the core; that in the penumbra varies inversely as the square of the radial distance, subject to defined limits and consistent with overall particle LET.

3.4.4d   Two Special Cases

In liquefied rare gases (LRG) the ejected electron has a long thermalization distance, because the subexcitation electrons can only be thermalized by elastic collisions, a very inefficient process predicated by the small mass ratio of the electron to that of the rare gas atom. Thus, even at a minimum of LET (for a ~1-MeV electron), the thermalization distance exceeds the interionization distance on the track, determined by the LET and the W value, by an order of magnitude or more (Mozumder, 1995). Therefore, isolated spurs are never seen in LRG, and even at the minimum LET the track model is better described with a cylindrical symmetry. This matter is of great consequence to the theoretical understanding of free-ion yields in LRG (see Sect. 9.6).

Recently, a biexcitonic quenching mechanism has been proposed to explain the variation of scintillation intensity in liquid argon (LAr) with the LET and quality of incident radiation (Hitachi *et al.*, 1992). According to this, quenching occurs mainly in the track core due to high-energy deposition density. This process is of importance in the operation of LAr calorimeters for high-energy heavy ions. For this purpose, energy partitions between the core and penumbra of Au, La, Fe, and Na ions have been calculated by Mozumder *et al.* (1995) over the energy interval 1–1000 MeV/amu. In the relativistic regime ($\geq$30 MeV/amu), kinematic and polarization corrections are required. A reinterpretation of the continuum electromagnetic theory of Fermi (see Mozumder, 1974) gives the core radius as $\lambda\beta$, where β is the ratio of incident particle speed to that of light and λ = 68 Å for LAr (Mozumder *et al.*, 1995).

For heavy ions at very high LET—that is, at lower energies—this implies multiple ionization in the core that very quickly degenerates into single ionizations by charge sharing with neighboring atoms. Thus, the core quickly expands to a radius $r_m = (S/2\pi NW)^{1/2}$, where S is the LET, N is the number density of Ar atoms, and W is the average energy needed for an ionization.

At still lower energies, Fermi's theory, based on continuum electrodynamics, becomes unrealistic and gives a core radius less than the interatomic distance in LAr. It is then replaced by the Bohr criterion with E_1 = 12 eV, the lowest excitation potential of Ar. The thusly determined core radius increases with particle energy, and in the relativistic regime, it is independent of particle quality (17–60 Å). Below ~30 MeV/amu, that radius decreases also with the atomic number of the incident particle (13–17 Å for Au, 9–17 Å for La, 4.2–17 Å for Fe, and 4.0–17 Å for Na). The energy deposition by the secondary electrons inside the core is estimated from their energy spectrum corrected for relativistic kinematics, their LET (Adams and Hansma, 1980), and their path detour (Bethe *et al.*, 1938). Thus, Mozumder *et al.* (1995) found that beyond ~40 MeV/amu, the percentage of energy within the core is independent of particle quality, reaching an asymptotic limit of 62–63%. Core expansion is clearly seen for Au and La ions below ~25 MeV/amu. Because of this effect and the increased electron LET at lower energies, the percentage of energy within the core increases at lower incident energies for these two ions, reaching ~72% and ~65% at 1 MeV/amu. The effect is not so evident for Fe and Na ions, in which cases the core percentage seems to level off at ~57%. These results have been used to calculate the luminescence quenching factor for relativistic Au ions in an allene-doped LAr calorimeter.

REFERENCES

Adams, A., and Hansma, P. K. (1980), Phys. Rev. B22, 4258.

Allen, A. O. (1961), *The Radiation Chemistry of Water and Aqueous Solutions*, ch. 5. Van Nostrand, Princeton, N.J.

Allison, S. K. (1958), Rev. Mod. Phys. 30, 1137.

Appleby, A. (1987), *Radiation Research*, v. 2, pp. 23–28, Taylor and Francis, London.

Appleby, A., and Gagnon, W. F. (1971), J. Phys. Chem. 75, 601.

Appleby, A., Christman, E. A., and Jayko, M. (1985), Radiat. Res. 104, 263.

Bethe, H. A., and Ashkin, J. (1953), in *Experimental Nuclear Physics*, v. 1. (Segrem, E., ed.), pp. 166–357, Wiley, New York.

Bethe, H. A., Rose, M. E., and Smith, L. P. (1938), Proc. Am. Phil. Soc. *78*, 573.

Bohr, N. (1913), Phil. Mag. *25*, 10.

Bohr, N. (1915), Phil. Mag. *30*, 581.

Bohr, N. (1948), Kgl. Danske Vied. Selskab. Mat-Fys. Medd. *18*, 8.

Boyd, A. W. (1963), J. Nucl. Mater. *9*, 1.

Boyle, J. W., Hochanadel, C. J., Sworski, T. J., Ghormley, J. A., and Kieffer, W. F. (1955), *Proc. 1st Int. Conf. on Peaceful Uses of At. Energy*, v. 7. p. 576, Geneva.

Burns, W. G., and Barker, R. (1965), in *Progress in Reaction Kinetics*, v. 3 (Porter, G., ed.), pp. 305–368, Pergamon, London.

Burns, W. G., and Reed, C. R. V. (1970), Trans. Faraday Soc. *66*, 2159.

Chatterjee, A., Maccabee, H. D.,and Tobias, C. A. (1973), Radiat. Res. *54*, 479.

Clifford, P., Green, N. J. B., and Pilling, M. J. (1982), J. Phys. Chem. *86*, 1318.

Clifford, P., Green, N. J. B., Oldfield, M. J., Pilling, M. J., and Pimblott, S. M. (1986), J. Chem. Soc., Faraday Trans. *1:82*, 2673.

Clifford, P., Green, N. J. B., Pimblott, S. M., and Burns, W. G. (1987), Radiat. Phys. Chem. *30*, 25.

Darwin, C. G. (1929), Proc. Roy. Soc. *A124*, 375.

Dexter, D. L. (1953), J. Chem. Phys. *21*, 836.

Falconer, J. W., and Burton, M. (1963), J. Phys. Chem. *67*, 1743.

Fano, U. (1963), Ann. Rev. Nucl. Sci. *13*, 1.

Fröhlich, H., and Platzman, R. L. (1953), Phys. Rev. *92*, 1152.

Forster, T. (1959), Discuss. Faraday Soc. *27*, 1.

Forster, T. (1960), Radiat. Res. Suppl. *2*, 326.

Franck, J., and Cario, G. (1922), Z. Physik. *11*, 161.

Ganguly, A. K., and Magee, J. L. (1956), J. Chem. Phys. *25*, 129.

Green, N. J. B., LaVerne, J. A., and Mozumder, A. (1988), Radiat. Phys. Chem. *32*, 99.

Heller, J. M., Hamm, R. N., Birkhoff, R. D., and Painter, L. R (1974), J. Chem. Phys *60*, 3483.

Herzenberg, A., and Mandl, F. (1962), Proc. Roy. Soc. *A270*, 48.

Hitachi, A., Doke, T., and Mozumder, A. (1992), Phys. Rev. *B46*, 11463.

Hummel, A., Allen, A. O., and Watson, F. H., Jr. (1966), J. Chem. Phys. *44*, 3431.

Imamura, M., Matsui, M., and Karasawa, T. (1970), Bull. Chem. Soc. Japan *43*, 2745.

Kubota, S., Nakamoto, A., Takahashi, T., Konno, S., Hamada, T., Miyajima, M., Hitachi, A., Shibamura, E., and Doke, T. (1976), Phys. Rev. *B13*, 1649.

Kuppermann, A., and Belford, G. G. (1962), J. Chem. Phys. *36*, 1427.

Lamerton, L. F. (1963), in *Proc. Second Intern. Cong. Radiation Res.* (Harrogate, England), p. 1, Yearbook Medical Publishing, Chicago.

LaVerne, J. A. (1988), *Bibliography of Studies of the Heavy Particle Radiolysis of Liquids and Aqueous Solutions*, Special Report SR-124 of the Notre Dame Radiation Laboratory, Notre Dame, Indiana.

LaVerne, J. A. (1996), Nucl. Instrum. Methods *B107*, 302.

LaVerne, J. A., and Schuler, R. H. (1983), J. Phys. Chem. *87*, 4564.

LaVerne, J. A., and Schuler, R. H. (1984), J. Phys. Chem. *88*, 1200.

LaVerne, J. A., and Schuler, R. H. (1986), J. Phys. Chem. *90*, 5995.

LaVerne, J. A., and Schuler, R. H. (1987a), J. Phys. Chem. *91*, 5770.

LaVerne, J. A., and Schuler, R. H. (1987b), *Radiation Research*, v. 2, pp. 17–22, Taylor and Francis, London.

Lemaire, G., and Ferradini, C. (1972), in *Proc. 3rd Tihany Symp. on Radiation Chemistry*, v. 2 (Dobo, J., and Hedvig, P., eds.), p. 1213, Akademiai Kiado. Budapest.

Magee, J. L., and Mozumder, A. (1973) in *Advances of Radiation Research* (Duplan, J. F., and Chapiro, A., eds.), v. 1, pp. 15–24, Gordon and Breach, New York.

Manion, J. P., and Burton, M. (1952), J. Phys. Chem. *56*, 560.

Mott, N. F. (1930), Proc. Roy. Soc. A126, 79.

Mozumder, A. (1969), Adv. Rad. Chem. 1, 1.

Mozumder, A. (1971), J. Chem. Phys. 55, 3020.

Mozumder, A. (1974), J. Chem Phys. 60, 1145.

Mozumder, A. (1995), Chem. Phys. Lett. 238, 143.

Mozumder, A., and LaVerne, J. A. (1987), in Radiation Research, v. 2 (Fielden, E. M., Fowler, J. F.,Hendry, J. H., and Scott, D., eds.), pp. 11–16, Taylor and Francis, London.

Mozumder, A., and Magee, J. L. (1966a), Radiat. Res. 28, 203.

Mozumder, A., and Magee, J. L. (1966b), J. Chem. Phys. 45, 3332.

Mozumder, A., Chatterjee, A., and Magee, J. L. (1968), Advan. Chem. Series 81, 27.

Mozumder, A., Doke, T., and Takashima, T. (1995), Nucl. Instru. & Meth. A365, 600.

Paretzke, H. G., Turner, J. E., Hamm, R. N., Ritchie, R. H., and Wright, H. A. (1991), Radiat. Res. 127, 121.

Pimblott, S. M. (1988), D. Phil. Thesis, Oxford University, Oxford, England.

Pimblott, S. M., and Mozumder, A. (1991), J. Phys. Chem. 95, 7291.

Pimblott, S. M., LaVerne, J. A., Mozumder, A., and Green, N. J. B. (1990), J. Phys. Chem. 94, 488.

Pimblott, S. M., Pilling, M. J., and Green, N. J. B. (1991), Radiat. Phys. Chem. 37, 337.

Pimblott, S. M., LaVerne, J. A., and Mozumder, A. (1996), J. Phys. Chem. 100, 8595.

Platzman, R. L. (1960), J. Phys. Radium 21, 853.

Platzman, R. L. (1962), Radiat. Res. 17, 419.

Platzman, R. L. (1967), in Radiation Research (Silini, G., ed.), pp. 20–52, North Holland, Amsterdam.

Samuel, A. H., and Magee, J. L. (1953), J. Chem. Phys. 21, 1080.

Schuler, R. H. (1967), J. Phys. Chem. 71, 3712.

Schuler, R. H., and Barr, N. F. (1956), J. Amer. Chem. Soc. 78, 5756.

Schwarz, H. A. (1969), J. Phys. Chem. 73, 1928.

Sowden, R. G. (1959), Trans. Faraday Soc. 55, 2084.

Stein, G. (1967), in The Chemistry of Ionization and Excitation (Johnson, G. R. A., and Scholes, G., eds.), p. 25, Taylor and Francis, London.

Stein, G. (1968), in Radiation Chemistry of Aqueous Systems, p. 83, Wiley, New York.

West, W., and Paul, B. (1932), Trans. Faraday Soc. 28, 688.

Wilson, C. T. R. (1923), Proc. Roy. Soc A104, 192

Zaider, M., Brenner, D. J., and Wilson, W. E. (1983), Radiat. Res. 95, 231.

CHAPTER 4

Ionization and Excitation Phenomena

4.1 GENERAL FEATURES

Excitation and ionization have a common origin–namely, raising the electronic level of an atom or a molecule from its ground state to a state of higher energy via the impact of charged particles or photons. Nevertheless, their chemical fates can be drastically different. In this chapter, we treat these phenomena descriptively.

Ionization normally means the removal of an electron from an atom or a molecule. The capture of an electron by a neutral entity may or may not result in a stable negative ion. When it does, the process is called an attachment. The inverse process—that is, the removal of an electron from a negative ion—should, in principle, be called detachment. However, chemists often also call this ionization.

4.1.1 IONIZATION AND APPEARANCE POTENTIALS

The minimum energy I required for ionization is called the (first) *ionization potential*. The W value is the average energy required to produce a pair of ions in the medium. Experimentally, this value is obtained by dividing the absorbed dose by the total number of collected ions. The W value depends primarily on the molecular nature of the medium and secondarily on the quality of the ionizing radiation. Normally, W greatly exceeds I, often by a factor between 2 and 3. The reasons for the excess energy requirement are the production of excited states, the kinetic energy of ionized electrons, and the removal of electrons with binding energy more than the minimum. The last-mentioned item sometimes results in an excited positive ion.

Franck and Hertz (1913) first demonstrated that an electron has to acquire a minimum energy before it can ionize. Thus, they provided an operational definition of the ionization potential and showed that it is an atomic or molecular property quite free from experimental artifacts. However, this kind of experiment does not tell anything about the nature of the positive ion; for this, one needs a mass spectrometric analysis. Although Thompson had demonstrated the existence of H^+, H_2^+, and H_3^+ in hydrogen discharge, it seems that Dempster (1916) was the first to make a systematic study of the positive ions.

The minimum electron energy necessary to observe a particular (positive) ion is called its *appearance potential* (AP). The concept is easily generalized to any molecular fragment. Knowledge of ionization potential and appearance potentials of fragments can be utilized to determine bond dissociation energies (D). Thus, applying energy–momentum conservation, one gets for the process

$$XY + e \rightarrow X^+ + Y + 2e,$$

the relation

$$D(XY) = AP(X^+) - I(X) - \frac{M}{M_X} T,$$

where M is the molecular mass of XY and M_X that of the fragment X. Extensive tabulation of appearance potentials for fragments of importance to radiation chemistry is available in the literature (Rosenstock *et al.*, 1977; Levin and Lias, 1982).

The adiabatic ionization potential (I_A) of a molecule, as shown in Figure 4.1, equals the energy difference between the lowest vibrational level of the ground electronic state of the positive ion and that of the molecule. In practice, few cases would correspond to adiabatic ionization except those determined spectroscopically or obtained in a threshold process. Near threshold, there is a real difference between the photoabsorption and photoionization cross sections, meaning that much of the photoabsorption does not lead to ionization, but instead results in dissociation into neutral fragments.

Most ionizations brought about by charged-particle impact are very fast processes governed by the Franck-Condon principle. According to this principle, the nuclear configuration of the molecule remains unchanged during fast electronic transitions. As seen in Figure 4.1, this means that the ionizing transitions from the lowest vibrational level of the molecular ground state must be vertically up. These transitions rarely lead to the lowest vibrational state of

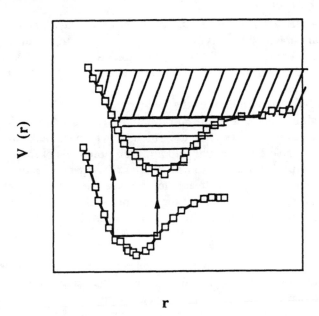

r

FIGURE 4.1 Illustration of adiabatic and vertical ionization potentials. Adiabatic I.P. refers to the energy difference between the lowest quantum states of the molecule and its positive ion. Often, Franck–Condon (vertical) transitions lead to a higher value, the vertical ionization potential.

the ion and, in general, would cost more energy than I_A. The term vertical ionization potential then corresponds to the Franck Condon process. It is usually not a precise value, but would cover a range with a lower limit equal to or somewhat greater than I_A. The difference between the two ionization potentials would be greater if the positions at minima and the shapes of the potential energy surfaces of the ion and the molecule differ significantly. For diatomic molecules, if the vertical process reaches above the dissociation limit of the parent ion, fragment ions would be observed, with the excess energy appearing as the relative kinetic energy of the fragments. For polyatomic molecules, the situation is complicated because the excess energy can be distributed among various modes of vibration.

The evaluation of ionization potentials by spectroscopic methods (adiabatic I_A), yields very accurate values. This amounts to identifying an appropriate Rydberg series in the absorption spectroscopy of complex atoms or molecules. The excitations leading to Rydberg-like series may be expressed as

$$E_i = RhZ^2[(n + a)^{-2} - (n_i + b)^{-2}], \tag{4.1}$$

where E_i is the ith transition energy, R is the Rydberg constant, h is Planck's constant, Z is the residual charge, and n, a, and b are adjustable parameters that remain fixed for a particular series. The energies for the different members of the series are obtained by varying the integer n_i. The appearance of the parameters a and b reflects the fact that the excited electron sees a somewhat sizable charge distribution. The parameters a and b are called the Rydberg correction factors.

The long-wavelength limit of the Rydberg series (see Eq. 4.1) gives a very accurate value of *one* of the ionization potentials of the molecule (Herzberg, 1950). It is frequently very difficult to identify which excited state the transition leads to. Table 4.1 gives a list of ionization potentials of some atoms, molecules, and radicals together with the W values (the *average* energy required to form one ion pair in a general irradiation of a specified type). Excellent tabulations of ionization potentials now exist together with the descriptions of methods of experimental determination or evaluation. For details, the reader is referred to the works by Christophorou (1971), Franklin *et al.* (1969), and Tabata *et al.* (1991).

The appearance potential (AP) of an ion refers to the experimentally determined minimum energy required to produce that specific ion from the ground state of the neutral atom. The appearance potentials of ions of specific types, such as the fragment ions, have been listed by Franklin *et al.* (1969). Some common appearance potentials of ions are given in Table 4.2.

4.1.1a   Primary and Secondary Ions

At a finite gas pressure and in the condensed phase, secondary ions are generated by reactions of primary ions with neutral gas molecules.

Dempster (1916) found that when extrapolated to zero pressure, the abundances of H^+ and H_3^+ in the ionization of hydrogen both go to zero, whereas that of H_2^+ goes to 100 percent (see, e.g., Smyth, 1931). Thus H_2^+ is the *dominant* primary ion and H^+ is mostly secondary. Of course, H_3^+ can only be a secondary ion, but a small amount of H^+ is always produced as a primary ion at a higher appearance potential (see Table 4.2). Cases similar to H_3^+ ($H_2^+ + H_2 \rightarrow H_3^+ + H$) are known in many diatomic molecules; such ion–molecule reactions play an important part in the radiation chemistry of liquid water, producing H_3O^+ and OH via the reaction $H_2O^+ + H_2O \rightarrow H_3O^+ + OH$ (Magee, 1964). In the following examples, the primary ions from a given molecule are shown within the parentheses: N_2 (N_2^+, N^+), O_2 (O_2^+, O^+), I_2 (I_2^+, I^+), K or K_2 (K^+, K_2^+), NO (NO^+, N^+, O^+), N_2O (N_2O^+, NO^+, O^+, N^+, N_2^+), CO_2 (CO_2^+, CO^+, O^+, C^+), H_2O (H_2O^+, OH^+, H^+), NH_3 (NH_3^+, NH_2^+, NH^+), CH_4 (CH_4^+, CH_3^+).

Secondary ions are also produced by charge or excitation transfer. Examples of the former are (1) $Ar^+ + H_2O \rightarrow Ar + H_2O^+$; (2) $He^+ + Ne \rightarrow He + Ne^+$; (3) $N^+ + NO \rightarrow N + NO^+$; (4) $O^+ + NO \rightarrow O + NO^+$; (5) $N_2^+ + N \rightarrow N_2 + N^+$; and so on. If sufficient energy is available, then sometimes charge transfer will be

TABLE 4.1 Ionization Potentials and W Values

	I (eV)	W (eV)[a]	W/I		I (eV)	W (eV)[a]	W/I
Atoms				*Other polyatomic molecules*			
H	13.6			NH_3	10.2	26.5	2.60
He	24.6	46.0	1.87	CH_4	12.7	29.1	2.29
Ne	21.6	36.6	1.69	C_2H_2	11.4	27.5	2.41
Ar	15.8	26.4	1.67	C_2H_4	10.5	28.0	2.67
Kr	14.0	24.0	1.71	C_2H_6	11.5	26.6	2.31
Xe	12.1	22.0	1.82	C_3H_6	9.7	27.1	2.79
F	17.3			C_3H_8	11.1	26.2	2.36
Cl	13.0			$n\text{-}C_4H_{10}$	10.6	26.1	2.46
Br	11.8			$n\text{-}C_6H_{12}$	9.90	25.0	2.53
I	10.6			C_6H_6	9.2	26.9	2.92
Diatomic molecules				$C_6H_5CH_3$	8.8		
H_2	15.4[b]	36.4	2.36	CH_3OH	10.8	25.0 (β)	2.31
N_2	15.6[b]	36.4	2.33	C_2H_5OH	10.5	32.6	3.1
O_2	12.1	32.2	2.66	CH_3I	9.5	24.8	2.61
CO	14.0	34.7	2.48	CH_3Br	10.5	34.6	3.30
HI	10.4			*Radicals*			
HCl	12.7	27.0	2.13	CH	11.1		
HBr	11.6	27.0	2.33	CH_2	10.4		
Triatomic molecules				CH_3	9.8		
H_2O	12.6	30.5	2.42	C_2H_5	8.4		
O_3	12.3						
CO_2	13.8	34.3	2.49				
N_2O	12.9	34.4	2.67				
H_2S	10.4	25.0 (γ)	2.40				

Further references: Meisels and Ethridge (1972); Miyajima *et al* (1974); Miller and Boring (1974); Jones (1973), Bichsel (1974); Takahashi *et al.* (1975) Jesse and Platzman, (1962); and Jesse (1964).

[a]W values reported are for α-radiation except where indicated in the parenthesis. In general, W values obtained with β-radiation is smaller.

[b]I determined spectroscopically.

accompanied by dissociation. Thus, for $Ne^+ + O_2 \rightarrow Ne + O + O^+$, the ionization potential of Ne (21.6 eV) exceeds the appearance potential of O from O_2 (19.0 eV).

Secondary ions formed in an excitation transfer process—namely, $A^* + B \rightarrow A + B^+ + e$—require that $E(A) \geq I(B)$, where $E(A)$ is the excitation

TABLE 4.2   Appearance Potentials of Some Parent and Fragment Ions

Ion	Produced from	AP (eV)	Ion	Produced from	AP (eV)
H_2^+	H_2	15.4	CH^+	C_2H_2	21.7
H^+	H_2	18.0	C_2H^+	C_2H_2	17.2
H_2^+	C_3H_8	17.0	$C_2H_2^+$	C_2H_2	11.4
H^+	C_3H_8	22.0	$C_2H_4^+$	C_2H_4	10.5
N_2^+	N_2	15.6	$C_2H_3^+$	C_2H_4	13.4
N^+	N_2	24.3	$C_2H_6^+$	C_2H_6	11.5
O_2^+	O_2	12.1	$C_2H_5^+$	C_2H_6	13.0
O^+	O_2	19.0	$C_3H_7^+$	$n\text{-}C_3H_7$	8.4
F_2^+	F_2	15.7	$C_3H_7^+$	$iso\text{-}C_3H_7$	7.5
F^+	CF_4	24.0	CH_4O^+	CH_3OH	11.0
Cl_2^+	Cl_2	11.5	CH_3O^+	CH_3OH	12.3
Cl^+	Cl_2	15.5	$C_2H_6O^+$	C_2H_5OH	10.5
Br_2^+	Br_2	10.5	$C_2H_5O^+$	C_2H_5OH	11.0
Br_2^+	Br_2	14.3	$C_3H_8O^+$	C_3H_7	10.1
I_2^+	I_2	9.3	$C_3H_7O^+$	C_3H_7	11.0
I^+	I_2	12.5	H_2O^+	H_2O	12.6
CH_4^+	CH_4	13.0	HO^+	H_2O	18.4
CH_3^+	CH_4	14.5	H^+	H_2O	19.2
CH_2^+	CH_4	15.3	O^+	H_2O_2	17.0

energy of A. This kind of ionization is generally called Penning ionization. When A is a rare gas atom, it gives rise to the Jesse effect—that is, the lowering of the measured W value of a rare gas (say, He) due to impurities (Jesse and Sadauskis, 1952, 1953, 1955; Jesse and Platzman, 1962). A process quite similar to Penning ionization is the associative (or collateral) ionization: $A^* + B \rightarrow AB^+ + e$. This process may occur even if $E(A) < I(B)$, provided that sufficient energy is available from the binding energy of AB and the excitation energy to overcome the ionization potential of AB. When B is a diatomic molecule X_2, the resultant ionization may be of the type $AX_2^+ + e$ or $AX^+ + X + e$; these are known as chemi-ionization processes or Hornbeck–Molnar processes.

4.1.2   NATURE OF EXCITED STATES

The ground state of most molecules is a singlet with exceptions such as O_2 and NO, which are triplets. Optically allowed transitions reach excited states of the same multiplicity as that of the ground state, whereas forbidden transitions reach states with a different multiplicity. In this chapter we will concern ourselves exclusively with excited singlets and triplets. Although states of higher multiplicity are in principle permissible, there is no great experimental evidence for their existence (Cundall, 1968).

The consideration of molecular excited states forms an integral part of the overall picture of radiation chemistry. For a period of time, a tendency developed among practicing radiation chemists to ignore the excited states. Partial justification for this came from the observation that, in many systems, luminescence from excited states and chemical product formation appeared to be mutually exclusive (Brocklehurst, 1970). Another important area of interdisciplinary interest is energy transfer from excited states. This is a vast field reviewed by several authors (Ausloos and Lias, 1967; Brocklehurst, 1970; Cundall, 1968, 1969; Forster, 1960, 1968; Kearwell and Wilkinson, 1969; Sieck, 1968; Singh, 1972; Stevens, 1968; Thomas, 1970). Even the early book by Pringsheim (1949) already covered some 800 pages, and no doubt important progress has been made since then.

In radiolysis, production of excited states is to a large extent indiscriminate, whereas in photolysis one can control the energy of the photon and thereby select excited states. With relatively small molecules, vacuum ultraviolet radiation is expedient. Various atomic radiations are routinely used for this purpose. Among these are hydrogen Lyman-α (10.2 eV), helium (21.2 eV, requires a windowless arrangement), Ar (11.6 and 11.8 eV), Kr (10.0 and 10.6 eV), and Xe (8.4 and 9.6 eV) radiations. For relatively low energy, one can use Hg (6.7 eV) or a Ar + Br_2 mixture (7.6 and 8.1 eV). Various lasers with frequency multiplication can also be used. Among the different methods for studying the excited state are light emission, energy transfer, and chemical reaction. The last item occasionally includes polymerization, isomerization, and the like.

4.2 IONIZATION EFFICIENCY: SUPEREXCITED STATES

The distinction between photoabsorption and photoionization is important, particularly near threshold, where the probability that ionization will *not* occur upon photoabsorption is significant. Thus, the ionization efficiency is defined by $\eta_i = \sigma_i/\sigma_a$, where σ_i is the photoionization cross section and σ_a, the photoabsorption cross section, is related to the absorption coefficient α by $\alpha = n\sigma_a$, n being the absorber density.

Platzman (1962a) has emphasized the implications of superexcited states in radiation chemistry. On the whole, his conjectures have been proved correct. Figure 4.2, using the data of Haddad and Samson (1986), shows the ionization efficiency in the gas phase of water. It shows that η_i starts with a value of 0.4 at the

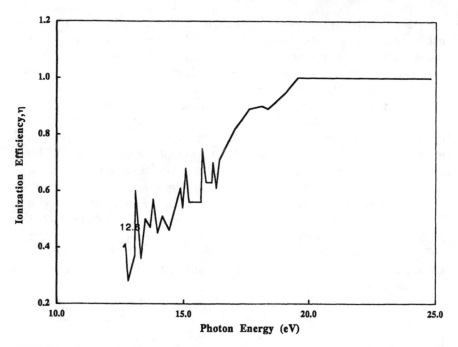

FIGURE 4.2   Ionization efficiency as a function of photon energy in the gas phase of water. Data from Haddad and Samson (1986), with permission of Am. Inst. Phys.©

ionization threshold (12.6 eV) and then gradually approaches unity at ~20 eV, showing some structure. It has also been found that in many cases, the photoionization efficiency is greater for the deuterated molecules. Thus, $\eta(CD_4) > \eta(CH_4)$, $\eta(C_6D_6) > \eta(C_6H_6)$, and so forth. The explanation of this phenomenon is found in the competition between dissociation and autoionization of the superexcited state. Making the plausible assumption that the dissociation rate is less for the deuterated molecule, the ionization rate is, by default, greater.

4.3   MECHANISMS OF EXCITED STATE FORMATION

Excited states may be formed by (1) light absorption (photolysis); (2) direct excitation by the impact of charged particles; (3) ion neutralization; (4) dissociation from ionized or superexcited states; and (5) energy transfer. Some of these have been alluded to in Sect. 3.2. Other mechanisms include thermal processes (flames) and chemical reaction (chemiluminescence). It is instructive to consider some of the processes generating excited states and their inverses. Figure 4.3 illustrates this following Brocklehurst (1970): luminescence $(1 \rightarrow 2)$

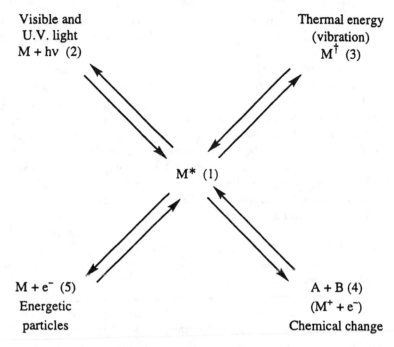

Visible and
U.V. light
M + hv (2)

Thermal energy
(vibration)
$M^{\dagger}$ (3)

M* (1)

M + e⁻ (5)
Energetic
particles

A + B (4)
($M^+ + e^-$)
Chemical change

FIGURE 4.3   Various processes leading to excited state formation and their inverses. See text for explanation. From Brocklehurst (1970).

vs. light absorption $(2 \rightarrow 1)$; collisional deactivation $(1 \rightarrow 3)$ vs. thermal process $(3 \rightarrow 1)$; autoionization or chemical reaction of the excited state $(1 \rightarrow 4)$ vs. ion neutralization or chemical change $(4 \rightarrow 1)$; and superelastic collision $(1 \rightarrow 5)$ vs. impact excitation $(5 \rightarrow 1)$. In the following, we will consider the principal mechanisms for excited state formation in some detail.

4.3.1 Light Absorption

Excited states formed by light absorption are governed by (dipole) *selection rules*. Two selection rules derive from parity and spin considerations. Atoms and molecules with a center of symmetry must have wavefunctions that are either symmetric (*g*) or antisymmetric (*u*). Since the dipole moment operator is of odd parity, allowed transitions must relate states of different parity; thus, $u \rightarrow g$ is allowed, but not $u \rightarrow u$ or $g \rightarrow g$. Similarly, allowed transitions must connect states of the same multiplicity—that is, singlet—singlet, triplet–triplet, and so on. The parity selection rule is strictly obeyed for atoms and molecules of high symmetry. In molecules of low symmetry, it tends to break down gradually; however, there may still be remnants of local symmetry, such as $n \rightarrow \pi$ transitions. Symmetry-forbidden transitions occur in molecules with intensities $\sim 10^{-1}$ to 10^{-3} compared with those of allowed transitions.

Spin-forbiddenness can be weakly violated by spin–orbit coupling. This gradually increases with the atomic number, until pure spin states (singlets or triplets) lose their significance. Spin–orbit coupling therefore depends significantly on the presence of heavy atoms in the molecule or in the surroundings (xenon bubbling promotion of spin-forbidden transition). McGlynn et al. (1964) cite an example of the former in the $S_0 \rightarrow T_1$ visible absorption in halonaphthalenes. The extinction coefficient in 1-chloronaphthalene is $\sim 10^{-2}$ l$\propto$M^{-1}cm^{-1}, but that in 1-iodonaphthalene is ~ 1. However, atomic number is not the only consideration. In aromatic hydrocarbons, cancellation of first-order terms in the spin–orbit interaction makes it very small (Hameka and Zahlan, 1967). On the other hand, the $^3P_1 \rightarrow {}^1S_0$ transition in neon has a lifetime only an order of magnitude larger than that of the $^1P_1 \rightarrow {}^1S_0$ transition.

These selection rules are affected by molecular vibrations, since vibrations distort the symmetry of a molecule in both electronic states. Therefore, an otherwise forbidden transition may be (weakly) allowed. An example is found in the lowest singlet-singlet absorption in benzene at 260 nm. Finally, the Franck-Condon principle restricts the nature of allowed transitions. A large number of calculated Franck-Condon factors are now available for diatomic molecules.

Of the different kinds of forbiddenness, the spin effect is stronger than symmetry, and transitions that violate both spin and parity are strongly forbidden. There is a similar effect in electron-impact induced transitions. Taken together, they generate a great range of lifetimes of excited states by radiative transitions, $\sim 10^{-9}$ to $\sim 10^3$ s. If nonradiative transitions are considered, the lifetime has an even wider range at the lower limit.

4.3.2   DIRECT EXCITATION BY CHARGED-PARTICLE IMPACT

Considerations leading to the Born–Bethe cross section are also applicable to impact excitation. Vibrational excitation in the ground electronic state has a very small cross section (Massey and Burhop, 1952), but the same is not true when attended by electronic excitation. Direct excitation may be (1) optically allowed, (2) forbidden, or (3) induced by electron exchange. These are characterized respectively by $E^{-1} \ln E$, E^{-1}, and E^{-3}, where E is the incident electron energy. Incident charged particles of high velocity excite the allowed states preferentially, but slow electrons, which inevitably accompany any ionizing radiation, tend to excite states more or less indiscriminately. Therefore, in radiolysis one cannot ignore forbidden transitions. To this, one should add the possibility of the formation of a solute excited state via the impact of electrons that have been rendered subexcitation in the solvent.

Often, the processes leading to a product from an excited state are very slow compared with the time scale of formation of the excited state. Thus, once the excited state is formed, its fate is independent of the manner of its formation. Polak and Slovetsky (1976) have described the theory of impact excitation via Born-Bethe theory. Using the Born-Oppenheimer approximation, this cross section is written as the product of the Franck-Condon factor and the square of a transition matrix element (see Sect. 4.5). For allowed transitions, Polak and Slovetsky find excellent agreement between these calculations and experiments for e-impact excitation in $H(1s \rightarrow 2p)$, $H_2(X^1\Sigma_g^+ \rightarrow B^1\Sigma_u^+, C^1\Pi_u, D^1\Pi_u)$, $CO(X^1\Sigma_g^+ \rightarrow A^1\Pi_g^+)$, and $N_2(X^1\Sigma_g^+ \rightarrow b)$. The excitation cross section increases with electron energy from threshold, reaches a broad peak at 3 to 4 times the excitation energy, and then decreases slowly, reminiscent of an impact-ionization cross section. As expected, excitation to forbidden states has a smaller cross section, and the energy dependence shows a sharp peak at a low energy characteristic of resonance in an exchange process.

Excited states often dissociate into neutral fragments. Sometimes, such an atomic or a diatomic radical fragment will form in an excited state, from which emission is observable. Lyman (α, β) and Balmer (α, δ) emissions have been observed in e-impact on H_2, H_2O, and various hydrocarbons in the gas phase. Polak and Slovetsky (1976) have discussed the total dissociation cross section giving neutral products under electron impact in H_2, N_2, CO, NO, NH_3, and CH_4. The dissociation threshold in these compounds ranges between 8 and 12 eV, the peak cross section is $\sim 10^{-16}$ cm^2, and the peak is observed variously between 15 and 100 eV electron energy. These cross sections are about an order of magnitude larger than the maximum cross sections measured in processes giving light emission (Vroom and deHeer, 1969a, b) or identified as excited neutral fragments (Polak and Slovetsky, 1976).

The fate of the remaining vast majority is only speculative. Considering Lyman emission by e-impact on hydrocarbons, the sum of photon energy (10.2 eV) and binding energy of the H atom exceeds the ionization potential of the molecule. The initial state produced, which is a neutral excited state, is in the ionization continuum. Such states were named superexcited states by Platzman (1962a, b). In the case of excited H atoms formed by e-impact on hydrocarbons, Vroom and deHeer (1969a) concluded that (1) the cross section for a given atomic state and electron energy is relatively independent of the hydrocarbon; and (2) the main molecular processes giving $n = 2$–6 atomic states on dissociation involve forbidden transitions. In contrast, excitation to the $n = 2$ atomic state of H_2 is mainly by the allowed process (Vroom and deHeer, 1969b), although higher states are progressively excited by forbidden transitions. These workers also found the isotope effect to be in the same direction as argued by Platzman (1962b) for the difference in the ionization yields in methane and deuterated methane. Dissociative excitation of H_2 was about 20% greater than in D_2, and similarly between CH_4 and CD_4. The reason simply is that ionization and dissociative excitation are competing channels for the decay of the superexcited state. The former is a mass-independent process. The latter is faster for the lighter isotope (Platzman, 1962b).

4.3.3   ION NEUTRALIZATION

In radiolysis, a significant proportion of excited states is produced by ion neutralization. Generally speaking, much more is known about the kinetics of the process than about the nature of the excited states produced. In inert gases at pressures of a few torr or more, the positive ion X^+ converts to the diatomic ion X_2^+ very rapidly. On neutralization, dissociation occurs with production of X^*. Apparently there is no repulsive He_2^* state crossing the He_2^+ potential curve near the minimum. Thus, without He_2^+ in a vibrationally excited state, dissociative neutralization does not occur; instead, neutralization is accompanied by a collisional radiative process. Luminescences from both He^* and He_2^* are known to occur via such a mechanism (Brocklehurst, 1968).

At the other extreme is ion-recombination luminescence observed in rigid systems, usually at low temperatures (Hamill, 1968). Two kinds of luminescence are known to occur: (1) thermoluminescence, observed on gentle warming, and (2) infrared stimulated emission (IRSE), in which the electron or the negative ion is bleached and then has a certain probability of recombining with the positive ion, resulting in light emission. In between these extremes lies the vast field of excited state production in liquids by ion neutralization. Both solvent and solute excited states are known to occur. The subject has been reviewed by Thomas (1971, 1976), Salmon (1976), Singh (1972), and others. From these studies, we conclude the following:

1. In alkanes, excited states are mainly produced by ion neutralization where the triplet contribution dominates.
2. In aromatics, both singlet and triplet excited states of the solvent are produced in comparable yields, with some contribution from direct excitation.
3. Energy transfer to the solute is readily observed in aromatics. When observed in alkanes, it is mainly a singlet-singlet transfer.
4. In alicyclic hydrocarbon solvents with aromatic solutes, energy transfer (*vide infra*) is unimportant and probably all excited solute states are formed on neutralization of solute cations with solute anions, which are formed in the first place by charge migration and scavenging in competition with electron solvent-cation recombination. The yields of naphthalene singlet and triplet excited states at 10 mM concentration solution are comparable and increase in the order cyclopentane, cyclohexane, cyclooctane, and decalin as solvents. Further, the yields of these

excited states decrease in the presence of charge scavengers following a normal scavenging pattern.

Lipsky and his group have extensively studied emission from a large variety of saturated hydrocarbons to determine the fraction δ of lowest excited states due to ion neutralization and the probability p of such an excited state being a singlet (Rothman *et al.*, 1973; Lipsky, 1974; Walter and Lipsky, 1975). Using bicyclohexyl as solvent and cyclic perflurocarbons as scavengers, and employing reasonable assumptions, Walter *et al.* (1976) found $\delta = 0.9\pm0.1$ and $p \geq 0.8$. Thus, *in this system*, almost all the lowest excited states are formed by ion neutralization, of which a very significant fraction are singlets.

For a spur containing n ion pairs and assuming (rather implausibly) that all ion pairs eventually neutralize, Magee and Huang (1972, 1974) give the probability of singlet excited state formation on ion neutralization as $p_n = (1/4)[1 + 3/(2n - 1)]$. Averaging p_n over the distribution of spur sizes given by Mozumder and Magee (1966) gives $\langle p \rangle = 0.7$. Using the energy distribution of Ore and Larsen (1964), Walter *et al.* (1976) compute $\langle p \rangle - 0.8$ and come to the conclusion that either theory comes close to the experimental value. Another formula for p_n, based on different theoretical assumptions, gives $p_n = (1/4)(1 + 3/n)$ (Higashimura *et al.*, 1972; Brocklehurst and Higashimura, 1974). In the present context, there is not much to choose between in these two formulations. Note, however, that theoretical calculations refer to *all* singlet excited states, whereas luminescence measurements only give the *first* excited one. Nevertheless, the comparison may be fair because of the high efficiency of internal conversion and the low polarization energy (preventing formation of the ground state directly).

4.3.4 DISSOCIATION FROM IONIZED OR SUPEREXCITED STATES

Not much is known about these processes, but they must be included to give a total picture. Emissions of Lyman and Balmer spectra of the H atom upon e-impact on hydrocarbons, H_2, and H_2O, discussed in Sect. 4.3.2, fall in this category. Similarly, many of the excited states observed in dissociated radicals via electron impact on stable molecules (Polak and Slovetsky, 1976) also belong to this category. It is known from the dipole oscillator spectrum of H_2O (Platzman, 1967) that most ionizations are accompanied by considerable excitation. Excitation transfer to the neighboring neutral molecule followed by fast dissociation cannot be ruled out.

4.3.5 ENERGY TRANSFER

Certain aspects of energy transfer were discussed in Sect. 3.2. Note that energy transfer involves the formation of an excited state of a molecule and the decay of another. Therefore, this subsection should be read together with Sect. 4.4. Among various phenomena depending on electronic energy transfer are the following: (1) sensitized fluorescence in which radiation absorbed by one component reappears as emission from another component at longer wavelengths; (2) scintillation under the impact of charged particles or photons; (3) chemical processes including isomerization and protection in radiation and photochemistry; (4) quenching; and (5) biological processes, as in photosynthesis. Energy transfer between neutral excited states may be broadly divided into resonant processes and collisional processes. These will be discussed in some detail.

Cario and Franck (1923) found that when a mixture of mercury and thallium was excited by the mercury resonance radiation, emission was observed from thallium also. In this, mercury was the sensitizer, or donor, and thallium the acceptor. Later, it was demonstrated that the transfer indeed occurred over long distances in a single step. Franck and Cario (1922) demonstrated that the excited state of mercury can transfer energy to a hydrogen molecule, resulting in chemical change. Perrin and Choucroun (1927, 1929) were the first to show luminescence resulting from energy transfer between dye molecules in the liquid phase. Kallmann and Furst (1950) invoked such processes in high-energy excitation of liquids. Burton and his co-workers studied various specific rates by measuring fast decay times of luminescence (Dresskamp and Burton, 1959; Yguerabide and Burton, 1962; Burton et al., 1964). Collisional transfer follows diffusion theory, whereas a theory of a long-range nonradiative transfer process has been advanced by Forster (1947) and extended by Dexter (1953) and Forster (1959).

4.3.5a   Resonance Energy Transfer

Forster's (1949a, b) experiments on resonance transfer with trypaflavine as donor and rhodamine-B as acceptor in solution established quenching of donor fluorescence and shortening of the lifetime of its excited state. On the basis of the latter, the trivial mechanism was ruled out and energy transfer over ~70 Å established. Later, Watson and Livingston (1950) demonstrated energy transfer from chlorophyll-b to chlorophyll-a by observing the fluorescence of both. Bowen and Brocklehurst (1953) carefully excluded other mechanisms in the energy transfer from 1-chloroanthracene to perylene in a benzene solution and established the resonance process. Keeping the donor-acceptor concentration ratio fixed at 5 : 1, they observed simultaneously a decrease in donor fluorescence and an increase in the total quantum yield of emission with increasing solute concentration.

Forster (1959) classifies the qualitative features based on which one can distinguish the various modes of energy transfer. Mainly, only collisional transfer depends on solvent viscosity (vide infra), whereas complexing between the donor and acceptor changes the absorption spectrum. On the other hand, the sensitizer lifetime decreases for the long-range resonant transfer process, whereas it should be unchanged for the trivial process.

Denoting the ground and excited states by the subscripts 0, 1, 2, ..., and indicating the donor first followed by the acceptor after the slash in the notation, Forster (1968) classifies excitation transfer into four groups: (1) $(S_1 \rightarrow S_0/S_0 \rightarrow S_1)$; (2) $(T_1 \rightarrow S_0/S_0 \rightarrow S_1)$; (3) $(S_1 \rightarrow S_0/T_1 \rightarrow T_2)$; and (4) $(T_1 \rightarrow S_0/S_0 \rightarrow T_1)$. Spin conservation is the most important factor in energy transfer. Processes (1), (3), and (4) conserve total spin, (1) and (3) conserve individual spins, whereas (2) violates total spin conservation. If the efficiency of the transfer process is taken as the ratio of the rate of transfer to that of intramolecular degradation, which is also spin dependent, then spin conservation in the acceptor is more important than in the donor. Process (3) involves excitation to the higher triplet state of the acceptor, which means that the acceptor must be first prepared in the T_1 state.

Forster (1968) further classifies the strength of coupling in resonant energy transfer as strong, weak, or very weak. The coupling is strong if the intermolecular interaction energy (IIE) exceeds the energy of interaction between electronic and nuclear motions in the individual molecules. Weak coupling refers to the reverse case under the condition IIE $> \Delta\varepsilon$, the width of the vibronic band in the individual molecules. Finally, coupling is very weak if IIE $< \Delta\varepsilon$. By this consideration, it may be said that most experimental work on the radiation chemistry of solutions is governed by the *weak coupling case*. Applying Fermi's *"golden rule,"* the transfer rate P or the probability per unit time of transition is given by

$$P = 2\pi\hbar^{-1}\rho\left(\int \phi_i^* H' \phi_f \, d\tau\right)^2, \tag{4.2}$$

where ρ is the density of final states, ϕ_i and ϕ_f are respectively the initial and final wavefunctions, and H' is the interaction Hamiltonian. Coulombic interaction between valence and core electrons of the donor with those of the acceptor constitute H'.

Developed into a power series in R^{-1}, where R is the intermolecular separation, H' exhibits the dipole–dipole, dipole–quadrupole terms in increasing order. When nonvanishing, the dipole–dipole term is the most important, leading to the Forster process. When the dipole transition is forbidden, higher-order transitions come into play (Dexter, 1953). For the Forster process, H' is well known, but ϕ_i and ϕ_f are still not known accurately enough to make an a priori calculation with Eq. (4.2). Instead, Forster (1947) makes a simplification based on the relative slowness of the transfer process. Under this condition, energy is transferred between molecules that are thermally equilibrated. The transfer rate then contains the same combination of Franck–Condon factors and vibrational distribution as are involved in the vibrionic transitions for the emission of the donor and the adsorptions of the acceptor. Forster (1947) thus obtains

$$P = \tau^{-1}\left(\frac{R_0}{R}\right)^6, \tag{4.3}$$

where τ is the true donor state lifetime and R_0 is a critical distance at which the energy transfer rate equals the spontaneous deactivation rate of the donor.

Equation (4.3) gives the inverse sixth power law of Forster,

$$R_0^6 = \frac{9\kappa^2 \eta_d}{128\pi^5 n^4 N'} \int f_d(v) \varepsilon_a(v) v^{-1} \, dv, \tag{4.4}$$

where v is the wavenumber, f_d (quanta per wavenumber) and ε_a (molar extinction coefficient) are obtained respectively from the donor emission and acceptor absorption spectra, η_d is the quantum efficiency of donor dipole emission, n is the refractive index, N' is Avogadro's number divided by 1000, and κ is a numerical factor representing the orientational dependence of the dipole-dipole interaction; in a random distribution, the average value of κ^2 is 2/3. The integral in Eq. (4.4) is to be performed over the range of significant overlap between donor emission and acceptor absorption.

Forster (1968) points out that R_0 is independent of donor radiative lifetime; it only depends on the quantum efficiency of its emission. Thus, transfer from the donor triplet state is not forbidden. The slow rate of transfer is partially offset by its long lifetime. The importance of Eq. (4.4) is that it allows calculation in terms of experimentally measured quantities. For a large class of donor-acceptor pairs in inert solvents, Forster reports R_0 values in the range 50–100 Å. On the other hand, for scintillators such as PPO (diphenyl-2,5-oxazole), pT (p-terphenyl), and DPH (diphenyl hexatriene) in the solvents benzene, toluene, and p-xylene, Voltz et al. (1966) have reported R_0 values in the range 15–20 Å. Whatever the value of R_0 is, it is clear that a moderate red shift of the acceptor spectrum with respect to that of the donor is favorable for resonant energy transfer.

4.3.5b   Collisional Energy Transfer

In liquids, collisional energy transfer takes place by multistep diffusion (the rate determining step) followed by an exchange interaction when the pair is very close. The bimolecular-diffusion-controlled rate constant is obtained from Smoluchowski's theory; the result, including the time-dependent part, may be written as

$$k = 4\pi N' a D \left[1 + \frac{a}{(\pi D t)^{1/2}} \right],$$

where a is the reaction radius and D is the mutual diffusion coefficient.

When the transient effect is negligible, k can then be determined by measuring the luminescence intensities under steady state conditions and the lifetimes of the decay in both the absence and the presence of a scavenger at concentration c. Indicating the intensities by I_0 and I, respectively, and the lifetimes by τ and τ_0, respectively, one gets the following from the Stern-Volmer relation:

$$\frac{I}{I_0} = \frac{\tau}{\tau_0} = (1 + k\tau c)^{-1}.$$

The value of k so determined could then be compared with the theoretical value of $4\pi N'aD$. However, when viscosity is considerable and/or for short lifetimes, the transient effect in diffusion is not negligible and ~30% of the transfer may be attributable to the transient phase. In such a case, the luminescence decay is not simply exponential (Sveshnikov, 1935). For a brief pulse excitation, a complicated decay ensures; on the other hand, for so prolonged an excitation as to generate a steady state, the resultant decay curve in many cases is indistinguishable from an exponential (Yguerabide et al., 1964).

The rate of energy transfer at a very short donor–acceptor separation R by the exchange mechanism has been given by Dexter (1953) as follows:

$$k(D^* \rightarrow A) = hW^2 \, \exp\left(\frac{-2R}{L}\right) \int f_D'(v)\varepsilon_A'(v) \, dv. \qquad (4.5)$$

In Eq. (4.5) the donor emission spectrum f_D' and the acceptor absorption spectrum ε_A' are separately normalized to unity, so that the transfer rate is independent of the oscillator strength of either transition. Unfortunately, the constants W and L are not easily determined by experiment. Nevertheless, an exponential dependence on the distance is expected. It should be noted that this type of transfer involves extensive orbital overlap and is guided by Wigner's (1927) spin rule.

4.4 DECAY OF EXCITED STATES

In the previous sections, we have encountered some of the fates of excited states due to energy transfer or dissociation. Other modes of decay include luminescence (fluorescence and phosphorescence), internal conversion and intersystem crossing, chemical change (unimolecular or bimolecular processes), and degradation to heat. In the following, we will discuss these briefly. Note that in a nonradiative process, there may or may not be an attendant chemical change. In the latter subgroup belongs radiationless decay to the ground state. The final degradation of all of the residual energy to heat involves energy sharing by many molecules, going from electronic to vibrational and then to translational and rotational modes. In condensed media and in gases at pressures of 10 torr or more, this is a significant process.

4.4.1   INTERNAL CONVERSION AND
INTERSYSTEM CROSSING

Internal conversion refers to radiationless transition between states of the same multiplicity, whereas intersystem crossing refers to such transitions between states of different multiplicities. The difference between the electronic energies is vested as the vibrational energy of the lower state. In the liquid phase, the vibrational energy may be quickly degraded into heat by collision, and in any phase, the differential energy is shared in a polyatomic molecule among various modes of vibration. The theory of radiationless transitions developed by Robinson and Frosch (1963) stresses the Franck–Condon factor. Jortner et al. (1969) have extensively reviewed the situation from the photochemical viewpoint.

Internal conversion and intersystem crossing from higher electronic states to the S_1 and T_1 states of the molecule are usually very fast processes when the excited states themselves are nondissociative. Triplet–triplet fluorescence is extremely rare. Even emission from higher singlet states is a rare process, occurring from the second excited states in azulene and benzpyrene. Together with this, one has to bear in mind that the quantum yield of fluorescence is nearly independent of the wavelength of excitation over a fairly wide range. For example, in anthracene the quantum yield of fluorescence remains unaltered at 0.3 for exciting wavelengths in the range 210–310 nm; in eosin, the quantum yield of fluorescence is 0.2 over excitation wavelengths in the range 210–600 nm. Representing the quantum yield of fluorescence as a product of the probability of transition to S_1 with the quantum yield from S_1 to S_0, one can argue that the stated probability must be close to unity.

To get a rough idea of the rate of radiationless transition from higher excited states, consider the quantum yield of fluorescence from that state as the ratio of nonradiative (τ_{nr}) to radiative (τ_r) lifetimes as follows:

$$\phi = \frac{\tau_{nr}}{\tau_r}.$$

Assuming plausible experimental values of $\tau_r \sim 1$ ns and $\phi \leq 10^{-3}$ from higher excited state, one gets $\tau_{nr} < 1$ ps. The rate of radiationless transition from higher to lower excited states is then $>10^{12}$ s^{-1}. In many compounds(e.g., aromatic hydrocarbons), however, the internal conversion to the ground state is far slower than this, so that fluorescence and intersystem crossing to T_1 may be competitive processes. Iet al., many systems, phosphorescence ($T_1 \rightarrow S_0$) is similarly competitive with intersystem crossing to the ground state.

Of course, in determining the various competitive processes relating S_0, S_1, and T_1, one also has to consider the types and symmetries of electronic states concerned. El-Sayed (1968) points out that the spin–orbit coupling between singlet and triplet states of the same type is much smaller than that in different types. Thus, in aromatic hydrocarbons and aliphatic ketones, intersystem crossing $(S_1 \rightarrow T_1)$ and fluorescence $(S_1 \rightarrow S_0)$ are competitive. One sees both fluorescence and phosphorescence in these systems. But in aromatic ketones at low temperatures, one only sees phosphorescence, because the intersystem crossing is much faster than fluorescence. The role of the symmetry of the electronic wavefunction is not well understood, but rules prohibiting the radiative process (e.g., $g \rightarrow g$) are expected.

There is some evidence that the crossing rate from T_1 to S_0 decreases rapidly with the energy gap. Presumably, the same is true for the S_1 to T_1 crossing, but in this case the energy gap is usually much smaller than in the former case. In the theory of Robinson and Frosch (1963), the energy gap is critical. There is no electronic state between S_0 and T_1; therefore, energy can only pass to highly excited vibrational states of S_0. Because of this, one may expect transfer to any of a large number of vibrational states that are in near resonance. But for the same reason of high vibrational energy, the wavefunctions have a large number of nodes that make the Franck-Condon overlap negligible for each of those states. This is a kind of interference effect and it apparently overcompensates for the effect of the density of states.

The importance of the energy gap and the Franck-Condon factor is also evident in another situation, in which there is a second triplet excited state T_2 close to S_1 (either lower or higher). This energy difference is usually smaller than the S_1-T_1 energy gap. If T_2 lies lower than S_1, energy goes to T_2 by intersystem crossing and then to T_1 by internal conversion. These processes are fast and temperature independent. On the other hand, if T_2 lies *slightly* above S_1, then energy likewise can go to T_1 via the same processes but, in addition to being somewhat slow, the processes are temperature dependent. Both these conclusions have been experimentally verified.

4.4.2 UNIMOLECULAR AND BIMOLECULAR CHEMICAL PROCESSES

Here we give a brief account of some unimolecular processes other than isomerization. No detailed description of bimolecular processes will be offered, except to remark that (1) the knowledge gained from the unimolecular processes is often useful in interpreting the bimolecular processes; and (2) in some cases, the bimolecular processes resemble normal diffusion-influenced reactions in the condensed phase.

Many molecules of interest to chemistry and biology contain H atoms. For the unimolecular dissociation of the excited states of such molecules, it is instructive to think of elimination of a particle or a group of particles. Thus, we can think of elimination of an electron, an H atom, an H_2 molecule, an alkane group (in suitable cases), and so forth. Other factors remaining the same, the rate of the process may be expected to decrease in that order. The first of these is simply ionization from a superexcited state. Some of the rest will now be discussed for some representative molecules such as H_2O, CO_2, NH_3, and certain hydrocarbons. At first, we should point out two experimental aspects—namely, scavenging and isotopic substitution. With their help, one can often distinguish whether the process is uni- or bimolecular, and whether the molecular H_2 elimination came from the same group.

4.4.2a   Water

At least seven modes of dissociation are theoretically possible below the ionization threshold, although their total yield in radiolysis is small (Platzman, 1967). The dissociation products are H, H_2, O, and OH, where the first two are in their ground (electronic) states but the last two may be either in ground or excited states. Only two modes of dissociation, $H_2O^* \rightarrow H_2 + O$ and $H_2O^* \rightarrow H + OH$, are possible for all excitation energies; UV photolysis indicates that the latter process is by far (90%) the most likely. Accordingly, in radiolysis there is a tendency to lump the decay of all excited states of the water molecule into H and OH.

4.4.2b   Carbon Dioxide

The decomposition of the excited state, $CO_2^* \rightarrow CO(^1\Sigma + O(^3P, {}^1D, \text{ and } {}^1S)$ is generally assumed simple. The state $O(^3P)$ has apparently not been seen, which is consistent with Wigner's (1927) rule. In any case, the reported quantum yield of unity for the production of CO is often used in far-UV chemical actinometry.

4.4.2c   Ammonia

Here at least nine dissociative channels are theoretically accessible below the ionization threshold. The dissociation products are $NH_3^* \rightarrow NH + H_2$, $NH_2 + H$, or $NH + 2H$ (final state), of which NH and NH_2 may exist either in the ground or an excited state. Production of molecular hydrogen is negligible at low excitation energies, but it can account for 15% or more of the dissociation processes when the excitation exceeds ~7 eV. Note that the lowest excited state of NH_3 (~4 eV) does dissociate into NH and H_2 but is spin forbidden.

4.4.2d   Hydrocarbons

We confine our discussion to alkanes and cyclic alkanes. Some of the features found in them are common to other hydrocarbons as well. An H atom is produced by the dissociation of *highly* excited alkane molecules:

$$C_nH_{2n+2}^* \rightarrow C_nH_{2n+1}^\bullet + H,$$

where the • indicates residual energy in the fragment. H atoms having a little excess kinetic energy can form H_2 by abstraction. Thus, there could be confusion with the following process:

$$C_nH^{\cdot}_{2n+2} \rightarrow C_nH_{2n} + H_2.$$

However, isotope and scavenging studies have shown that in the vapor phase, epithermal H atoms only account for about 15% of hydrogen yield at 1470 Å. But this contribution rapidly increases to ~50% at 1225 Å. The energy-rich intermediate can undergo successive fragmentation, which makes the overall process difficult to distinguish from the one-step C—C cleavage generating free radicals:

$$C_nH^{\cdot}_{2n+2} \rightarrow C_{n'}H_{2n'+1} + C_{n''}H_{2n''+1} \qquad (n = n' + n'').$$

In addition, a higher alkane can eliminate a lower one as follows:

$$C_nH^{\cdot}_{2n+2} \rightarrow C_{n'}H_{2n'+2} + C_{n''}H_{2n''} \qquad (n = n' + n'').$$

In cyclic alkanes, analogously, one has three dissociative processes: (1) H_2 elimination, (2) C—H cleavage with retention of cyclic structure, and (3) C—C cleavage with opening of the ring. There are some simple yet useful generalizations about the cyclic alkanes. The ring opening requires considerable energy. Therefore, at low excitation energies, it occurs in the molecules already under strain e.g., cyclopropane and cyclobutane. In these cases, H_2 formation, which is the major channel in cyclopentane and cyclohexane, is small. However, even in the latter two cases, a certain amount of ring opening will occur at high excitation energy, and it seems to be correlated with the degree of excitation.

4.4.3 Luminescence

Certain features of light emission processes have been alluded to in Sect. 4.4.1. Fluorescence is light emission between states of the same multiplicity, whereas phosphorescence refers to emission between states of different multiplicities. The Franck-Condon principle governs the emission processes, as it does the absorption process. Vibrational overlap determines the relative intensities of different subbands. In the upper electronic state, one expects a quick relaxation and, therefore, a thermal population distribution, in the liquid phase and in gases at not too low a pressure. Because of the combination of the Franck-Condon principle and fast vibrational relaxation, the emission spectrum is always red-shifted. Therefore, oscillator strengths obtained from absorption are not too useful in determining the emission intensity. The theoretical radiative lifetime in terms of the Einstein coefficient, $\tau = A^{-1}$, or $(\Sigma A_l)^{-1}$ if several lower states are involved, requires the dipole moment matrix element for its evaluation. However, in most cases, a priori calculation is not possible except for the simplest atoms. The oscillator strength for absorption can be obtained from experimental measurement of the extinction coefficient, by which the following expression may be obtained (Kearwell and Wilkinson, 1969):

$$\tau^{-1} = \frac{18,424\pi c}{n} v^2 \int \varepsilon(v)\, dv. \qquad (4.6)$$

Equation (4.6) is strictly applicable for line spectra (atoms and diatomic molecules), but in these cases the extinction coefficient is not easily measured as a function of the wavenumber. Out of the several modifications of the concept applied to the polyatomic molecules, that owing to Strickler and Berg (1962) involves the refractive index of the solvent as well as the fluorescence spectrum $I(v)$ expressed as photons per unit wavenumber interval. Their formula is

$$\tau^{-1} = \frac{18,424\pi c}{n} v^2 \gamma \int \varepsilon(v)\, \frac{dv}{v}, \qquad (4.7)$$

where n is the refractive index and γ is given by

$$\gamma^{-1} = \frac{\int v^{-3} I(v)\, dv}{\int I(v)\, dv}.$$

For large, rigid aromatic compounds, agreement between values calculated from Eq. (4.7) and those directly determined by experiment is good; when the quantum yield for fluorescence is close to unity, the agreement is excellent (Kearwell and Wilkinson, 1969). But for small, flexible molecules, the agreement is poor. In some triatomics such as NO_2, the calculated value is two orders of magnitude less than measured (Brocklehurst, 1970). Kearwell and Wilkinson (1969) have given the following equation for estimating the order of magnitude of allowed radiative lifetimes:

$$\tau \text{ (in μs)} \approx \frac{100}{\varepsilon_{max}},$$

where ε_{max} is the extinction coefficient at the absorption maximum. This formula would predict radiative lifetimes between 1 μs and 10 ns for most allowed transitions. It should be remembered that experiments measure the overall lifetime, not only the lifetime from the final emitting state. Thus, in temperature-dependent fluorescence, there may be a long delay in populating the emitting state by intersystem crossing from a triplet with thermal assistance (see Sect. 4.4.1). In such cases, the rate determining step is the rate of intersystem crossing, and measured values will often greatly exceed calculated values. A similar situation exists in some luminescence processes induced in radiolysis (*vide infra*).

If the shapes of the potential energy surfaces are similar in the excited and ground electronic states of a molecule, then one can expect a mirror image relationship between absorption and emission spectra. Actually, this is more nearly true in phosphorescence versus singlet-triplet absorption perturbed by oxygen. Phosphorescence lifetimes lie in the range 10 ms to 10 s or longer. Considering a photostationary situation, the following relations can be obtained connecting the quantum yields of fluorescence (ϕ_f) and phosphorescence (ϕ_p) with the rates of radiative and nonradiative processes:

$$\phi_f = \frac{k_f}{k_f + k + k'}, \tag{4.8a}$$

$$\phi_p = \phi_t \Gamma_p. \tag{4.8b}$$

In Eqs. (4.8a, b), k_f is the rate of fluorescence, k is the intersystem crossing rate $(S_1 \rightarrow T_1)$, k' is the radiationless decay rate to the ground state $(S_1 \rightarrow S_0)$, ϕ_t is the quantum yield for producing the triplet, and Γ_p is the probability of phosphorescence from that state. Note that in many hydrocarbons, $\phi_t \approx 1 - \phi_f$, and for a group of carbonyl compounds, $\phi_t \approx 1$.

According to Ludwig (1968), there is a some similarity between UV- and high-energy-induced luminescence in liquids. In many cases (e.g., p-terphenyl in benzene), the luminescence decay times are similar and the quenching kinetics is also about the same. However, when a mM solution of p-terphenyl in cyclohexane was irradiated with a 1-ns pulse of 30-KeV X-rays, a long tail in the luminescence decay curve was obtained; this tail is absent in the UV case. This has been explained in terms of excited states produced by ion neutralization, which make a certain contribution in the radiolysis case but not in the UV case (cf. Sect. 4.3). Note that the decay times obtained from the initial part of the decay are the same in the UV- and radiation-induced cases. Table 4.3 presents a brief list of luminescence lifetimes and quantum yields.

4.5 IMPACT IONIZATION AND PHOTOIONIZATION

Photoionization refers to the ionization of a neutral entity. Its inverse is the *radiative* capture of a free electron. By virtue of detailed balancing, the cross section of radiative capture, σ_r, and that of photoionization, σ_i, involving the same initial and final states, i and f, are related by

$$\frac{\sigma_r}{\sigma_i} = \frac{\omega_i}{2\omega_f} \frac{h^2 v^2}{mc^2 T}. \tag{4.9}$$

TABLE 4.3 Luminescence Lifetimes and Quantum Yields (qy) of Some Selected Compounds

Compound	Solvent	qy	τ	Process[a]
Anthracene	EtOH	0.3		f
Naphthalene[b]	EtOH/ether	0.3	63 s	p
1-Iodonaphthalene[b]	EtOH/ether	0.2	10 ms	p
Fluorene	EtOH/ether	0.5		f
Phenol	Water	0.2		f
Fluorescein	0.1 M NaOH solution	0.9	4 ns	f
Rhodamine B	EtOH ml	~1	6 ns	f
Perylene	Benzene	0.9	4.8 ns	f
Benzene	Argon (84 K)		14 s	p
Phenanthrene	EPA (77 K)		2.3 s	p
Pyrene	EPA (77 K)		0.5 s	p
Biphenyl	EPA (77 K)		4.2 s	p
p-Terphenyl	EPA (77 K)		2.6 s	p

[a]Unless otherwise specified, both qy and τ refer to the same process, indicated by f for fluorescence and p for phosphorescence. For naphthalene, qy is for fluorescence and τ is for phosphorescence.
[b]Note heavy atom effect in phosphorescence.

In Eq. (4.9), v is the frequency of radiation and ω_i and ω_f are the statistical weights of the initial and final states. It should be remembered that Eq. (4.9) refers to the photoionization cross section, not the total photoabsorption cross section (see Sect. 4.2).

Ionization produced by charged-particle impact is of great importance, because it provides a basis for dosimetry. In principle, a *fivefold* differential cross section is needed for the complete description of impact ionization: one in ejected electron energy, and two each for the angular distributions of the ejected electron and the incident particle. In radiation chemistry, one frequently is concerned only with the total number of ionizations, and such details are unnecessary. Even for track structure, a cross section doubly differential in ejected electron energy and angle is sufficient. In the case of electron-impact ionization at high energy, Mott (1930) incorporated the indistinguishability of the outgoing electrons into the Rutherford formula (see Sect. 2.3.3) as follows:

$$\left(\frac{d\sigma}{dE}\right)_M = \Gamma \sum_i n_i [Q_1^{-2} - Q_1^{-1}(T - E)^{-1} + (T - E)^{-2}]. \tag{4.10}$$

In Eq. (4.10), $\Gamma = 4\pi a_0^2 R^2/T$ and the second term within the brackets represents interference between direct and exchange scatterings. A corresponding relativistic treatment has been given by Møller (1931).

The quantum-mechanical ionization cross section is derived using one of several approximations—for example, the Born, Ochkur, two-state, or semi-classical approximations—and numerical computations (Mott and Massey, 1965). In some cases, a binary encounter approximation proves useful, which means that scattering between the incident particle and individual electrons is considered classically, followed by averaging over the quantum-mechanical velocity distribution of the electrons in the atom (Gryzinski, 1965a–c). However, Born's approximation is the most widely used one. This is discussed in the following paragraphs.

Consider inelastic scattering of an incident ion when the impinging velocity is large compared with the orbital velocities of the electrons. [The description here is nonrelativistic. Relativistic modifications have been reviewed by Inokuti (1971), who has also given a critical appraisal.] In the first Born approximation, the incident particle is described by a plane wave and the scattered particle by a slightly perturbed plane wave. The differential cross section for scattering into the solid angle $d\Omega$ attended by excitation (or ionization) to the state n with energy ε_n is then given by

$$\frac{d\sigma_n}{d\Omega} = (2\pi)^{-2} \ \hbar^{-4} \ M^2 \ \frac{k'}{k} \ |\gamma|^2, \tag{4.11}$$

where

$$\gamma = \int \psi_n^* \ \exp[i\vec{r} \cdot (\vec{k} - \vec{k}')] \, U\psi_0 \, d\tau.$$

In the foregoing, U is the interaction potential, M is the reduced mass of the colliding system, $\hbar\vec{k}$ and $\hbar\vec{k}'$ are respectively the momentum of the projectile before and after the collision, ι_0 and ι_n are respectively the wavefunctions of the atom (or molecule) in the ground and nth excited states, and the volume element $d\tau$ includes the atomic electron and the projectile. Since U for charged-particle impact may be represented by a sum of coulombic terms in most cases, Eq. (4.11) can be written as (Bethe, 1930; Inokuti, 1971)

$$\frac{d\sigma_n}{d\Omega} = 4z^2 \ |\vec{k} - \vec{k}'|^{-4} \ \frac{k'}{k} \left(\frac{Me^2}{\hbar^2}\right)^2 |\alpha_n|^2,$$

where ze is the charge of the incident particle and $\alpha_n(\vec{k} - \vec{k}')$ is the matrix element of the operator

$$\sum_j \exp[i\vec{r}_j \cdot (\vec{k} - \vec{k}')]$$

between the ground and excited states of the atom called the *form factor* for inelastic collision. The summation is over the coordinates of the electrons of the atom.

Barring exceptional cases, the matrix element α_n can be taken as a function of the magnitude of the momentum transfer ($K = |\mathbf{k} - \mathbf{k}'|$). Thus, integrating over the angle and converting the solid angle into momentum transfer, one gets

$$\frac{d\sigma_n}{d(K^2)} = 4\pi z^2 K^{-4} k^{-2} \left(\frac{Me^2}{\hbar^2}\right)^2 |\alpha_n(K)|^2. \tag{4.12}$$

Further, one may write $K^2 = 2mQ/\hbar^2$, where Q is the energy imparted *if* the momentum were transferred to a free electron (Q is not the actual energy transfer, which is ε_n), and obtain

$$\frac{d\sigma_n}{d(\ln Q)} = \frac{2\pi z^2 e^4}{mv^2} \frac{|\alpha_n(K)|^2}{Q}. \tag{4.13}$$

Bethe (1930) defined *the generalized oscillator strength* in terms of the form factor as

$$F_n(K) = \frac{\varepsilon_n}{Q} |\alpha_n(K)|^2 \tag{4.14}$$

and proved the sum rule

$$\sum_n F_n(K) = Z,$$

where Z is the total number of electrons in the target. The generalized oscillator strength is analogous to the dipole oscillator strength, which determines the cross section for optical transitions (see Sect. 4.7). For present purposes, the latter can be written as

$$f_n = \frac{\varepsilon_n}{R} |x_n|^2 \ (ea_0)^{-2},$$

where a_0 is the Bohr radius, R is the Rydberg energy (13.6 eV), and x_n is the dipole moment matrix element. Expanding the exponential, it is readily shown that

$$\lim_{k \to 0} F_n(K) = f_n.$$

This limit is actually nonphysical for inelastic collisions, as an excitation to a state with energy ε_n cannot occur with zero momentum transfer. Nonetheless, it is an important limit, which serves to check on experiments on inelastic collisions (Lassettre and Francis, 1964), and it shows clearly the similarity between optical transitions and collision with fast charged particles for which K is small. From (4.13) and (4.14), one can write

$$\frac{d\sigma_n}{d[\ln(Ka_0)^2]} = \frac{4\pi a_0^2 z^2 R^2}{T} \frac{F_n(K)}{\varepsilon_n}, \tag{4.15}$$

where $T = mv^2/2$. Equation (4.15) shows that the cross section for the inelastic collision is essentially proportional to the factor $F_n(K)/\varepsilon_n$.

To further reduce of the cross section formula (4.11), we note that it is proportional to the area of the curve of $F_n(K)/\varepsilon_n$ plotted against $\ln(Ka_0)^2$ between the maximum and minimum momentum transfers. Since T is large and the generalized oscillator strength falls rapidly with the momentum transfer, the upper limit may be extended to infinity. In addition, the minimum momentum transfer decreases with T in such a manner that the limit $F_n(K)$ may be replaced by f_n, the dipole oscillator strength for the same energy loss. This implies that a mean momentum transfer can be defined independently of T such that the relevant area of the curve of $F_n(K)/\varepsilon_n$ is equal to $(f_n/\varepsilon_n)[(\ln \overline{K}a_0)^2 - (\ln Ka_0)^2]$. Thus, by definition (Bethe, 1930; Inokuti, 1971),

$$\ln(\overline{K}a_0)^2 = \int_0^\infty \frac{F_n(K)}{f_n}\, d[\ln(Ka_0)^2] - \int_{-\infty}^0 \left[1 - \frac{F_n(K)}{f_n}\right] d\,[\ln(Ka_0)^2].$$

With this value of $\overline{K}$ one gets, from (4.11),

$$\sigma_n = \frac{4\pi a_0^2 z^2 R^2}{T}\, \frac{f_n}{\varepsilon_n} \left\{ \ln\left[\frac{4RT}{\varepsilon_n^2}\, (\overline{K}a_0)^2\right] + O\!\left(\frac{\varepsilon_n}{T}\right) \right\} \qquad (4.16)$$

where the second term on the right-hand side represents a correction of the order of ε_n/T for finite T. In Eq. (4.16), we have used the kinematic expression for K_{min} for large T—namely, $(Ka_0)_{min}^2 = \varepsilon_n^2/4RT + O(\varepsilon_n/T)$. Bethe (1930) defines a T-independent number through

$$\ln c_n = \ln\left[\frac{(\overline{K}a_0)^2 R^2}{\varepsilon_n^2}\right].$$

From this relation and Eq. (4.6), we get, ignoring the small correction,

$$\sigma_n = \frac{4\pi a_0^2 z^2 R^2}{T}\, \frac{f_n}{\varepsilon_n} \ln\left(\frac{4c_n T}{R}\right) \qquad (4.17)$$

which is the fundamental Born–Bethe formula for the total inelastic cross section σ_n for the energy loss ε_n. Note that basically this formula depends on the dipole oscillator strength, although the generalized oscillator strength appears (weakly) through c_n.

To convert Eq. (4.13) to the ionization cross section, we have to replace f_n by df/dQ_i and ε_n by Q_i, where $Q_i = E + I_i$ is the energy transferred to the electron in the ith orbital of ionization energy I_i, and E is the secondary electron energy. Summing over all the occupied orbitals consistent with the same E, we get (see Kim, 1975a)

$$\frac{d\sigma}{dE} = \frac{4\pi a_0^2 z^2 R^2}{T} \sum_i Q_i^{-1} \frac{df}{dQ_i} \ln\left(\frac{4c_c T}{R}\right). \tag{4.18}$$

In Eq. (4.18), it is implicitly assumed that the ionization is a direct, one-electron process; that is, the contribution of superexcited states to ionization is *not* included. The latter process is indirect and essentially of a two-electron nature. When the energy loss is much larger than the ionization potential, however, ionization is almost a certainty. For high energies of the secondary electron, Eq. (4.18) approaches the Rutherford cross section, or the Mott cross section if the incident particle is an electron.

The total ionization cross section is found from (4.18):

$$\sigma = \int_0^{E_{max}} \frac{d\sigma}{dE}\, dE, \tag{4.19}$$

where E_{max} is the maximum energy of the ejected electron. For incident heavy particles, $E_{max} \approx 2mv^2/(1 - \beta^2)$ with $\beta = v/c$. In the case of incident electrons, it is customary to designate the more energetic of the two outgoing electrons as the primary; thus, in this case, $E_{max} = (1/2)(mv^2/2 - I)$, where I is the ionization potential of the target.

Platzman suggested the plotting of the function

$$Y(T, Q) = \left(\frac{Q}{R}\right)^2 \frac{T}{4\pi a_0^2 z^2} \frac{d\sigma}{dE},$$

which is the ratio of the ionization cross section to the Rutherford cross section for ionization for the least bound electron. When experimental values are used for $d\sigma/dE$, the curve of Y vs. R/Q has an area proportional to the total ionization cross section (see Eq. 4.19), the constant of proportionality being $T/(4\pi a_0^2 R)$. Kim (1975a) has advocated the use of the Platzman plot in analyzing experimental data. Some advantages of using the Platzman plot as emphasized by Kim (1975a) are (1)

$$\lim_{Q\to\infty} Y = Z,$$

the total number of electrons in the system; (2) for E less than about $10R$, the shape of the Y curve resembles that of $Q(df/dQ)$; (3) autoionization and Auger processes show up as dips or peaks at appropriate energies; and (4) in proton impact, a peak is seen at ejected electron speed equal to the proton speed representing the continuum state of the H atom. Kim (1975a, b) has made repeated use of the Platzman procedure in the careful analysis of the energy distribution of the secondary electrons.

4.6   PHOTOIONIZATION
AND PHOTODETACHMENT

The simplest case of photodetachment—that is, ionization of a negative ion—is that of H⁻. Chandrasekhar (1945a, b) studied this case theoretically for its astrophysical interest. For this ion, the experimentally determined spectrum of photodetachment is nearly in perfect agreement with theory (Branscomb, 1962). Other examples of atomic negative ions are found in C⁻, O⁻, and so forth. Due to the nature of the short-range forces responsible for the binding of an extra electron, it is not expected that every atom or molecule will have a negative ion. For the same reason, when a negative ion exists, one generally expects very few bound states, usually one.

The threshold dependence of cross section with energy is somewhat different for photodetachment and photoionization. In the latter case, the angular momentum does not play a dominant role. Thus, photoionization cross sections at threshold are expected to be finite and relatively independent of energy (in practice, this is not always true). This is not true for photodetachment, where the cross section σ starts with zero at threshold and then gradually increases with energy. In this case, the outgoing electron sees a short-range potential and the angular momentum barrier becomes an important problem. The transition in H⁻ is of the s–p type; at threshold, $\sigma \propto (\varepsilon + A)\varepsilon^{3/2}$ or $\varepsilon^{3/2}$ if ε, the kinetic energy of the outgoing electron, is much less than A, the electron affinity. The transitions in O⁻, C⁻, and so forth, are of the p–s type, thus giving $\sigma \propto \varepsilon^{1/2}$ at threshold, which is a milder form of variation. In some respects, photodetachment from molecular negative ions resembles the atomic case. Of course, there is a complication due to the Franck–Condon effect. Thus, it is conceivable that the vertical detachment energy may be different from the electron affinity and even of opposite sign in cases involving large differences between the equilibrium positions of the ion and the neutral.

Table 4.4 lists some threshold wavelengths and cross sections for photoionization. For atoms, the cross section refers to that at spectral head; for molecules, the maximum cross section is given. Photoionization cross sections for simple atoms and molecules are often hard to measure, mainly for lack of suitable sources. Usual optical sources go up to the beginning of vacuum-UV, whereas soft X-rays come down to ~1 KeV, leaving a gap in the most important region for the photoionization of small molecules. Synchrotron sources are being used in this region, but these are not easily available. An alternative procedure is to determine the generalized oscillator strength from fast electron scattering as a function of momentum transfer (k) at fixed excitation energy ε. The calculated (admittedly unrealistic) limit

$$\lim_{k \to 0} \frac{df(k, \varepsilon)}{d\varepsilon}$$

TABLE 4.4   Threshold Wavelength and Cross Section
for Photoionization

	λ_T (nm)	σ_1 (Mb)
Atoms		
H	91.2	6.3
He	50.4	7.4
C	110.0	11
N	85.2	9
O	91.0	2.6
F	71.3	6
Na	241	0.1
Ne	57.5	4.0
Ar	78.7	35
Kr	88.5	—
Xe	102.0	—
Molecules		
H_2	80.5	7.4
CO	86.8	16.5
N_2	79.2	26
O_2	77.0	22
H_2O	96.0	35
NH_3	121.0	10
CH_4	96.7	56

equals the density of optical oscillator strength $df/d\varepsilon$. A test is usually applied to verify that the thusly stated limit is independent of incident electron energy. This procedure gives a "photoabsorption" cross section by electron impact (Inokuti, 1971; Lassettre *et al.*, 1969). Also, with a coincidence measurement of two outgoing electrons, one can determine the "photoionization" cross section. Once the DOSD (differential oscillator strength distribution) is obtained, the cross sections for the photoabsorption and the fast-charged-particle impact processes are given basically by $df/d\varepsilon$ and $\varepsilon^{-1} df/d\varepsilon$, respectively, apart from determinable parameters (see Sects. 2.3.4 and 2.5.2).

We have already mentioned the threshold behavior of photodetachment. Morrison (1957) postulated the rule that at or near threshold, the cross section for a process should vary as $\sigma \propto (\varepsilon - \varepsilon_0)^n$, where ε is the energy transferred, ε_0 is the minimum energy required for the process, and n equals the number of outgoing electrons minus 1. Thus, as shown schematically in Figure 4.4, the threshold cross section should vary linearly with energy for

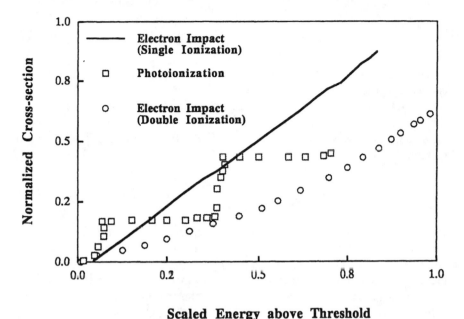

Scaled Energy above Threshold

FIGURE 4.4   Schematic of threshold behavior of ionization processes. Under ideal conditions, one expects a step function for photoionization, a linear variation with energy under electron impact, and a parabolic dependence for double ionization by electron impact.

electron impact ionization ($n = 1$), should show a step behavior for photoionization ($n = 0$), and should vary parabolically with energy for double ionization by electron impact ($n = 2$). Experiments with electron impact on He (both single and double ionizations) support this rule. So does photoionization in some simple molecules, although there are significant exceptions. Electron attachment to neutral molecules sometimes exhibits a resonance character, which also somehow is supposed to be covered by the Morrison rule. However, the rule should not be taken too seriously. It has not been rigorously derived, and the range of validity never has been appropriately defined. Detailed discussion of experimental photoionization and photoabsorption cross sections of atoms and molecules has been made by Christophorou (1971), to which the interested reader is referred.

4.7   OSCILLATOR STRENGTH
AND SUM RULES

Closely related to the absorption coefficient is the concept of the *oscillator strength* for the transition, or the *f*-number. It is given by

$$f = \frac{mc}{\pi n e^2} \int \alpha(v)\, dv, \qquad (4.20)$$

where m, c, and e have their usual meanings, and the integration in Eq. (4.20) is over the range of frequencies associated with the transition [see Eq. (4.6) in Sect. 4.4.3]. This concept already has been alluded to in Sects. 4.5 and 4.6, with which the present section should be read. The connection between the oscillator strength and the radiative lifetime τ or the Einstein A coefficient may be given as follows:

$$f = \frac{mc\lambda^2}{8\pi^2 e^2 \tau} \frac{g_j}{g_i} = \frac{mc\lambda^2}{8\pi^2 e^2} A_{ji} \frac{g_j}{g_i}, \qquad (4.21)$$

where λ is the wavelength of radiation, and g_i and g_j are respectively the statistical weights of the lower (i, absorbing) and upper (j, emitting) states. This shows that f is an atomic (or molecular) property independent of extraneous factors. To relate the A coefficient to the properties of an atom (or molecule), one uses semiclassical theory where the matter is treated quantum-mechanically but the radiation classically (see Schiff, 1955). The result is

$$A_{ji} = \frac{64\pi^4 v^3}{3hc^3} |R_{ij}|^2, \qquad (4.22)$$

where $|R_{ij}|^2 = |x_{ij}|^2 + |y_{ij}|^2 + |z_{ij}|^2$. The matrix element for the x component of the dipole moment operator is defined as

$$x_{ij} = \left\langle \psi_j^* \left| e \sum_k x_k \right| \psi_i \right\rangle$$

where x_k is the x coordinate of the kth electron and Ψ_j and Ψ_i are the wavefunctions corresponding to the states j and i; similarly for y_{ij} and z_{ij}. From (4.21) and (4.22), one can write the oscillator strength in terms of the dipole moment matrix elements as follows:

$$f = \frac{8\pi^2 mv}{3e^2 h} \frac{g_j}{g_i} |R_{ij}|^2.$$

The dipole oscillator strength is the dominant factor in dipole-allowed transitions, as in photoabsorption. Bethe (1930) showed that for charged-particle impact, the transition probability is proportional to the matrix elements of the operator $\exp(i\vec{k} \cdot \vec{r})$, where $\hbar\vec{k}$ is the momentum transfer. Thus, in collision with fast charged particles where $\vec{k} \cdot \vec{r}$ is small, the process is again controlled by dipole oscillator strength (see Sects. 2.3.4 and 4.5).

Therefore, fast-charged-particle impact resembles optical transition to some extent. The oscillator strength introduced in Chapter 2 corresponds to this kind of transition, whereas that for the entire operator $\exp(i\mathbf{\bar{k}} \cdot \mathbf{\bar{r}})$ is called the generalized oscillator strength, which also has some interesting properties (Inokuti, 1971).

It is not easy to calculate oscillator strengths from first principles except in some very simple cases. On the other hand, the oscillator strength distribution must fulfill certain sum rules, which in some cases help to unravel their character. Referring the (dipole) oscillator strength for the transition from the ground state with excitation energy ε_n to state n as f_n, a sum may be defined by

$$S(i) = \sum_n \varepsilon_n^i f_n ,$$

where integration over continua is implied. The properly normalized $S(i)$ gives the mean value of the ith power of excitation energy over the excitation probabilities. While $S(i)$ diverges for large $|i|$, for $-4 < i < 2$, it is related to important physical properties of the system (Nicholls and Stewart, 1962). $S(0) = Z$, the total number of electrons in the system; this relation, called the *Thomas–Kuhn sum rule*, is also satisfied by the generalized oscillator strength, as shown by Bethe (1930). $S(-4)$ can be obtained experimentally from data on the refractive index and the Verdet constant, the latter referring to the rotation of the plane of polarization per unit thickness per unit magnetic field parallel to the propagation direction. $S(-2) = \alpha/4$, where α is the polarizability. $S(2) = (16\pi Z/3)$ times the average electron concentration at the center of the system. $S(-1)$ and $S(1)$ are not so easily interpreted, but they are related, apart from a constant ground state energy, to position and momentum correlations, respectively (Nicholls and Stewart, 1962).

Another class of sum rules can be obtained by formally differentiating $S(i)$ with respect to I, giving

$$L(i) = \sum_n \varepsilon_n^i f_n \ln \varepsilon_n = S(i) \frac{d}{di} S(i).$$

The quantity $L(0) = \ln I$, where I is the mean excitation potential introduced by Bethe, which controls the stopping of fast particles (see Sect. 2.3.4); $L(2) = \ln K$, where K is the average excitation energy, which also enters into the expression for Lamb shift (Bethe, 1947). Various oscillator sum rules have been verified for He and other rare gases to a high degree of accuracy. Their validity is now believed to such an extent that doubtful measurements of photoabsorption and electron-impact cross sections are sometimes altered or corrected so as to satisfy these.

4.8   THE W VALUE

If an incident particle with energy E generates on the average n_i ions of either sign while being completely absorbed, then the *integral* W value of the medium is defined by

$$W = \frac{E}{n_i}. \tag{4.23a}$$

Ionizations of all generations are included in the definition of W, which is therefore a gross average. Closely related to the *integral* W value, there exists a more theoretically significant *differential* ω value defined over a small track segment such that the inelastic collision cross sections are sensibly constant. The relationship between these quantities is given by (Dalgarno, 1962)

$$\frac{1}{\omega} = \frac{d}{dE}\left(\frac{E}{W}\right), \tag{4.23b}$$

where E is the incident particle energy. Obviously, if W is independent of E, so is ω, and vice versa.

Table 4.1 gives the experimental W values and ionization potentials (I) for some atoms, molecules, and radicals. Christophorou (1971) lists the W values of many more compounds and mixtures, from which we can draw the following conclusions: (1) except at low energies, W is insensitive to particle energy; (2) it is also nearly independent of the radiation quality; and (3) generally speaking, the ratio W/I is smaller for rare gases (1.7–1.9) than for molecules (mostly 2.3–3.0). This may be attributed to additional nonionizing losses in molecules such as vibrational excitation, dissociation of superexcited states into neutral fragments, and so forth. The insensitivity of W to particle energy means that the measurement of ionization is equivalent to the measurement of relative energy loss. It also implies that the curve of specific ionization—that is, the number of ions created per unit path length, or the Bragg curve—has the same shape as the energy-loss curve. These factors are extremely important for radiation dosimetry. Fano's (1946) qualitative explanation for the near independence of ω from E is based on the insensitivity of the relative (note: *not* absolute) cross sections on energy (e.g., σ_{ex}/σ_{ion}). That the assumption is plausible follows from the Born approximation.

Dalgarno and Griffing (1958) made a detailed theoretical analysis of the ionization produced by a beam of protons penetrating a gas of H atoms. They find that the W value remains constant at around 36 eV, to within 2.5 eV, for proton energies of 10 KeV and up. However, below about 100 KeV, the near constancy of the W value is also partially due to the fact that the beam is a near equilibrium composition of protons and H atoms because of charge exchange. Therefore, at lower energies, the increase of the W value specific to the penetrating protons is largely canceled by the smaller W value specific to H atoms. This interesting effect extends the range of the constancy of W, and it should be considered along with Fano's argument presented in the previous paragraph.

Platzman (1961) gave the ratio of the W value to the ionization potential (I) as

$$\frac{W}{I} = \frac{\overline{E}_{\mathrm{ion}} + v_{\mathrm{ex}}\overline{E}_{\mathrm{ex}} + \overline{E}_{\mathrm{s}}}{I}, \qquad (4.23c)$$

where $\overline{E}_{\mathrm{ion}}$, $\overline{E}_{\mathrm{ex}}$, and $\overline{E}_{\mathrm{s}}$ are respectively the average energies for ionization, excitation, and subexcitation electrons and v_{ex} is the relative number of excitations to ionizations. For helium, Platzman takes $\overline{E}_{\mathrm{ion}}/I = 1.06$ to account for a small production of doubly charged ions and for the production of excited ions. The value of v_{ex} was estimated to be 0.4 and $\overline{E}_{\mathrm{ex}}/I$ and $\overline{E}_{\mathrm{s}}/I$ were taken to be 0.85 and 0.31, respectively. Thus, Platzman obtained $W = 42.3$ eV and $W/I = 1.71$ for He, in agreement with the main experimental results. Although the exact relationship between W and I remains somewhat elusive, many empirical rules are known. For molecules, as an example, Christophorou (1971) obtained by the method of least squares $W = 9.8 + 1.67I$ for α-particles and $W = 5.9 + 1.82I$ for β-particles and high-speed electrons.

A semianalytical theory of the W value was invented by Spencer and Fano (1954). It consists of writing an integral equation for *the degradation spectrum* $y(E, T)$, defined so that the total path length of all electrons in the medium between energies E and $E + dE$ is $y(E, T)\, dE$ when initially there is a unit flux of electrons at energy T. Although y is not easily obtained, the advantage is that, once it is known, the yield of any primary species is obtained merely by an integration as follows:

$$N_x(T) = N \int_{I_x}^{T} \sigma_x(E)y(E, T)\, dE. \qquad (4.23d)$$

In Eq. (4.23d), N is the number density of molecules, σ_x is the cross section for production of any primary species x at electron energy E having an energetic threshold I_x for its production, and $N_x(T)$ is the total number of such species formed by the complete absorption of an electron of kinetic energy T. In this sense, y can be thought of as a distribution function. Specifically, if x refers to ionization, I_x is simply the ionization potential I, σ_x is the ionization cross section σ_i, and $N_x = n_i$, the total number of ionizations.

Another method for finding the W value, called the Fowler equation approach (Inokuti, 1975), is based on three assumptions, some of which can be relaxed. These are (1) that the incident particle is an electron; (2) that there is only one ionization potential; and (3) that the ionization efficiency is unity—that is, any energy loss $E > I$ results in an ionization with a primary of energy

$T - E$ and a secondary electron with energy $E - I$. One can then write the following *integral equation* by classifying the first inelastic event:

$$n_1(T) = p_i(T) + \int_I^{(T+I)/2} [n_1(T - E) + n_1(E - I)] \frac{dp_i(E, T)}{dE}$$

$$+ \sum_n p_n(T)n_1(T - E_n). \qquad (4.23e)$$

Equation (4.23e), which is the *Fowler equation* of the problem, is self-explanatory. Here $n_1(T)$ is the total number of ionizations created by the complete absorption of a primary electron of energy T. The first term on the right-hand side represents the probability of first primary ionization. It creates two electrons, of energies $T - E$ and $E - I$. The second term gives the *total* number of ionizations produced by the complete absorption of these two electrons. Finally, the last term represents the excitation of all n levels short of ionization, leaving the electron with energy $T - E_n$, whence by definition the number of further ionizations caused by the subsequent complete absorption of the electron is given by $n_1(T - E_n)$. The probabilities $p_i(T)$ and $p_n(T)$ are given respectively by $\sigma_i(T)/\sigma_t$ and $\sigma_n(T)/\sigma_t$, where σ_i and σ_n are respectively the cross sections of ionization and excitation to the nth level with

$$\sigma_t = \sigma_i + \sum_n \sigma_n.$$

Given the relevant cross sections, one can solve Eq. (4.23e) numerically with $n_1 = 0$ for $T < I$ and $n_1 = p_i$ for $I \leq T < I + E_1$, where E_1 is the first excitation potential. Numerical computation for He shows that for $T \gg I$ (say, above 100 eV), a simple linear solution may be obtained:

$$n_1(T) \approx \frac{T - U}{W_\infty}, \qquad (4.23f)$$

where W_∞ is the asymptotic approximation to the W value (at $T \to \infty$) and U is a parameter having the dimension of energy. Detailed calculation gives $U = 14$ eV for He and less for other atoms and molecules (Inokuti, 1975). Substituting Eq. (4.23f) into Eq. (4.23e) and rearranging, one gets

$$W_\infty = I + U + \sum_n \frac{E_n p_n(T)}{p_i(T)},$$

which is slightly inconsistent in the sense that now W_∞ is a mild function of T. However, this inconsistency is not serious, since $p_n(t)/P_i(T)$ is generally insensitive to T. Inokuti suggests a first-order correction by using Eq. (4.23f) for $E \geq 2I$, but taking $n_1(E) = p_i(E)$ for $I \leq E < 2I$ and $n_1(E) = 0$ for $E < I$. This leads, via a self-consistent procedure, to the following result:

$$U + I = E' \equiv \frac{\int_{I}^{2I} [E \, dp_i(E, T)/dE] \, dE}{\int_{I}^{2I} [dp_i(E, T)/dE] \, dE}. \tag{4.23g}$$

Equation (4.23g) may be interpreted as E' being equal to the average energy transfer in an ionizing collision where the ejected electron is *subionization*—that is, incapable of causing further ionization. The energy U is then the excess of E' over the ionization potential I. Therefore, it is expected that $E_1 \leq U \leq I$. Values of U calculated for typical cases using Eq. (4.23g) and realistic cross sections give results in good agreement with numerical solutions. From Eqs. (4.23a) and (4.23f), an energy-dependent $W(T)$ may be obtained as follows:

$$W(T) = W_\infty \left(1 - \frac{U}{T}\right)^{-1}. \tag{4.23h}$$

The significance of Eq. (4.23h) is that when $T \gg U$, $W(T) = W_\infty$, a constant as indeed observed. The energy dependence of $W(T)$ is given by Eq. (4.23h), where the details of the cross sections are represented in the single parameter U. Note that Eq. (4.23h) cannot be valid for very small T unless $U = I$, since it will give $W(T) \rightarrow \infty$ as $T \rightarrow U$, whereas the proper limit is reached at $T \rightarrow I$. This is not a great difficulty, since for small values of T either a direct Monte Carlo scheme or a numerical solution of the Fowler equation can be found without too much difficulty. Inokuti (1975) relaxed some of the assumptions inherent in the Fowler equation and concluded that the basic solution (Eq. 4.23h) remains valid under fairly general conditions.

Cole (1969) has measured the W value in air for electrons of energy 5 KeV down to 20 eV. In his experiment, monoenergetic electrons are completely absorbed in air. In the ionization chamber, the central electrode, when negatively charged, carries a current J^+ due to positive ions. When the polarity is reversed, it carries the total current $J = J^+ + i$, where i is the input electron beam current. By difference, i is calculated and the W value is given by $W = T/(J^+/i)$. Later, Combecher (1980) measured energy-dependent W values in several gases including water vapor. It is noteworthy that W is consistently larger than ω. The trend of the variation of W and ω with particle energy is believed to be of general validity.

Another procedure for calculating the W value has been developed by LaVerne and Mozumder (1992) and applied to electron and proton irradiation of gaseous water. Considering a small section Δx of an electron track, the energy loss of the primary electron is $S(E) \Delta x$, where $S(E)$ is the stopping power at electron energy E. The average number of primary ionizations produced over Δx is $N\sigma_i \Delta x$ where σ_i is the total ionization cross section and N is the number density of molecules. Thus, the W value *for primary ionization* is $\omega_p = S(E)/N\sigma_i(E)$. If the differential ionization cross section for the production of a secondary electron at energy ε is denoted by $d\sigma_i/d\varepsilon$, then the total number of secondary electrons produced over Δx in *all generations* will be given by

$$N\Delta_x \int_I^{\varepsilon_m} \frac{d\sigma_i}{d\varepsilon} \frac{\varepsilon}{W(\varepsilon)} d\varepsilon,$$

where I is the ionization potential, $\varepsilon_m \approx (E - I)/2$ is the maximum energy of the secondary electron, and $W(\varepsilon)$ is the *integral W* value for an electron of energy ε. Counting all ionizations, primary and secondary, the differential ω value at primary electron energy E is given by

$$\omega(E) = \frac{S(E)}{N\left[\sigma_i + \int_I^{\varepsilon_m} \frac{d\sigma_i}{d\varepsilon} \varepsilon W^{-1}(\varepsilon)\, d\varepsilon\right]} = \frac{\omega_p(E)}{1 + \int_I^{\varepsilon_m} \frac{d\rho_i}{d\varepsilon} \varepsilon W^{-1}(\varepsilon)\, d\varepsilon},$$

where *the relative secondary electron spectrum* is defined by $d\rho_i/d\varepsilon$.

To calculate the differential ω value using the preceding equation, and hence the integral W value from Eq. (4.23b), requires the knowledge of stopping powers, differential ionization cross sections, and integral W values for electrons up to energy ε_m. Clearly, some *aufbau principle* is needed for which the electron energy span is divided into the intervals $I - 3I, 3I - 7I, 7I - 15I$, and so forth. In the first interval, no secondary ionization is possible and $\omega(E) = \omega_p(E)$; nevertheless, Eq. (4.23b) must be used to compute $W(E)$. Ionization cross sections and stopping powers are obtainable from experiments or compilations. It should be noted that at low electron energies (≤ 40 eV), extraelectronic processes, mainly vibrational, contribute significantly to stopping power (Hayashi, 1989). In the second and subsequent intervals, the integral W values are available from earlier intervals. The procedure for incident protons is quite similar, except that the stopping power, ionization cross section, and ε_m must be appropriate for protons. In the high-energy limit, ω and W often are indistinguishable. Further making the plausible assumption that most secondary ionizations are created by higher-energy electrons, one gets in this limit ($W = W_\infty$)

$$W = \frac{S(E)}{N\left[\sigma_i + W^{-1}\int_I^{\varepsilon_m} \varepsilon \frac{d\sigma_i}{d\varepsilon}\, d\varepsilon\right]} = \frac{\omega_p}{1 + \langle\varepsilon\rangle W^{-1}} \omega_p - \langle\varepsilon\rangle,$$

where

$$\langle\varepsilon\rangle \equiv \sigma_i^{-1} \int_I^{\varepsilon_m} \varepsilon \frac{d\sigma_i}{d\varepsilon}\, d\varepsilon$$

is the mean energy of the secondary electrons between I and ε_m. The preceding equation may be interpreted intuitively—namely, the energy requirement for overall ionization is reduced from that of primary ionization by an amount equal to the mean energy of secondary electrons that are capable of further ionization.

By detailed analysis of available experiments and compilations, LaVerne and Mozumder (1992) concluded that

1. Various experiments on the electron energy dependence of ionization cross sections in gaseous water are basically consistent with each other. Rudd's (1989) model computes these cross sections quite well.

2. Hayashi's (1989) compilation highly overestimates the total inelastic cross section below ~100 eV. These are inconsistent with measured W values. Although the total cross section is reasonably well determined, uncertainties in the elastic cross section might have led Hayashi to overestimate the inelastic cross section. Figure 4.5 shows these cross sections. It is seen, however, that one theoretical calculation is consistent with W value measurement (Pimblott et al., 1990). In any case, Hayashi's values for total inelastic cross section are much greater than all major calculations, and the discrepancy is directly traceable to overestimates of excitation cross sections.

3. At incident electron energy ≥ 1 KeV, asymptotic W_∞ is reached. The calculated value (34.7 eV) is greater than measured (29.6 eV; Christophorou, 1971) by 15%, which is attributable to relative errors in total and inelastic cross sections. At 1 KeV, $\omega_p = 43.5$ eV is calculated, which shows the importance of secondary ionization.

4. For proton irradiation, asymptotic W_∞ is reached at 500 KeV. The computed integral W value, 28.9 eV, compares well with experimental determination (30.5 eV; Christophorou, 1971); ω_p at this energy is calculated to be 53.5 eV, showing great importance of secondary ionization.

4.9   SOME SPECIAL CONSIDERATIONS

4.9.1   CONDENSED PHASE EFFECTS

Ionization in the condensed phase presents a challenge due to the lack of a precise operational definition. Only in very few cases, such as the liquefied rare gases (LRG), where saturation ionization current can be obtained at relatively low fields, can a gas-phase definition be applied and a W value obtained (Takahashi et al., 1974; Thomas and Imel, 1987; Aprile et al., 1993).

Operationally, a procedure may be based on measuring the yield of a reaction traceable to ionization, usually giving a lower limit to the ionization yield. Thus, in the radiation chemistry of hydrocarbon liquids, the product of an electron scavenging reaction (for example, $C_2H_5 \cdot$ radical from the scavenger C_2H_5Br)

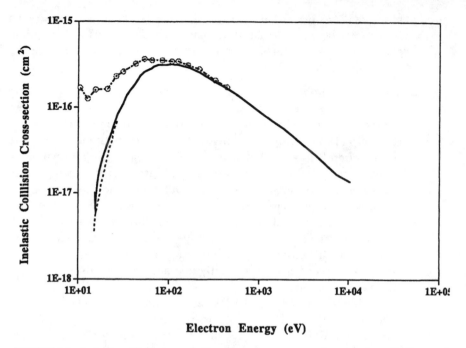

FIGURE 4.5   Inelastic collision cross section of water vapor versus electron energy (LaVerne and Mozumder, 1992). Circles: compilation of Hayashi (1989); dashed line: unmodified theoretical formula (Pimblott *et al.*, 1990); dot-dashed line: theoretical formula scaled to match compilation; full curve: theoretical formula scaled to match experimental W values.

may be measured as a function of scavenger concentration and extrapolated to infinite concentration. This procedure is somewhat uncertain because of the fairly long extrapolation. Nevertheless, in some cases it results in a W value comparable to that in the gas phase. For example, in cyclohexane, the thusly determined W value for ^{60}Co-γ radiation is about 26 eV.

For highly polar media, the yield of the solvated electron can serve as a lower limit to the ionization yield. This method needs short-time measurement and may work for liquid water and ammonia. Farhataziz *et al.* (1974) determined the G value—that is, the 100-eV yield—of solvated electrons in liquid NH_3 to be about 3.1 at ~50 ns. This corresponds to a W value of 32 eV, compared with the gas-phase value of 26.5 eV. The difference may be attributed to neutralization during the intervening time. In liquid water, it has been found that $G(e_h)$ increases at short times and has a limiting value of 4.8 (Jonah *et al.*, 1976; Sumiyoshi *et al.*, 1985). This corresponds to W_{liq} = 20.8 eV compared with W_{gas} = 30 eV (Combecher, 1980). Considering that the yield of e_h can only be a lower limit of the ionization yield, suggestions have been made that the ionization potentials in polar liquids should be smaller than that in the gas phase (and, therefore, also the W value). It is much more likely that superexcited states will preferentially ionize in polar liquids. In the gas phase, superexcited states mainly dissociate into neutral fragments. In polar liquids, such dissociations may be frustrated by the relatively short dielectric relaxation time. If the yield of nonionizing superexcited state in the gas phase of water (0.9) is added to the ionization yield (3.3), the result (4.2) still falls short of the ionization yield in the liquid (4.8) by about 0.6 G-unit. The difference could be attributed to the lowering of the ionization potential in the liquid phase.

Biphotonic ionization has been observed in liquid water by Nikogosyan *et al.* (1983). Ionization begins to occur at ~6.5 eV with very small quantum yield. It rises very slowly with energy up to ~8 eV, beyond which the rise is much faster. The extrapolation of the higher-energy ionization efficiency (η) data to $\eta = 0$ gives an ionization potential $I_{liq} = 8.3$ eV (Pimblott and Mozumder, 1991), which is close to the value suggested by Goulet *et al.* (1990) and by Kaplan *et al.* (1990). Thus, it is important to distinguish between the appearance potential (6.5 eV) and the ionization potential (8.3 eV) in liquid water. This point has often been confused in the literature. Pimblott and Mozumder (1991) have used $I_{liq} = 8.3$ eV, $W_{liq} = 20.8$ eV, and a statistical method to give an ionization distribution in spurs for water radiolysis (*vide infra*).

Scavenging experiments in hydrocarbon liquids (Rzad *et al.*, 1970; Kimura and Fueki, 1970) tend to give low *observed* ionization yield, although the primary yield may be greater. The situation is similar for free-ion yield measurement under a relatively large external field. Both processes require large extrapolations to obtain the W value.

4.9.1a   Yields of Excited States in the Radiolysis of Liquids

Table 4.5 lists some of the yields of the lowest excited states (singlets and triplets) observed in the radiolysis of liquids. From this table, it seems that these yields vary widely, and it is hard to make theoretical estimates. One uncertain feature is the intersystem crossing ratio. Another is the possibility of spin-exchange (triplet) excitations by slow electrons. Magee and Huang (1972) considered this parametrically, but it is not very clear-cut. Walter and Lipsky (1975) suggest that the yield of singlet states measured by energy transfer tends to give seriously underestimated values. If this is true, then many of those yields will need revision.

In alkanes, excited states are produced (not exclusively) by ion neutralization. Singlet states are also produced, but exact pathways are debatable. In alkane solutions, excited states of solutes, produced by ion neutralization, are mainly triplets, but some singlets are also formed with biphenyl as a solute. Mechanisms of formation and decay of triplet states have been treated in detail

TABLE 4.5   Excited State Yields in the Radiolysis of Liquids

Compound	Solvent/Concentration	G-Value[a,b]	Reference
Singlets			
Benzene	Cyclohexane/.01 M	0.3	Dainton *et al.* (1973)
Cyclohexane	—	1.2[c]	Beck and Thomas (1972)
Toluene/*p*-xylene	Cyclohexane/.01 M	0.4	Dainton *et al.* (1973)
Anthracene	Cyclohexane/1 mM	0.2	Dainton *et al.* 1973)
Biphenyl	Cyclohexane/.01 M	0.44	Baxendale and Wardman (1971)
Anthracene	Cyclohexane/.01 M	0.4	Baxendale and Wardman (1971)
Benzene	—	1.6[c]	Thomas (1976)
Toluene	—	2.1[c]	Thomas (1976)
Mesitylene	—	1.6[c]	Thomas (1976)
p-xylene	—	2.0[c]	Thomas (1976)
Benzyl alcohol	—	0.7[c]	Thomas (1976)
Dimethylaniline	—	0.9[c]	Thomas (1976)
Napthalene	Liquid	2.4	Holroyd and Capellos (1972)
Napthalene	Cyclohexane/0.1 M	1.4	Dainton *et al.* (1972)
Napthalene	Cyclopentane/0.01 M	0.3	Salmon (1976)
Napthalene	Decalin/0.01 M	1.6	Salmon (1976)
1-Methyl-naphthalene	Liquid	2.7	Holroyd and Capellos (1972)
Toluene[f]	Cyclohexane/0.2 M	0.8	Miyazaki *et al.* (1975)
Toluene[f]	3-MP/0.2 M	0.7	Miyazaki *et al.* (1975)
Cyclohexane	Liquid	1.4–1.7	Walter and Lipsky (1975)
Methyl-cyclohexane	Liquid	1.9–2.2	Walter and Lipsky (1975)
Dodecane	Liquid	3.3–3.9	Walter and Lipsky (1975)
Bicyclohexyl	Liquid	3.5	Walter and Lipsky (1975)

(Continued)

by McGlynn *et al.* (1969), and their participation in pulse radiolysis has been reviewed by Singh (1972).

With many aromatic hydrocarbons as solutes, excited state yields in alkane solutions are nearly equally divided between singlets and triplets, and these yields increase with solute concentration until ~0.1 M (Salmon, 1976; Thomas *et al.*, 1968). In these systems, both the solute anion and the solute excited state yields increase similarly with solute concentration. With anthracene as a solute, the rate of growth of anthracene triplet matches that of the decay of the anthracene anion. With aromatic solvents, on the other hand, solute ions play

TABLE 4.5   (Continued)

Compound	Solvent/Concentration	G-Value[a]	Reference
Triplets			
Biphenyl	Cyclohexane/0.7 M	11,000[d]	Thomas *et al.* (1968)
Biphenyl	Polystyrene/0.2 M	1.3	Ho and Siegel (1969)
Anthracene	Cyclohexane/.01 M	1.0	Land and Swallow (1968)
Anthracene	Benzene/.01 M	2.0	Land and Swallow (1968)
Anthracene	Toluene/.03 M	4.0	Cooper and Thomas (1968)
Benzophenone	Benzene/1 M	3.2	Land and Swallow (1968)
Benzophenone	Neat	2.2	Holroyd *et al.* (1970)
Biacetyl	Benzene/.04 M	2.7	Cundall *et al.* (1968)
Benzene	—	3.8[c]	Thomas (1976)
Toluene	—	2.4[c]	Thomas (1976)
Mesitylene	—	1.8[c]	Thomas (1976)
p-Xylene	—	2.4[c]	Thomas (1976)
Benzyl alcohol	—	1.1[c]	Thomas (1976)
Dimethylaniline	—	3.1e	Thomas (1976)
Naphthalene	Benzene/1 M	2.3	Land and Swallow (1968)
Naphthalene	Acetone/.05 M	1.2	Hayon (1970)
Naphthalene	Cyclohexane/.01 M	1.0[c]	Dainton *et al.* (1972)
Naphthalene	Cyclohexane/.1 M	2.2[c]	Dainton *et al.* (1972)
Naphthalene	Cyclopentane/.01 M	0.58	Salmon (1976)
Naphthalene	Decalin/.01 M	1.2	Salmon (1976)

[a]100-eV yield.

[b]A short list is given here. A more extended list has been given by Singh (1972).

[c]About 40–50% of these are estimated to form by intersystem crossing (see Salmon, 1976).

[d]Product value of $G\varepsilon$.

[e]*G* values refer to the neat liquids, although in most cases energy transfer to solutes is used to establish the values

[f]Solid phase at 77 K.

a minor role. Excited singlet and triplet states of the solvent are first produced, followed by energy transfer (Thomas, 1976). It is suspected that in nonpolar aromatic systems, the initial ion yield is large but excited states are produced quickly on neutralization. Gangwer and Thomas (1975) set a lower limit of 1.0 for the yield of excited states in liquid alkyl benzenes by ion neutralization. There is some effect of the solvent on the yields of solute excited states; this effect is not clearly understood, although in some cases it is systematic (Salmon, 1976).

4.9.2   EXCESS ENERGY IN IONIZATION

Platzman (1967) has emphasized that most direct ionizations in molecules leave the positive ions in an excited state. Based on crude DOSD, he estimated that in water the average positive ion will have about 8 eV excitation energy. Later, the less approximate calculation of Pimblott and Mozumder (1991) reduced that figure to about 4 eV. The chemical role of this excitation energy is unknown, although it may have some effect in the radiolysis of highly concentrated solutions.

4.9.3   THE AUGER EFFECT AND K-PROCESSES

Normally, in impact ionization, outer electrons are removed. Infrequently, however, an inner electron may be ejected or a K-process may occur such as an orbital electron capture or β-decay. In such cases, the result is an electronic rearrangement, in preference to emission. Since enough energy is available, frequently the resultant ion is multiply charged. The cross section for this process follows the usual Bethe-type variation $\sim T^{-1} \ln(BT)$, where B is a constant (Fiquet-Fayard et al., 1968). In charged particle irradiation, the amount of energy lost in the K-processes is very small, usually much less than 1%. On the other hand, some specific effect may be attributable to that; that is, experiments can be so designed.

   One interesting feature of the Auger effect in molecules is that, contrary to the expectation based on the availability of a large amount of energy, the resultant molecular dissociation is rather infrequent. One tends to get more multiply charged molecules, at least initially. The reason is that the Auger process is much faster than molecular dissociation. Before the excess energy can be localized on bonds, it is largely dissipated as the kinetic energy of ejected electrons and in sharing of charge with neighboring molecules.

4.9.4   IONIZATION AND EXCITATION DISTRIBUTION IN SPURS

In Sect. 4.9.1, experimental rationalization was provided for the W value of ionization in gaseous and liquid water, giving respectively 30.0 and 20.8 eV. The corresponding ionization potentials are respectively 12.6 and 8.3 eV. For the purpose of diffusion and stochastic kinetics, one often requires the statistical distribution $P(i, j)$ of the number of ionizations i and excitations j, conditioned on i ionizations for a spur of energy ε. Pimblott and Mozumder (1991) write $P(i, j) = \Gamma(i)\Omega(j; i)$ where $\Gamma(i)$ is the probability of having i ionizations and $\Omega(j; i)$ is the probability of having j excitations conditioned on i ionizations. These probabilities are separately normalized to unity.

The first ionization gives a *progenitor electron* with energy $\varepsilon - J$, where $J = 1 +$ mean excitation energy in the positive ion, E^+. Subsequent ionizations cost, on an average, an energy $I' = J + T$, where T is the mean subexcitation electron energy. Thus, practically the maximum number of ionizations for spur energy ε is given by $i_m = 1 + \text{int}[(\varepsilon - J)/I']$, where $\text{int}(x)$ denotes the nearest integer smaller than x.

Pimblott and Mozumder consider each ionization subsequent to the first as a random walk of the progenitor electron with probability $q = \langle\sigma_i\rangle/(\langle\sigma_i\rangle + \langle\sigma_{ex}\rangle)$ of giving ionization at each stage. Here $\langle\sigma_i\rangle$ and $\langle\sigma_{ex}\rangle$ are respectively the mean cross sections of ionization and excitation. $\Gamma(i)$ is then given by the Bernoulli distribution

$$\Gamma(i) = \frac{(i_m - 1)! \; q^{i-1}(1 - q)^{i_m - i}}{(i_m - i)! \; (i - 1)!}, \tag{4.24}$$

With i ionizations, the maximum number of excitations is given by $j_m = \text{int}[(\varepsilon - iI' + T)/E']$, corresponding to conversion of all the $i_m - i$ deficits of ionization into excitation, where E' is the average excitation energy. The relative probability of having one less excitation is the same as converting the excitation energy E' into $\delta[\equiv E'/(v, r)']$ inferior energy losses such as vibration and rotation, where $(v, r)'$ is the average energy of such a process. Thus, $\Omega(j - 1; i)/\Omega(j; i) = (1 - p)^\delta \equiv \phi$, where $p = \langle\sigma_{ex}\rangle/(\langle\sigma_{ex}\rangle + \langle\sigma_v\rangle + \langle\sigma_r\rangle + \cdots)$ and $\langle\sigma_v\rangle, \langle\sigma_r\rangle, \ldots$ are the average cross sections for vibration, rotation, and so forth. This results a geometric series of excitation probabilities and gives, subject to normalization at each i,

$$\Omega(j; i) = \frac{\phi^{j_m - j}(1 - \phi)}{1 - \phi^{j_m + 1}}. \tag{4.25}$$

From (4.24) and (4.25), one gets

$$P(i, j; \varepsilon) = \frac{(i_m - 1)!}{(i - 1)! \; (i_m - i)!} q^{i-1}(1 - q)^{i_m - i} \frac{\phi^{j_m - j}(1 - \phi)}{1 - \phi^{j_m + 1}}. \tag{4.26}$$

Pimblott and Mozumder (1991) used Eq. (4.26) for both gaseous and liquid water, utilizing experimental information on ionization potentials, W values, ionization efficiencies, and the relevant cross sections. Their findings are briefly summarized as follows:

1. *Integral W values of ionization* for incident electron energies E, as measured in Combecher's (1980) experiments on gaseous water, can be well fitted by the equation $W(E) = W(\infty)(1 - I/E)^{-1}$, where $W(\infty) = 30.0$ eV is the value in the high-energy limit. A similar equation is assumed for liquid water. In contrast, *the entity-specific* W_i *value of ionization*, defined for a certain energy deposition in a spur, shows a minimum at ~20 eV in

the gas phase and at ~17.5 eV in the liquid. From the ionization threshold to the minimum, the decrease of W_i is due to the general increase in ionization efficiency. Beyond the minimum until a second ionization can occur, W_i increases because of energy wastage in electronic and vibrational excitation. After the occurrence of the second ionization, W_i steadily approaches the asymptotic limit $W(\infty)$ for spur energy 100 eV.

2. When averaged over the distribution of energy loss for a low-LET radiation (e.g., a 1-MeV electron), the most probable event in liquid water radiolysis generates one ionization, two ionizations, or one ionization and excitation, whereas in water vapor it would generate either one ionization or an excitation. *In liquid water, the most probable outcomes for most probable spur energy (22 eV) are one ionization and either zero (6%) or one excitation (94%); for the mean energy loss (38 eV), the most probable outcomes are two ionizations and one excitation (78%), or one ionization and three excitations (19%).* Thus, it is clear that a typical spur in water radiolysis contains only a few ionizations and/or excitations.

3. The calculated ionization yields agree well with experiment in both gaseous and liquid water, with the $W(\infty)/I.P.$ ratio being about the same in both phases. However, the computed yield of excited states ($g_{exc} = 3.1$) is greatly in excess of the experimental value in liquid water (0.8). A similar conclusion was arrived at by Kaplan *et al.* (1990) and interpreted as cage recombination. If this is true, then ~74% of the products of initial excitation undergo cage recombination, leaving only ~26% available for experimental detection.

REFERENCES

Aprile, E., Bolotnikov, A., Chen, D., and Mukherjee, R. (1993), Phys. Rev. *A48*, 1313.

Ausloos, P., and Lias, S-G. (1967), in *Action Chimiques et Biologiques des Radiations (Haissinsky, M., ed.), Masson, Paris, p. 61 et seq.*

Baxendale, J. H., and Wardman, P. (1971), *67*, 2997.

Beck, G., and Thomas, J. K. (1972), *76*, 3856.

Bethe, H. A. (1930), Ann. d. Physik. *5*, 325.

Bethe, H. A. (1947), Phys. Rev. *72*, 339.

Bichsel, H. (1974), in *Proc. 4th Symp. on Microsocimetry* (Booz, J., ed.), p. 1015 (review), Comm. European Communities, Luxembourg.

Bowen, E. J., and Brocklehurst, B. (1953), Trans. Faraday Soc. *49*, 1131.

Branscomb, L. M. (1962), in *Atomic and Molecular Processes* (Bates, D. R., ed.), p. 100, Academic, New York.

Brocklehurst, B. (1968), Radiation Res. Revs. *1*, 223.

Brocklehurst, B. (1970), Radiation Res. Revs. 2, 149.

Brocklehurst, B., and Higashimura, T. (1974), J. Phys. Chem. 78, 309.

Burton, M., Ludwig, P. K., Kennard, M. S., and Povinelli, R. J. (1964), J. Chem. Phys. 41, 2563.

Cario, G., and Franck, J. (1923), Z. Physik *17*, 202.

Chandrasekhar, S. (1945a), Astrophys. J. *102*, 223.

Chandrasekhar. S. (1945b), Astrophys. J. *102*, 395.

Christophorou, L. G. (1971), *Atomic and Molecular Radiation Physics*, Wiley-Interscience, London.

Cole, A. (1969). Radiat. Res. *38*, 7.

Combecher, D. (1980). Radiat. Res. *84*, 189.

Cooper, R., and Thomas, J. K. (1968), J. Chem. Phys. *48*, 5097.

Cundall, R. B. (1968), in *Energetics and Mechanisms in Radiation Biology* (Phillips, G. O., ed.), p. 227. Academic, London

Cundall, R. B. (1969), in *Energy Transfer and Storage by Molecules* (Burnett, G. M., and North, A. M., eds.), p. 1, Wiley-Interscience, London.

Cundall, R. B., Evans, G. B., Griffiths, P. A., and Keene, J. P. (1968), J. Phys. Chem. *72*, 3871.

Dainton, F. S., Morrow, T., Salmon, G. A., and Thompson, G. F. (1972), Proc. Roy. Soc. (London) *A328*, 457.

Dainton, F. S., Ledger, M. B., May, R., and Salmon, G. A. (1973), J. Phys. Chem. *77*, 45.

Dalgarno, A. (1962) in *Atomic and Molecular Processes* (D. R. Bates ed.), p. 622, Academic, New York.

Dalgarno, A., and Griffing, G. W. (1958), Proc. Roy. Soc. *A248*, 415.

Dempster, A. J. (1916), Phil. Mag. *31*, 438.

Dexter, D. L. (1953), J. Chem. Phys. *21*, 836.

Dresskamp, H., and Burton, M. (1959), Phys. Rev. Lett. *2*, 45.

Finstein, A. (1917), Physik. Z. *18*, 121.

El-Sayed, M. A. (1968), Accts. Chem. Res. *1*, 8.

Fano, U. (1946), Phys. Rev. *70*, 44.

Farhataziz, Perkey, L. M., and Hentz, R. R. (1974), J. Chem. Phys. *60*, 717.

Fiquet-Fayard, F., Chiari, J., Muller, F. and Ziezel, J-P.(1968), J. Chem. Phys. *48*, 478.

Forster, Th. (1947), Ann. Physik 2, 55.

Forster, Th. (1949a), Z. Electrochem. *53*, 93.

Forster, Th. (1949b), Z. Naturforsch. *4a*, 321.

Forster, Th. (1959). Diss. Faraday Soc. 27, 7.

Forster, Th. (1960), in *Comparative Effects of Radiation* (Burton, M., Kirby-Smith, J., and Magee, J. L., eds), p. 300, Wiley, New York.

Forster, Th (1968), in *Energetics and Mechanisms in Radiation Biology* (G. O. Phillips ed.), p. 183, Academic Press, London.

Franck, J., and Cario, G. (1922), Z. Physik *11*, 161.

Franck, J., and Hertz, G. (1913), Ver. der Phys. Ges. *15*, 373; see also Lenard (1902).

Franklin, J. L., Dillard, J. G, Rosenstock, H. M., Herron, J. T., Draxl, K., and Field, F. H. (1969). NSRDS-NBS 26, National Bureau of Standards, Washington, D.C.

Gangwer, T. E., and Thomas, J. K. (1975), Int. J. Radiat. Phys. and Chem. 7, 305.

Goulet, T., Bernas, A., Ferradini, C., and Jay-Gerrin, J.-P. (1990). Chem. Phys. Lett. *170*, 492.

Gryzinski, M. (1965a), Phys. Rev. *A138*, 305.

Gryzinski, M. (1965b), Phys. Rev. *A138*, 322.

Gryzinski, M. (1965c), Phys. Rev. *A138*, 336.

Haddad, G. N., and Samson, J. A. R. (1986), J. Chem. Phys. *84*, 6623.

Hameka, H. F., and Zahlan, A. B., eds. (1967), *The Triplet State*, p. 1, Cambridge Univ. Press, London.

Hamill, W. H. (1968), in *Radical Ions* (Kaiser, E. T., and Kevan, L., eds.), p. 321, Interscience, New York

Hart, J. F., and Herzberg, G. (1957), Phys. Rev. *106*, 79.

Hayashi, M. (1989), in *Atomic and Molecular Data for Radiotherapy*, pp. 193–199, International Atomic Energy Agency, Vienna.

Hayon, E. (1970), J. Chem. Phys. *53*, 2353.

Herzberg, G. (1950), *Molecular Spectra and Molecular Structure I. Spectra of Diatomic Molecules*, 2nd Ed., p. 388, Van Nostrand, Toronto.

Higashimura, T., Hirayama, K., and Katsuura, K. (1972), Ann. Rep. Res. Reactor Inst. Kyoto Univ. 5, 11.

Ho, S. K., and Siegel, S. (1969), J. Chem. Phys. 50, 1142.

Holroyd, R. A., and Capellos, C. (1972), J. Phys. Chem. 76, 2485.

Holroyd, R. A., Theard, L. M., and Peterson, F. C. (1970), J. Phys. Chem. 74, 1895.

Inokuti, M. (1971), Revs. Mod. Phys. 43, 297.

Inokuti, M. (1975), Radiat. Res. 64, 6.

Jesse, W. P. (1964), J. Chem. Phys. 41, 2060.

Jesse, W. P., and Sadauskis, J. (1952), Phys. Rev. 88, 417.

Jesse, W. P., and Sadauskis, J. (1953), Phys. Rev. 90, 1120.

Jesse, W. P., and Sadauskis, J. (1955), Phys. Rev. 100, 1755.

Jesse, W. P., and Platzman, R. L. (1962), Nature 195, 790.

John, T. W. (1960), Astrophys. J. 131, 743.

Jonah, C. D., Hart, E. J. and Matheson, M. S. (1973), J. Phys.Chem. 77, 1838.

Jonah, C. D., Matheson, M. S., Miller, J. R., and Hart, E. J. (1976), J. Phys. Chem. 80, 1267.

Jones, W. M. (1973), J. Chem. Phys. 59, 5688.

Jortner, J., Rice, S. A., and Hochstrasser, R. M. (1969), Advances in Photochemistry, v. 7 (Pitts, J. N., Jr., Hammond, G. S., and Noyes, W. A., Jr., eds.), p. 149, Interscience, New York.

Kallmann, H., and Furst, M. (1950), Phys. Rev. 79, 857.

Kaplan, I. G., Miterev, A. M., and Sukhonosov, V. Ya. (1990), Radiat. Phys. Chem. 36, 493.

Kearwell, A., and Wilkinson, F. (1969), in Energy Transfer and Storage by Molecules (Burnett, G. M., and North, A. M., eds.), p. 94, Wiley-Interscience, London.

Kim, Y.-K. (1975a), Radiat. Res. 64, 96.

Kim, Y.-K. (1975b), Radiat. Res. 64, 205.

Kimura, T., and Fueki, K. (1970), Bull. Chem. Soc. Japan 43, 3090.

Ladenburg, R. (1921), Zeits. f. Physik 4, 451.

Land, E. J., and Swallow, A. J. (1968), Trans. Faraday Soc. 64, 1247.

Lassettre, E. N., and Francis, S. A. (1964), J. Chem. Phys. 40, 1208.

Lassettre, E. N., Skerbele, A., and Dillon, M. A. (1969), J.Chem. Phys. 50, 1829.

LaVerne, J. A., and Mozumder, A.(1992), Radiat. Res. 131, 1.

Lenard, P. (1902), Ann. d. Phys. (Lpz.) 8, 149.

Levin, R. D., and Lias, S. G. (1982), NSRDS-NBS 71, National Bureau of Standards, Washington, D.C.

Lipsky, S. (1974), in Chemical Spectroscopy and Photochemistry in the Vacuum Ultraviolet (Sandorfy, C., Ausloos, P. J., and Robin, M. B., eds.), p. 495, D. Reidel, Boston.

Ludwig, P. K. (1968), Molecular Crystals 4, 147.

Magee, J. L. (1964), Radiat. Res. Supplement 4, 20.

Magee, J. L., and Huang, J-T. J. (1972), J. Phys. Chem. 76, 3801.

Magee, J. L., and Huang, J-T. J. (1974), J. Phys. Chem. 78, 310.

Magee, J. L., and Mozumder, A. (1973), in Advances in Radiation Research, v. 1 (Duplan, J. F., and Chapiro, A., eds.), p. 15, Gordon and Breach, New York.

Massey, H. S. W., and Burhop, E. H. S. (1952), Electronic and Ionic Impact Phenomena, Oxford University Press, Oxford.

McGlynn, S. P., Smith, F. J., and Cilento, G. (1964), Photochem. Photobiol. 3, 269.

McGlynn, S. P., Azumi, T., and Kinoshita, M. (1969), Molecular Spectroscopy of the Triplet State, Prentice Hall, New Jersey.

Meisels, G. G., and Ethridge, D. R. (1972), J. Phys. Chem. 76, 3842.

Miller, M. S., and Boring, J. W. (1974), Phys. Rev. A9, 242.

Miyajima, M., Takahashi, T., Konno, S., Hameda, T., Kubota, S., Shibamura, H., and Doke, T. (1974), Phys. Rev. A9, 1438.

Miyazaki, T., Tanaka, T., and Kuri, Z. (1975), Int. J. Radiat. Phys. Chem. 7, 627.

Møller, C. (1931), Zeits. f. Physik 70, 786.

Morrison, J. D. (1957), J. Appl. Phys. 17, 419.

Mott, N. F. (1930), Proc. Roy. Soc. (London) A126, 259.

Mott, N. F., and Massey, H. S. W. (1965), The Theory of Atomic Collisions, Oxford University Press, Oxford.

Mozumder, A. (1971), J. Chem. Phys. 55, 3026.

Mozumder, A., and Magee, J. L. (1966), J. Chem. Phys. 45, 3332.

Nicholls, R. W., and Stewart, A. L. (1962), in Atomic and Molecular Processes (Bates, D. R. ed.), p. 47, Academic, New York.

Nikogosyan, D. N., Oraevsky, A. A., and Rupasov, V. I.(1983), Chem. Phys. 77, 131

Ore, A., and Larsen, A (1964), Radiat. Res. 21, 331.

Perrin, J., and Choucroun, Mlle. (1927), Compt. Rend. 184, 1097.

Perrin, J., and Choucroun, Mlle. (1929), Compt. Rend. 189, 1213.

Pimblott, S. M., LaVerne, J. A., Mozumder, A., and Green, N. J. B. (1990), J. Phys. Chem. 94, 488.

Pimblott, S. M., and Mozumder, A. (1991), J. Phys. Chem. 95, 7291.

Platzman, R. L. (1961), Int. J. Appl. Rad. and Isotopes 10, 116.

Platzman, R. L. (1962a), Radiat. Res. 17, 419.

Platzman, R. L. (1962b), Vortex 23, 372.

Platzman, R. L. (1967), in Radiation Research (Silini, G., ed.), p. 20, North Holland, Amsterdam.

Polak, L. S., and Slovetsky, D. I. (1976), Int. J. Radiat. Phys. and Chem. 8, 257.

Pringsheim, P. (1949), Fluorescence and Phosphorescence, Interscience, New York.

Robinson, G. W., and Frosch, R. P. (1963), J. Chem. Phys. 38, 1187.

Rosenstock, H. M., Draxl, K., Steiner, B. W., and Herron, J. T. (1977), J. Phys. Chem. Ref. Data, 6, suppl. 1.

Rothman, W., Hirayama, F., and Lipsky, S. (1973), J. Chem. Phys. 58, 1300.

Rudd, M. E. (1989), Nucl. Tracks Radiat. Meas. 16, 213.

Rzad, S. J., Schuler, R. H., and Hummel, A. (1969), J. Chem. Phys. 51, 1369.

Rzad, S. J., Infelta, P. P., Warman, J. M., and Schuler, R. H. (1970), J. Chem. Phys. 52, 3971.

Salmon, G. A. (1976), Int. J. Radiat. Phys. and Chem. 8, 13.

Schiff, L. I. (1955), Quantum Mechanics, 2nd Ed., p. 261, McGraw-Hill, New York.

Sieck, L. W. (1968), in Fundamental Processes in Radiation Chemistry (Ausloos, P., ed.), p. 119, Interscience, New York.

Singh, A. (1972), Radiat. Res. Revs. 4, 1.

Smyth, H. D. (1931), Revs. Mod. Phys. 3, 347.

Spencer, L. V., and Fano, U. (1954), Phys. Rev. 93, 1172.

Stevens, B. (1968), in Energetics and Mechanisms in Radiation Biology (Phillips, G. O., ed.), p. 253, Academic, London

Strickler, S. J., and Berg, R. D. (1962), J. Chem. Phys. 37, 814.

Sumiyoshi, T., Tsugaru, K., Yamada, T., and Katayama, M. (1985), Bull. Chem. Soc. Jpn., 58, 3073.

Sveshnikov, B. L. (1935), Acta Physicochem. USSR 3, 257.

Tabata, Y., Ito, Y., and Tagawa, S. (1991), CRC Handbook of Radiation Chemistry, CRC Press, Boca Raton, Fla.

Takahashi, T., Konno, S, and Doke, T. (1974), J. Phys. C7, 230.

Takahashi, T., Konno, S., Hamada, T., Miyajima, M., Kubota, S., Nakamoto, A., Hitachi, A., Shibamura, E., and Doke, T. (1975), Phys Rev. A12, 1771.

Thomas, J., and Imel, D. A. (1987), Phys. Rev. A36, 614.

Thomas, J. K. (1970), Annual Rev. of Phys. Chem. 21, 17.

Thomas, J. K. (1971), Record of Chemical Progress 32, 145.

Thomas, J. K. (1976), Int. J. Radiat. Phys. and Chem. 8, 1.

Thomas, J K., Johnson, K., Klippert, T., and Lowers, R. (1968), J. Chem. Phys. 48, 1608.

Voltz, R., Klein, J., Heisel, F., Lauri, H., Laustriat, G., and Coche, A. (1966), J. Chem. Phys. 63, 1259

Vroom, D. A. and deHeer, F. J. (1969a), J. Chem. Phys. *50*, 573.
Vroom, D. A. and deHeer, F. J. (1969b), J. Chem. Phys. *50*, 580.
Walter, L., and Lipsky, S. (1975), Int. J. Radiat. Phys. Chem. *7*, 175.
Walter, L., Hirayama, F., and Lipsky, S. (1976), Int. J. Radiat. Phys. Chem. *8*, 237.
Warman, J. M., Asmus, K.-D., and Schuler, R. H. (1969), J. Phys. Chem. *73*, 931.
Watanabe, K., and Jursa, A. S. (1964), J. Chem. Phys. *41*, 1650.
Watson, W. F., and Livingston, R. (1950), J. Chem. Phys. *18*, 802.
Wigner, E. (1927), Nachr. Ges. Wiss. Gottingen, Math.-Physik. *K1*, 375
Yguerabide, J., and Burton, M. (1962), J. Chem. Phys. *37*, 1757.
Yguerabide, J., Dillon, M., and Burton, M. (1964), J. Chem. Phys. *40*, 3040.

CHAPTER 5

Radiation Chemistry of Gases

5.1 Mechanisms of Reactions in the Gas Phase
5.2 Radiolysis of Some Common Gases and Mixtures
5.3 Some Theoretical Considerations

The main characteristics of gas-phase radiation chemistry may be stated as follows (see Ausloos and Lias, 1967):

1. At low densities, the molecules are far apart. Therefore, there is little LET effect.

2. Ionization yield is readily measured; this offers relatively simple dosimetry, as the W value for ionization is nearly independent of the quality and energy of the incident particle (see Sect. 4.8).

3. Primary processes are easier to delineate, as these refer to isolated molecules. Thus, fragmentation patterns of positive ions and, to some extent, of neutral excited states can be studied and compared with mass spectrometric results.

4. The effect of an electric field on the ensuing chemistry can be studied experimentally, although interpretations of the results are not always straightforward. The main disadvantage of the gas-phase work is complications due to the wall effect: At low densities, some reactions such as neutralization take place mainly on the walls; these are not easy to account for.

Since this book is primarily concerned with condensed-phase studies, we do not go into the details of gas-phase radiation chemistry. We will briefly outline some important mechanisms of gas-phase reactions, followed by a presentation of some specific examples and certain theoretical considerations. In Chapter 4, we considered ionization and excitation in some detail. Many of these considerations apply to the gas phase; these will not be repeated. We stress that the measurement of the W value is of utmost importance in the gas phase, as yields of reactions, molecules created or destroyed, are often normalized per unit energy deposition. For a more complete study of the radiation chemistry of gases, the reader is referred to the works of Mund (1956), Lind (1961), Anderson (1968), Ausloos and Lias (1967, 1968), Dixon (1970), Freeman (1968), Meisels (1968), Spinks and Woods (1976), and Swallow (1973). Additional useful references are Noyes and Leighton (1941) for the photochemistry of gases, Firestone and Dorfman (1971) for pulse radiolysis of gases, and Christophorou (1971) for electron attachment in the gas phase.

5.1   MECHANISMS OF REACTIONS IN THE GAS PHASE

As for condensed phases, two important aspects of gas-phase radiolysis are the analysis of molecular products and the effect of added scavengers on these products. For hydrocarbons, the main molecular products are hydrogen and other hydrocarbons. Scavenger studies have dual purpose of inhibiting free radical reactions that yield molecular products and identifying a specific radical. For saturated hydrocarbons, trace impurities or accumulated products will react effectively with radicals, as the hydrocarbon itself is relatively inert toward them. Thus, good purification and a low dose are required. In the case of unsaturated hydrocarbons, the radicals, especially H atoms, tend to attach to the unsaturated hydrocarbon in preference to reacting with radiation-produced impurities. At the same time, parent molecules are reactive toward ions and ionic fragments, both primary and secondary, so that elucidation of reaction mechanisms is often difficult due to a long sequence of ion-molecule reactions. For the present discussion, the various reaction mechanisms will be classified as follows: ion-molecule reactions, fragmentation, dissociation of excited molecules, neutralization, and chain reactions. We will now describe these briefly.

5.1.1   ION-MOLECULE REACTIONS

Ion-molecule reactions involve a positive ion and a neutral molecule, frequently the parent molecule. Historically, there has been a dichotomy in the interpretation of the radiation-chemical yields in hydrocarbon gases. Early work by Lind (1961) and by Mund (1956) indicated the involvement of ion clustering, exemplified in the radiation-induced polymerization of acetylene as follows:

$$C_2H_2 \rightarrow C_2H_2^+ + e \quad \text{(ionization)},$$
$$C_2H_2^+ + nC_2H_2 \rightarrow (C_2H_2)^+_{n+1} \quad \text{(clustering)},$$
$$e + (C_2H_2)^+_{n+1} \rightarrow \quad \text{(polymer product)},$$

Later, H. Eyring *et al.* (1936) emphasized the involvement of free radicals produced in excitation, ionization, and secondary reaction processes. This theory gained a great deal of currency while maintaining analogy with photochemistry and free-radical chemistry. In the example of acetylene, first H and C_2H radicals will be formed according to this scheme. These radicals, reacting with parent molecules, will form C_2H_3 and C_4H_3 as intermediate products, which on further reaction may give the polymer. The role of the ion–molecule reactions was once again emphasized and clarified by later works, Talroze and Lyubimova (1952) and Franklin *et al.* (1956), among others.

Ion-molecule reactions are ubiquitous. A simple example in hydrogen is given by

$$H_2^+ + H_2 \rightarrow H_3^+ + H, \tag{5.I}$$

a reaction that is known to occur in the (high-pressure) mass spectrometer. In an H_2–D_2 mixture, one may have the following additional reactions giving HD:

$$H_3^+ + D_2 \rightarrow H_2 + HD_2^+ \quad \text{(proton transfer)},$$
$$HD_2^+ + H_2 \rightarrow HD + H_2D^+ \quad \text{(deuteron transfer)}.$$

Similarly, in water vapor one has the reaction

$$H_2O^+ + H_2O \rightarrow H_3O^+ + OH, \tag{5.II}$$

which is believed to be the main source of OH radicals in the liquid phase.

In the case of hydrocarbons, following Ausloos and Lias (1967) we can look at the various ion-molecule reactions in terms of transfer of H^- and H_2^- from the molecule to the ion, transfer of H, H_2, and H^+ from the ion to the molecule, and condensation reactions. Some examples are now given.

5.1.1a   Hydride Transfer Reaction

The hydride transfer reaction is

$$C_2H_5^+ + C_3H_8 \rightarrow C_2H_6 + C_3H_7^+.$$

According to H. Eyring *et al.* (1936), the relative rate of the hydride transfer reaction, which is proportional to the collision cross section, should increase monotonically with α/m_y, where α is the polarizability and m_y is the reduced mass. Using $C_2D_5^+$, $C_3D_7^+$, and $C_4D_9^+$ as ions, Ausloos *et al.* (1966) have found confirmation of the theory for a number of alkanes and cycloalkanes.

5.1.1b H_2^- Transfer Reactions

In the radiolysis of cyclopentane, propane is believed to be formed by H_2^- transfer:

$$C_5H_6^+ + C_5H_{10} \rightarrow C_5H_8 + C_5H_8^+.$$

Generally speaking, the occurrence of this type of reaction is universal in all saturated hydrocarbons:

$$C_nH_m^+ + RH_2 \rightarrow C_nH_{m+2} + R^+.$$

That propane is indeed formed by H_2^- reaction is known by observing the distribution of yields of various isotopic compositions of propane from the radiolysis of an equimolar mixture of cyclopentane and deuterated cyclopentane. Further evidence is provided by the facts that (1) propane is not formed by photolysis below the ionization threshold, and (2) an electric field has no effect on the yield.

5.1.1c   H Transfer Reactions

A typical example of an H transfer reaction is

$$C_3H_6 + C_3H_6^+ \rightarrow C_3H_7 + C_3H_5^+.$$

This reaction competes with the H_2 transfer reaction (*vide infra*). Note that in reactions (5.I) and (5.II), an H atom is transferred from the molecule to the ion.

5.1.1d   H$_2$ Transfer Reactions

Symbolically, the H_2 transfer reaction can be written as

$$C_nH_m + RH_2^+ \rightarrow C_nH_{m+2} + R^+,$$

where the reactant molecule is some unsaturated hydrocarbon. Another example in cyclopentane is provided by

$$C_5H_{10}^+ + C_3D_6 \rightarrow C_5H_8^+ + C_3D_6H_2 .$$

5.1.1e   Proton Transfer Reactions

Reactions of this type are important in some gases and certainly in many gaseous mixtures. The basic proton donors are H_3^+, CH_5^+, and ions of the types $C_nH^+_{2n+1}$ and $C_nH_{2n-1}^+$ in addition to certain others. In some cases, such as in cyclopropane, the product of the proton transfer reaction is a stable ion:

$$CH_5^+ + C_3H_6 \rightarrow CH_4 + C_3H_7^+.$$

In many other cases, the resultant ion decomposes (dissociative proton transfer). Examples are

$$CH_5^+ + C_4H_{10} \rightarrow CH_4 + C_4H_{11}^+ \rightarrow 2CH_4 + C_3H_7^+,$$
$$H_3^+ + C_2H_6 \rightarrow C_2H_7^+ + H_2 \rightarrow C_2H_5^+ + 2H_2 .$$

Similarly, when CH_4 containing a little n-C_5D_{12} is irradiated the immediate product of the proton-transfer reaction, $C_5D_{12}H^+$ will dissociate in several different ways. It is believed that (1) the relative probability of dissociative proton transfer reaction increases with exothermicity; (2) at high pressures, the different modes of dissociation on proton transfer from H_3^+ are about equally competitive; and (3) there is little or no H atom reshuffling on proton transfer (Ausloos and Lias, 1967). The last-mentioned item may be established by isotope studies.

5.1.1f Condensation Reactions

In condensation reactions, one eventually obtains a product ion of greater molecular weight than the reactant ion through partial stabilization. In the extreme case, complete stabilization may even occur (collisional stabilization). An example is

$$C_2H_4^+ + C_2H_4 \rightarrow C_4H_8^+ \rightarrow C_3H_5^+ + CH_3.$$

The formation of polymers upon irradiation of hydrocarbons through the ionic mechanism, of course, will belong to this class of reactions.

5.1.2 FRAGMENTATION

The process of fragmentation refers both to the parent and product ions. Each fragmentation mode has a critical energy requirement, called the appearance potential (see Chapter 4). The fragmentation pattern depends critically on the excess energy over the highest relevant appearance potential. However, the role of molecular collisions in the deactivation of an excited parent (or secondary) ion in a high-pressure system is not clearly understood. If it is assumed that the fragmentation pattern in a high-pressure system is similar to unimolecular fragmentation except for the shortening of the time scale, then one can take the fragmentation pattern at 10–100 ps as standard. But Ausloos and Lias (1966) demonstrated that, in the case of photolysis of n-butane by 105-nm UV radiation, this assumption is not reasonable. In this case, there is only about 0.6 eV excess energy above the fragmentation process giving a propyl ion, $C_4H_{10}^+ \rightarrow C_3H_7^+ + CH_3$. These authors used inert gases, He, Ne, and N_2, to achieve pressures up to 700 torr, and the photon energy (11.7 eV) was insufficient to ionize the inert gases. Even so, they found that the parent ion deactivation, measured empirically as the plot of reciprocal yield, $(M/N)^{-1}$, versus pressure, was different for different inert gases.

To gain knowledge of the fragmentation pattern of positive ions, mass spectrometric studies are considered relevant. The identity of resultant ions can be deduced from such studies. The relative abundances usually depend on the electron energy; however, at ~100 eV or more, they vary little with energy and thus can be relied on. Taking propane as an example, the following scheme has been established for the fragmentation of the parent ion:

$$C_3H_8^+ \rightarrow C_3H_7^+ + H$$
$$\rightarrow C_3H_6^+ + H_2 \text{ (or 2H)}$$
$$\rightarrow C_2H_5^+ + CH_3$$
$$\rightarrow C_2H_4^+ + CH_4 .$$

In this case, all these fragment ions carry excess energy and, in principle, can undergo further fragmentation in almost all conceivable ways, ultimately yielding simpler stable ions such as $C_2H_3^+$, $C_2H_2^+$, $C_3H_3^+$, $C_3H_5^+$, $C_2H_4^+$, and $C_3H_4^+$. As in the case of mass spectra, one can invoke the quasi-equilibrium theory of H. Eyring and associates (see Rosenstock et al., 1952) for the relative abundance of the fragment ions, although the basis of that theory may not be entirely acceptable (see Sect. 6.3).

5.1.3   DISSOCIATION OF EXCITED MOLECULES

In Sect. 4.4.2, we discussed certain aspects of the dissociation process of excited molecules that are applicable to gas-phase radiation chemistry. The dissociation processes ensuing from excited alkane molecules may be classified as elimination of H, H_2, CH_2, or entire alkane molecules. In that sense, there is some analogy with the ionic fragmentation process. In another sense, also, there is an analogy—namely, in certain cases the dissociation product has enough energy to undergo further dissociation. For example, in the photolysis of CH_3CD_3 by 147-nm UV radiation, $CHDCD_2^*$ is formed in one mode of dissociation:

$$CH_3CD_3^* \rightarrow H_2 + CHDCD_2^*.$$

The excited product $CHDCD_2^*$ can be collisionally stabilized, but can it otherwise dissociate as follows:

$$CHDCD_2^* \rightarrow HD + CDCD.$$

In the case of $CH_3CD_3^*$, various other dissociation modes exist, including C—C and C—H bond breakage, either separately or in various combinations (Freeman, 1968). A few other examples follow of dissociation of excited inorganic and organic molecules:

$$CO_2^* \rightarrow CO + O,$$
$$N_2O^* \rightarrow N_2 + O,$$
$$N_2O^* \rightarrow N + NO,$$
$$NH_3^* \rightarrow NH + H_2 \text{ (or 2H)}$$

$$\rightarrow NH_2 + H,$$
$$H_2O^* \rightarrow H + OH$$
$$\rightarrow H_2 + O$$
$$\rightarrow 2H + O,$$
$$CH_4^* \rightarrow CH_2 + H_2$$
$$\rightarrow CH + H + H_2,$$
$$C_2H_6^* \rightarrow C_2H_4^* + H_2 \text{ (or 2H)}$$
$$\rightarrow CH_4 + CH_2,$$
$$C_4H_{10}^* \rightarrow C_2H_4 + C_2H_6$$
$$\rightarrow C_3H_6 + CH_4,$$
$$C_6H_{12}^* \rightarrow 3C_2H_4 \text{ (overall)}$$
$$\rightarrow 2C_3H_6.$$

Generally speaking, the dissociation modes are established by photolysis, isotope studies, the electric field effect, and, to some extent, by special mass spectrometric methods. In addition, polymerization and isomerization studies have been helpful.

5.1.4 NEUTRALIZATION

Many investigators consider neutralization as a poorly understood process in the gas phase. While in the best possible cases it may be considered as a homogeneous second-order process, detailed experimentation in specific cases sometimes fails to establish that (Freeman, 1968; Meisels, 1968). At low dose rates and low gas pressures, wall effects can be seen as a major inhibiting factor, as most neutralizations would then be expected to occur on the walls. Coating the wall with specific chemicals has not lead to a uniform conclusion. On the other hand, wall effects are also present at high dose rates. In such cases, and with gas pressure greater than about 0.1 atm, normal positive ions cannot reach the walls if the size of the vessel is ~10 cm or more (Freeman, 1968). Even for electrons, it is hard. Large-scale convection is supposed to be the chief transport mechanism; this, however, is difficult to establish experimentally.

There are basically two kinds of neutralization processes for the cation, reaction with the electron and with a negative ion. In each case, it may be assumed that neutralization will occur with the parent or fragment ion of lowest energy. It is believed that the various degradation processes for the cation-fragmentation, ion-molecule reaction, and so forth—are much faster than the neutralization process. In addition, one considers charge transfer, without decomposition, from the cation formally as a neutralization of that species. To effect that, of course, one must have a molecule with a lower ionization potential introduced as a scavenger. Meisels (1965) established such a process by studying the formation of butene (C_4H_8) in the radiolysis of ethylene. The normally small yield of the product is greatly enhanced by adding compounds of lower ionization potential than butene, whereas adding compounds of higher ionization potential nearly eliminates butene formation. The explanation of the enhancement process is charge transfer neutralization of $C_4H_8^+$ before it can react with C_2H_4 as follows:

$$C_2H_4^+ + C_2H_4 \rightarrow C_4H_8^+,$$
$$C_4H_8^+ + A \rightarrow C_4H_8 + A^+,$$
$$C_4H_8^+ + C_2H_4 \rightarrow C_6H_{12}^+.$$

In general, neutralization of the cation by the electron, represented schematically by

$$RH^+ + e \rightarrow R + H$$
$$\rightarrow \text{other unspecified neutral products,}$$

has rate constants $\sim 10^{14}$ liter/(mole•s). Neutralization by the negative ion has a somewhat smaller rate constant, typically by a factor of 10 or so. However, in the latter case the negative ion originally has to be formed, presumably by the reaction of an electron with a scavenger, the parent molecule, or a highly reactive radical. Such a reaction may simply be an attachment, or it may result in a dissociation (Christophorou, 1971). Neutralization by anions usually involves popular electron scavengers such as SF_6, CCl_4, and N_2O. There is evidence that neutralization by SF_6^- or Cl^- leads to less fragmentation. It is also reasonably clear that ionic neutralizations do not generate H atoms; if they did, then adding an electron scavenger would reduce H_2 yield in a C_3H_8–C_3H_6 mixture to a point below that scavenged by propylene, which is not observed. On that basis, it has been concluded that in that system about two-thirds of the electron neutralization results in the formation of H atoms.

Finally, it was observed by Woodward and Back (1963) that in the radiolysis of propane at 800 torr pressure, the H_2 yield decreased with the external field and also decreased upon decreasing the dose rate. Both effects are explainable on the assumption of a competition between the gas-phase electron neutralization and ionic neutralization at the wall. The latter process presumably is accompanied by at least a partial stabilization of the excited state produced.

5.1.5　CHAIN REACTIONS

Gas-phase radiolysis can sometimes result in chain reactions involving H atoms or other radicals. As in other cases with chain reactions, termination is due to either recombination or reaction with other radicals. Typical chain length is ~1000 or more. Some specific examples will be considered in Sect. 5.2.

5.2 RADIOLYSIS OF SOME COMMON GASES AND MIXTURES

5.2.1 DIATOMIC GASES AND MIXTURES

5.2.1a Hydrogen

In very pure hydrogen, there can be hardly any permanent chemical change produced by irradiation. However, the ion-molecule reaction (5.I) does occur in the mass spectrometer, and it is believed to be important in radiolysis. The H_2 molecule can exist in the ortho (nuclear spin parallel) or para (antiparallel) states. At ordinary temperatures, equilibrium should favor the ortho state by $3:1$. However, the rate of equilibration is slow in the absence of catalysts but can be affected by irradiation. Initially, an H atom is produced either by the reaction (5.I) or by the dissociation of an excited molecule. This is followed by the chain reaction (H. Eyring *et al.*, 1936)

$$H + H_2(para) \rightarrow H + H_2(ortho).$$

The chain length, using α-radiation is estimated to be $\sim 10^3$, and termination occurs through recombination, presumably at the wall, $2H \rightarrow H_2$.

A closely similar situation prevails for HD production in the irradiation of a mixture of H_2 and D_2 (or HT in a $H_2 + T_2$ mixture). The chain length in a nominally pure mixture is $\sim 10^3$, but in mixtures carefully freed of rare gases, chain lengths $\sim 10^4$ have been observed. Originally, a free-radical mechanism was proposed as follows:

$$H + D_2 \rightarrow HD + D,$$
$$D + H_2 \rightarrow HD + H.$$

Termination was through reaction with trace oxygen impurities, $H(D) + O_2 \rightarrow H(D)O_2$. Later, it was realized that most chain carriers are ions, and the following mechanism was established after the initial reaction (5.I) or its D equivalent:

$$H_3^+ + D_2 \rightarrow HD_2^+ + H_2,$$
$$HD_2^+ + H_2 \rightarrow HD + H_2D^+,$$
$$H_2D^+ + D_2 \rightarrow HD + HD_2^+.$$

In the presence of a rare gas atom A, there may be inhibition of initiation $(H_2^+ + A \rightarrow AH^+ + H)$ or of chain propagation $(H_3^+ + A \rightarrow AH^+ + H_2)$. Termination is usually by neutralization such as $H_3^+ + B^- \rightarrow H_2 + H + B$, where B is either H or an impurity; other modes of neutralization are also possible.

5.2.1b Nitrogen

As in the case of hydrogen, no net chemical change is expected upon irradiation of pure N_2, but an exchange reaction is known to take place between

$^{14,14}N_2$ and $^{15,15}N_2$ isotopes with a G value of about 8. That N atoms are not entirely responsible for the exchange is proved by the facts that (1) the dissociation energy of N_2 is high, and (2) in the presence of NO, which scavenges N atoms very efficiently (N + NO→N_2 + O), there is still a residual exchange yield of about 3. Two alternative mechanisms are known (Anderson, 1968): one is exchange reaction between ground state N_2 and excited state N_2 (excepting the lowest triplet), and the other is an ionic mechanism involving N_4^+ given as follows:

$$^{14,14}N_2^* + {}^{15,15}N_2 \rightarrow {}^{14,15}N_2 + {}^{14,15}N_2^*$$

and

$$^{14,14}N_2 + {}^{15,15}N_2^+ \text{ (or vice versa)} \rightarrow N_4^+,$$
$$N_4^+ + e \rightarrow {}^{14,15}N_2 + {}_{14}N^+ + {}^{15}N.$$

5.2.1c   Oxygen

Production of ozone by irradiation or electric discharge in oxygen is readily revealed by the characteristic odor. The yield, though, is not easy to establish. In a closed system, it is very small; in a flow system ,it increases with the flow rate and decreases with the dose rate. These have been explained by taking the formation reaction as a three-body combination with O atom, $O + 2O_2 \rightarrow O_3 + O_2$ (no chain), where the back reaction is indeed a chain process (Magee and Burton, 1951). In the mass spectrometer, O_2^+ is the main ion but the total yield of excited states is about the same as that of ionization (about 3.1), giving the following initial species:

$$O_2 \rightarrow O_2^*$$
$$\rightarrow 2O$$
$$\rightarrow O_2^+ + e$$
$$\rightarrow O^+ + O + e.$$

The O atoms so produced may be in excited states, too. Unless the pressure is very low, the electron invariably attaches to O_2 in a three-body process, $e + 2O_2 \rightarrow O_2^- + O_2$, and neutralization occurs through the reaction $O_2^+ + O_2^- \rightarrow 2O + O_2$.

In careful experiments by pulse radiolysis, the maximum G value of ozone production is 13.8, of which 6.2 comes from ionization and eventual neutralization, each such sequence giving two O atoms. If the remaining yield is attributed to the dissociation of excited states, either directly or indirectly, then the total yield of excitation will be about the same as that of ionization, 3.8 in this case, because each dissociation also gives two O atoms.

When a mixture of N_2 and O_2 is irradiated, various oxides of nitrogen are produced in addition to ozone if the mixture is dry. In presence of moisture, nitric acid is produced to the exhaustion of the moisture. The actual mechanisms of the forward and backward reactions are very complex and probably not too well known. However, it is believed that an equilibrium, depending on the composition and the dose rate, is finally established. Several attempts have been made to utilize atmospheric nitrogen in making nitrate fertilizer via these reactions.

5.2.1d   Carbon Monoxide

$G(-CO)$ is about 8. In the absence of oxygen, the main products are CO_2 ($G = 2$), a suboxide of carbon that is solid, and various gaseous compounds. In the presence of oxygen, the suboxide is inhibited but a chain reaction occurs, ultimately giving CO_2, probably through an ionic mechanism.

5.2.1e   Hydrogen-Chlorine Mixture

HCl is formed by a chain mechanism with a very high yield (M/N value ~10^5), presumably because of the high reactivity of the Cl atom. The chain is initiated by the production of H and Cl atoms. Dissociative electron capture by Cl_2 requires about 1.6 eV energy. Therefore, presumably both kinds of ions are initially produced by excitation and ionization. The chain is propagated simply as follows:

$$H + Cl_2 \rightarrow HCl + Cl,$$
$$Cl + H_2 \rightarrow HCl + H.$$

Termination is by the recombination of H and Cl. There is complete corroboration of this mechanism from photochemical studies. This is undoubtedly a free-radical chain propagation mechanism.

5.2.2   POLYATOMIC GASES

5.2.2a   Water Vapor

Despite its obvious importance, the interpretation and even the measured yields in the radiolysis of water vapor were doubtful until the sixties. It was not because of lack of experimental data; rather, it was because of difficulties of comparing measurements of different workers due to artifacts and sheer experimental problems (Anderson, 1968). The greatest discrepancy is in the reported hydrogen yields, which varied between the extremes by a factor of ~10^4 (Dixon, 1970; Anderson, 1968). It is now agreed that $G(H_2)$ in water vapor at ~10^{21} eV/g varies around 10^{-3}. But, as pointed out by Dixon, the absolute yield of hydrogen in pure water vapor is not a very meaningful quantity, because a steady state is achieved and a consistent steady state concentration of H_2 and O_2 may be obtained, which varies as the square root of the dose rate. At high dose rates, the $[H_2]/[O_2]$ ratio approaches 2, but the ratio decreases at lower dose rates.

In the presence of scavengers of electron and H atoms, the situation is clearer; most workers agree on an unscavengable hydrogen yield of ~0.5. To determine $G(-H_2O)$, Firestone (1957) studied the isotope exchange reaction $D_2 + H_2O \rightarrow HD + HDO$ under tritium-β irradiation. Below 150°, he obtained a yield $G(HD) = 11.7$ and equated it to $G(H_2O)$, which was not easily reconciled with theoretical concepts. Since then, several workers (see Baxendale and Gilbert, 1964) measured hydrogen yield in the presence of H atom scavengers and reported $G(-H_2O) = 7.5 \pm 1$ at ~1 atm and below 150°, conditions that are comparable to Firestone's. This value, which is closer to the theoretical concept, is now accepted. Electron yield measurement by suppression of H_2 in presence of e-scavengers and by the consideration of the effect of temperature on total yields gives $G(e) = 3 \pm 10\%$, which agrees well with the W value for ionization. Similar considerations give $G(H) = 3.5 \pm 10\%$.

Of the various modes of production and decay of the excited states of H_2O, only two can occur at any excitation energy, giving $H_2 + O$ and $H + OH$ respectively as products (see Chapter 4). Of these, photochemistry indicates that the latter process is by far the dominant one. Mass spectrometric studies reveal that the main ions are H_2O^+ (~77%) and OH^+ (~18%), with minor positive ions such as H^+, O^+, and so forth, making up the rest. Under normal conditions of water vapor radiolysis, both of these major ions may be converted into H_3O^+ by reaction (5.II) and the reaction $OH^+ + H_2O \rightarrow H_3O^+ + O$. The fate of the O atom is not known clearly. It may attach with water molecules, giving two OH radicals in an endothermic reaction. Alternatively it may undergo scavenging reactions. The H_3O^+ ion so produced almost certainly undergoes clustering before neutralization produces an H atom. Based on the previous discussion, the following scheme may be presented for water vapor radiolysis:

$$H_2O \rightarrow H_2O^*, e, H_2O^+, OH^+, H, \text{etc.},$$
$$H_2O^+ + H_2O \rightarrow H_3O^+ + OH,$$
$$OH^+ + H_2O \rightarrow H_3O^+ + O,$$
$$H_3O^+ + nH_2O \longleftrightarrow H^+(H_2O)_{n+1},$$
$$H^+(H_2O)_{n+1} + e \rightarrow H + (n+1)H_2O,$$
$$e + H_3O^+ \rightarrow H + H_2O$$
$$\rightarrow 2H + OH,$$
$$H_2O^* \rightarrow H + OH$$
$$\rightarrow H_2 + O.$$

Clustering following the first ion-molecule reaction is apparently a reversible reaction. The equilibrium of this reaction is of some importance, as one can get two H atoms by the electron neutralization of H_3O^+.

5.2.2b   Carbon Dioxide

Like certain other permanent gases, CO_2 is remarkably radiation resistant. This, together with its advantageous heat transfer properties, has made it attractive for use as a coolant in graphite-moderated reactors. On the other hand, there is not a complete absence of chemical change. It only indicates efficient back reaction, and a small amount of permanent change does occur either with additives or on the walls. From photochemical studies, we know that the excited states dissociate into CO and O, and occasionally into C atoms. In the mass spectrometer, CO_2^+ is the main ion and the minor ions CO^+, O^+, and C^+ amount only to a few percent each. In the operating conditions in a graphite reactor, a certain amount of the reaction $C + CO_2 \rightarrow 2CO$ does take place, but it is not known clearly if that is due to the combined effect of temperature (about 650°) and neutron irradiation on the graphite. However, it is known that considerable graphite loss can occur through this process.

5.2.2c   Ammonia

In the initial stages of the radiolysis of NH_3, the products are H_2, N_2, and N_2H_4 (hydrazine), with the decomposition yield being ~3. On prolonged irradiation, back reactions almost completely exhaust the hydrazine, and an equilibrium is established among N_2, H_2, and NH_3 with about 90% decomposition of ammonia. Apparently, the same equilibrium is established starting with a $H_2 + N_2$ mixture. Photochemical evidence indicates the production of H, NH, and NH_2. Some hydrogen is probably produced by the recombination of H atoms at the wall. The recombination of $2NH_2$ radicals gives $N_2 + 2H_2$, but in the presence of a third body, it can also give N_2H_4. Mass spectrometric evidence gives NH_3^+ and NH_2^+ as main ions, both of which finally give NH_4^+ by ion-molecule reactions:

$$NH_2^+ + NH_3 \rightarrow NH_3^+ + NH_2,$$
$$NH_3^+ + NH_3 \rightarrow NH_4^+ + NH_2.$$

The neutralization reaction gives $H + NH_3$. Again, dimerization of NH_2 gives hydrazine, but the latter is susceptible to radical attack, which is the beginning of the back reactions:

$$H + N_2H_4 \rightarrow H_2 + N_2H_3,$$
$$NH_2 + N_2H_4 \rightarrow NH_3 + N_2H_3.$$

Further reactions ensue until only H_2 and N_2 are left.

5.2.2d   Methane

Photochemistry gives the primary reactive species as CH_3, CH_2, and CH, and mass spectrometry indicates the main ions as CH_4^+ (46%), CH_3^+ (40%), and

CH_2^+ (8%), with some minor ions such as CH^+ and C^+ accounting for the rest. The radical reactions are mainly either recombination in the presence of a third body or insertion reactions with the parent molecule from which, all in all, a great variety of reactions can ensue (Meisels, 1968). Some examples are

$$2CH_3 + M \rightarrow C_2H_6,$$
$$CH + CH_4 \rightarrow C_2H_5^* \rightarrow C_2H_4 + H,$$
$$CH_2 + CH_4 \rightarrow C_2H_6^* \rightarrow 2CH_3.$$

The H atom can react with products or with the parent gas metathetically:

$$H + CH_4 \rightarrow H_2 + CH_3.$$

The parent ion and the fragment ion CH_3^+ undergo fast ion-molecule reactions:

$$CH_4^+ + CH_4 \rightarrow CH_5^+ + CH_3,$$
$$CH_3^+ + CH_4 \rightarrow C_2H_5^+ + H_2.$$

Neutralization of these product ions and of CH_2^+ generates H_2, H, and other free radicals, which then undergo a large variety of free-radical reactions (Meisels, 1968). On balance, the main products of methane radiolysis by MeV electrons or γ-rays are H_2, C_2H_6, and C_3H_8 with approximate G values of 6, 2, and 0.3, respectively. In addition, there are large groups of products of smaller yield, and a liquid polymer accounting for a G value of conversion of CH_4 of about 2.

5.2.2e   Ethane

There is some similarity in the radiolysis of ethane and higher paraffins with that of methane in the sense that the main products are hydrogen, various dimeric combinations, and unsaturated compounds. In addition, there are some products arising out of C—C bond fission. From photochemistry, one knows that the excited states dissociate in various ways, giving H, H_2, CH_2, CH_4, C_2H_4, and C_2H_5 as immediate products, where the last two species may be in excited states. Among further reactions of these radicals, the excited ethyl radical can dissociate into ethylene and an H atom. Similarly, acetylene and H_2 can be formed by the dissociation of C_2H_4.

Unlike the case for methane, the parent ion is not the major component in the mass spectroscopy of ethane. Various ions are seen, with the following accounting for most: $C_2H_4^+$ (45%), $C_2H_3^+$ (15%), $C_2H_6^+$ (12%), $C_2H_2^+$ (10%), and $C_2H_5^+$ (10%). It appears that in ethane, the ion–parent molecule reactions are not as efficient as in methane. The main products in the high-energy radiolysis of ethane are H_2, CH_4, C_3H_8, and n-C_4H_{10}, with approximate yields of 7, 0.6, 0.5, and 1.0. In addition, a number of minor products and a liquid polymer are formed. In reporting yields in hydrocarbon gases, we should stress that these are the yields when the pure gas is irradiated. A small amount of additive (e.g., NO) may materially affect the yields.

5.2.2f   Ethylene

The photochemistry of ethylene is fairly well understood, but not the radiation chemistry. UV-photolysis shows that the excited states dissociate mainly by elimination of an H atom or a H_2 molecule as follows:

$$C_2H_4^* \rightarrow C_2H_3 + H$$
$$\rightarrow C_2H_2 + H_2 .$$

The vinyl (C_2H_3) radical is sometimes in an excited state; if so, it can eliminate an H atom also. Both H and C_2H_3 will undergo addition reactions with the parent molecule:

$$H + C_2H_4 \rightarrow C_2H_5 ,$$
$$C_2H_3 + C_2H_4 \rightarrow C_4H_7 .$$

Butenyl (C_4H_7) mainly disappears by reacting with ethyl radicals. The leftover ethyl radicals either recombine (faster process) or undergo disproportionation (slower process):

$$2C_2H_5 \rightarrow n\text{-}C_4H_{10}$$
$$\rightarrow C_2H_4 + C_2H_6 .$$

Under mass spectrometric conditions, the chief ions are $C_2H_4^+$ (38%), $C_2H_3^+$ (23%), and $C_2H_2^+$ (22%). Various ion-molecule reactions do ensue, but the (plausible) involvement of long-lived complexes is not clear. Some illustrations of these reactions are

$$C_2H_4^+ + C_2H_4 \rightarrow [C_4H_8^+] \rightarrow C_3H_5^+ + CH_3$$
$$\rightarrow C_4H_7^+ + H,$$

where the long-lived complex $[C_4H_8^+]$ as well as the intermediate product ions can undergo further ion-molecule reactions. A similar situation prevails for reaction of the ions $C_2H_3^+$ and $C_2H_2^+$, with the parent molecule producing the complexes $[C_4H_7^+]$ and $[C_4H_6^+]$, respectively. The acetylene ion can also undergo simple charge transfer, giving acetylene: $C_2H_2^+ + C_2H_4 \rightarrow C_2H_2 + C_2H_4^+$.

In general, unsaturated hydrocarbons are more sensitive to radiation. They react vigorously with free radicals and thereby compete for them with radiation-produced intermediates and products. Therefore, to get to the true yields, low doses are required so that conversion is limited to ~1%. In ethylene, the yield of decomposition is ~15 and the main products are H_2 , C_2H_2 , butanes, butenes, C_2H_6 , and C_3H_6 , with G values for high-energy radiation given respectively as 1.2, 1.5, 0.5, 0.4, 0.3, and 0.2. In addition, polymers are formed. At STP, the polymer is a liquid with a G value for the converted monomeric molecules being ~10. However, both the yield and the phase of the polymer (liquid or solid) are sensitive to pressure and temperature. At STP, the polymerization is believed to proceed by a free-radical mechanism.

The radical can either be an H atom or an organic free radical. Indicating the radical as R, the reaction is formally written as

$$R + C_2H_4 \rightarrow RC_2H_4 \rightarrow C_2H_4 \text{ polymer.}$$

At the liquid nitrogen temperature, a lower molecular weight branched-chain polymer is obtained, which is believed to be produced in an additive ion-molecule chain reaction of the type

$$C_nH^+_{2n} + C_2H_4 \rightarrow C_{n+2}H^+_{2(n+2)}.$$

5.2.2g   Acetylene

Photoexcitation of an acetylene molecule results in either dimerization or dissociation of the molecule ($C_2 + H_2$ or $C_2H + H$). In the mass spectrometer, the major positive ions are $C_2H_2^+$ (75%) and C_2H^+ (15%). However, at STP no gaseous products are seen under radiolysis. There are only two major products, benzene and a polymer, cuprene with an empirical formula $C_{40}H_{40}$. The detailed mechanisms are still debatable. However, the following remarks may be made:

1. Probably the mechanisms of formation of benzene and cuprene are different.
2. Benzene probably is formed from the excited state:

$$C_2H_2 \rightarrow C_2H_2^*,$$
$$C_2H_2^* + C_2H_2 \rightarrow (C_2H_2)_2^*,$$
$$(C_2H_2)_2^* + C_2H_2 \rightarrow C_6H_6.$$

3. Formation of cuprene is either by a free-radical chain reaction or by clustering around the parent ion (cluster size ~20) followed by neutralization, which is not a chain process. The M/N value for decomposition of acetylene is about 20, giving the corresponding G value as 70–80, which is very large. The G value of benzene production is ~5, whereas the G of conversion of monomers into the polymer is ~60.

5.3   SOME THEORETICAL CONSIDERATIONS

In this section, we will briefly consider two useful theoretical aspects in studying the radiation chemistry of gases.

5.3.1   THE QUASI-EQUILIBRIUM THEORY (QET)

The quasi-equilibrium theory (QET) is the most widely used theoretical framework for the discussion of the fragmentation pattern of the parent ion in a unimolecular process. Although other unimolecular theories (see Levine, 1966) have been subsequently proposed, the QET has traditionally been applied for mass spectrometric studies and its discussion is relevant to radiation chemistry. It is a combination of the absolute reaction rate theory of H. Eyring (Glasstone *et al.*, 1941) and statistical mechanical principles in which the excess energy in the ion plays a crucial role. Instead of going into mathematical details, we will here give the physical basis of the theory.

Before we do this, though, we point out that for a simple diatomic molecule, assuming ideal conditions, one can in principle calculate the rate of the unimolecular process. This is so because the lower excited states of the ion are (relatively) few and well separated. If the potential curves are then given, the value of the rate can be provided. For a polyatomic molecule, two great complications immediately arise: (1) the number of lower excited states increases tremendously; and (2) multidimensional potential energy surfaces make trajectory calculations intractable.

The situation calls for a statistical treatment, which was employed by Rosenstock *et al.* (1952), taking advantage of the following facts: (1) fragment ions are usually formed with little kinetic energy; (2) fragmentation pattern is sensitive to structure; (3) the probability of a particular bond scission is independent of finding that bond in the molecule; and (4) all outer electrons participate more or less equally in the ionization process (Vestal, 1968). These considerations imply that the fragmentation process does not follow immediately upon ionization; rather, time must be allowed for several vibrations to occur. Thus, neither a diatomic-type model nor a model based on ionization of individual bonds would be adequate. Instead, energy must be reshuffled between the degrees of nuclear motion, vibration and rotation, before modes of fragmentation are established.

The QET envisages a particular mode of fragmentation as a transition from the reactant to the products going through the activated complex along a reaction coordinate. If the activation energy for that mode is indicated by ε_0, then the theory asserts that the probability of the ion attaining *that* activated complex is proportional to the number of available quantum (i.e., quasi-stationary) states above ε_0 consistent with available energy E. In terms of H. Eyring's absolute reaction rate theory (Glasstone *et al.*, 1941), one writes the velocity $k(E)$ of reaction in *that* channel as

$$k(E) = \frac{S}{h} \rho(E)^{-1} W^{\neq}(E - \varepsilon_0),$$

where h is Planck's constant, $\rho(E)$ is the density of states of the parent ion at energy E, and S is a degeneracy factor giving the number of reactions with the identical activated complex; $W^{\neq}$ is the total number of states available for the activated complex, given in terms of the corresponding density of states $\rho^{\neq}$ by

$$W^{\neq} = \int_0^{E-\varepsilon_0} \rho^{\neq}(E, \varepsilon_0, \varepsilon) \, d\varepsilon,$$

where ε is the kinetic energy associated with the reaction coordinate. Originally, QET considered classical motion at the activated complex saddle point. This gave a discontinuity at $E = \varepsilon_0$, since quantum-mechanical reflection and tunneling were ignored. One would then get $k = 0$ for $E < 0$, and $k = k_{min} = S/h\rho(\varepsilon_0)$ at $E = \varepsilon_0$ because at that energy there is just one state for the activated complex. A quantum-mechanical treatment, of course, removes this discontinuity.

In QET, one usually ignores the contribution of the motion of electrons. The nuclear motion is described in terms of oscillators for vibration with a certain number of free rotors. In enumerating the available state number W^*, QET originally treated the entire problem classically, with a somewhat confusing outcome. To get agreement with experiment, one had to use an effective number of oscillators that was a good deal less than the actual number. Later, Vestal et al. (1962) realized that the main problem was the discrete quantum states of the oscillators. When the necessary quantum correction was made, good agreement was obtained between theory and experiment, as exemplified in the case of propane (see Ausloos and Lias, 1967). In the actual application of QET (or, for that matter, any unimolecular theory), one has to know the fragmentation pattern at a given energy, and then an averaging may be necessary over the distribution of available energy. Fortunately for radiation chemistry, the fragmentation pattern stabilizes at $E \sim$ 2–3 times the ionization potential. Following E. Eyring and Wahrhaftig (1961), this situation is shown for the fragmentation of the parent ion of propane in Figure 5.1.

5.3.2   ION-MOLECULE REACTION

At low gas pressures, the parent ion will continue to fragment until stable ions are formed, and then these ions will undergo ion-molecule reactions. At higher gas pressures, ion clustering may occur, and this might effectively stabilize the ion against fragmentation. Magee and Funabashi (1959) considered the possibility of ion clustering during the time scale τ of neutralization, but actually the scale should involve all other reactions, too. They argue that

1. Clustering should be proportional to the square of the pressure at low pressures (three-body process) and proportional to the pressure at high pressures (two-body process).
2. An equilibrium of the type $A^+ + nB \longleftrightarrow A^+B_n$ is quickly established in times much less than τ, where A and B may or may not be identical.
3. Simplified considerations are applicable to calculating ion energies classically and for also the partition functions Γ.

The equilibrium constants are given by

$$K_n = \frac{\Gamma_n}{\Gamma_0 \Gamma_B^n} \exp\left(\frac{-E_n}{kT}\right),$$

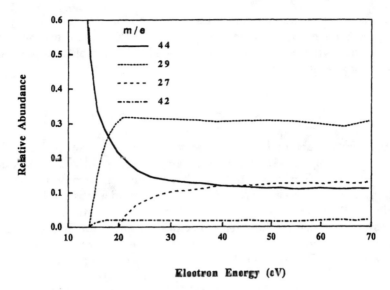

FIGURE 5.1 Major fragmentation pattern of the parent ion in propane. Numbers on the curves denote *m/e* values. The relative abundance of these, as well as of the minor ions, stabilizes beyond ~40 eV. Adapted from Eyring and Wahrhaftig (1961), with permission from Am. Inst. Phys.©

where E_n is the energy change in clustering reaction of order n and the subscripts refer to cluster of a given size.

Assuming that the concentration of the gas molecules C_B is little affected by clustering, the following two equations determine the equilibrium concentrations of clusters of various sizes C_n , $n = 0, 1, 2, \ldots$:

$$\sum K_n C_B^n C_0 = C(\text{total cluster concentration}) = \sum C_n .$$

The problem then reduces to the determination of the partition functions and the energy. For the former, the Magee and Funabashi apply simple statistical mechanical considerations using moments of inertia, mean frequencies, and symmetry numbers. Energy change in the clustering reaction E_n is taken as the sum of change of the electrostatic energy, the chemical bond energy, and the energy of intermolecular attraction of the neutrals in the cluster.

A detailed calculation would be very difficult, but classical arguments are used to arrive at an approximation. The chemical bond energy is hard to guess, but it is noted that it saturates quickly with n, so that it can mostly be treated as an additive parameter (at least when $n \gg 1$). The change in the electrostatic energy is simply taken as the difference in the potential energy of a sphere of radius a (size of A^+) and that of a sphere of radius b (cluster size) in a medium of dielectric constant κ. This energy also saturates—that is, tends to a finite value as $n \to \infty$. The intermolecular attraction is proportional to the cluster size for large n, being given approximately by $-n\varepsilon_H + 4\pi b^2 T$, where T is the surface tension and ε_H is the heat of vaporization for the liquid form of B. Thus, the total energy change is given by

$$E_n = 4\pi b^2 T - n\varepsilon_H + \frac{e^2}{2}(1 - \kappa^{-1})(b^{-1} - a^{-1}) - \text{const.}$$

Magee and Funabashi distinguish the cases of large and small clustering by $n \gg 1$ and $n \sim 1$, respectively, with no clustering being a nontrivial special case of small clustering. Their analysis shows that the cluster distribution, for large clusters, is expected to be a gaussian centered around a most probable value. For the small clusters they proceed, using charge conservation, with the plausible assumption that the clustered molecules are within the first coordination shell with approximately equal binding energy. Detailed comparison with experiment is difficult to make, but the authors find qualitative agreement in the case of Li+ ion clustering in rare gases.

It is common knowledge that, in the absence of clustering, the ion-molecule reactions often have large cross sections. They are typically much larger than gas-kinetic cross sections for neutral molecules. Langevin (1905) first described this process for structureless particles in terms of orbiting collisions between an ion and a polarizable molecule. Large cross sections are calculated in the Langevin theory, but even larger cross sections have been reported experimentally when the neutral molecule has a permanent dipole moment (see Theard and Hamill, 1962). Considering first the case with no dipole moment, one can write the equations of motion referring to the center-of-mass coordinates as follows:

$$\frac{d}{dt}(mr^2\dot{\theta}) = 0, \tag{5.1}$$

$$\frac{d}{dt}(m\dot{r}) - mr(\dot{\theta})^2 = f(r) \equiv \frac{\partial V}{\partial r}. \tag{5.2}$$

In Eqs. (5.1) and (5.2), m is the reduced mass of the colliding system, V is the interaction potential at ion-molecule separation r, θ is the angle between the direction of r and the center-of-mass velocity, and the dot indicates differentiation with respect to time. Integration of (5.1) just gives the angular momentum L, which is conserved in the collision. Substitution in (5.2) gives

$$m\ddot{r} = F(r) \equiv -\frac{\partial}{\partial r}V_{\text{eff}},$$

where $V_{eff} = V_r + L^2/2mr^2$ is the effective potential including the centrifugal force. Taking the ion-molecule interaction as given solely by the polarizability α, assumed as a scalar, the effective potential appears as

$$V_{eff} = -\frac{\alpha e^2}{2r^4} + \frac{L^2}{2mr^2}. \qquad (5.3)$$

The potential given by Eq. (5.3) has a maximum of $V_{max} = L^4/8\alpha e^2 m^2$ at $r_{max} = (e/L)(2\alpha m)^{1/2}$. The relative kinetic energy E at large distance is simply $E = (1/2)mv_0^2$, where v_0 is the relative velocity at that distance. When $E < V_{max}$, the distance of closest approach is greater than r_{max}, so that the particles are reflected always experiencing repulsion. When E is somewhat greater than V_{max}, the particles experience (mostly centrifugal) repulsion for $r > r_{max}$, but then they experience (mostly by induced polarization) attraction for $r < r_{max}$ and orbiting might ensue. In terms of the impact parameter b, the distance of closest approach if there were *no* interaction, the angular momentum is also given by $L = mbv_0$. Eliminating the angular momentum between this equation and the critical energy for orbiting, $E_c = V_{max}$, one gets the critical impact parameter as $b_c = (4\alpha e^2/mv_0^2)^{1/4}$. The cross section for orbiting collisions is then given by ($b < b_c$)

$$\sigma = \pi b_c^2 \sim E^{-1/2} \qquad (5.4)$$

For the potential given by (5.3), it is easy to show that when $b > b_c$ the distance of closest approach is $b_c/2^{1/2}$, whereas for $b < b_c$ the only thing preventing interpenetration is a repulsive core potential, which is not explicitly considered here. Equation (5.4) is actually the classical collision cross section for the problem. To translate this into a reaction cross section, we may assume that there is another critical separation r_0 such that when $r < r_0$ chemical forces complete the reaction and no reaction takes place if $r > r_0$. If r_0 is less than $b_c/2^{1/2}$, then Eq. (5.4) is also the reaction cross section, since reaction definitely takes place if $b < b_c$ and it definitely does not take place if $b > b_c$. According to this modification, the high-energy limit of the reaction cross section is πr_c^2 rather than zero as given by (5.4). One therefore has

$$\sigma(\text{reaction}) = \pi\left(\frac{2\alpha e^2}{E}\right)^{1/2} \qquad \text{for } E < \frac{\alpha e^2}{2r_c^4},$$

$$\sigma(\text{reaction}) = \pi r_c^2\left(1 + \frac{\alpha e^2}{2r_c^4 E}\right) \qquad \text{for } E > \frac{\alpha e^2}{2r_c^4}.$$

The reaction complex that is formed has angular momentum about the center of mass, which is conserved during the entire reaction. Now the products can have different polarizabilities, reduced masses, and so on. Therefore, it is conceivable that some of the complexes may not give products (i.e., decrease of reaction cross section). That in practice this does not generally happen must be interpreted as the products being left rotationally excited. That is, the conversion into the rotational angular momenta of the products must be efficient.

To explain the extraordinarily large ion—polar molecule reaction cross section, Theard and Hamill (1962) included the charge—dipole interaction as

$$V(r) = -\frac{\mu e}{r^2} - \frac{\alpha e^2}{2r^4},$$

where μ is the dipole moment. This amounts to assuming instantaneous dipole alignment with the field. It is doubtful whether that assumption is realistic; especially at higher energies, there is not enough time for alignment. Even so, the authors calculated an enhanced collision cross section, given by

$$\sigma = \pi \left[\frac{\mu e}{E} + \left(\frac{2\alpha e^2}{E} \right)^{1/2} \right]$$

Dugan and Magee (1967) and Dugan et al. (1968, 1969) have made extensive numerical calculations on the trajectories of ion-molecule collisions and defined capture collisions for polar molecules. Their major findings may be summarized as follows:

1. The Langevin cross section is a lower limit of computed cross sections.
2. Orbiting collisions, in the sense that the polar angle changes $>\pi$ and the azimuth changes $>2\pi$, are not found. These types of collisions are not found with molecules that are only polarizable and also they are unimportant when the dipole term dominates.
3. In the case of ion–polar molecule collisions, extraordinarily long-lived complexes have been seen in computed plots. Some of them live as long as $\sim 10^3$ times specular reflection periods. However, they arise via a different phenomenon. The rotating dipole changes the potential for outward motion, thereby introducing or altering the turning points and causing multiple reflections.

REFERENCES

Anderson, A. R. (1968), in *Fundamental Processes in Radiation Chemistry* (Ausloos, P., ed.), p. 281, Interscience, New York.
Ausloos, P., and Lias, S.-G. (1966), J. Chem. Phys. 45, 524.
Ausloos, P., and Lias, S.-G. (1967), in *Actions Chimiques et Biologiques des Radiations*, v. 11 (Haissinsky, M., ed.), p. 1, Masson, Paris.

Ausloos, P., and Lias, S.-G. (1968), Radiat. Res. Revs. *1*, 75.

Ausloos, P., Lias, S.-G., and Scala, A. A. (1966), Adv. Chem. Ser. *58*, 264.

Baxendale, J. H., and Gilbert, G. P. (1964), J Am. Chem. Soc. *86*, 516.

Christophorou, L. G. (1971), *Atomic and Molecular Radiation Physics*, chs. 6 and 7, Wiley-Interscience, London.

Dixon, R. S. (1970), Radiat. Res. Revs. *2*, 237.

Dugan, J. V., and Magee, J. L. (1967), J. Chem. Phys. *47*, 3103.

Dugan, J. V., Rice, J. H., and Magee, J. L. (1968), Chem. Phys. Lett. *2*, 219.

Dugan, J. V., Rice, J. H., and Magee, J. L. (1969), Chem. Phys. Lett. *3*, 323.

Eyring, E. M., and Wahrhaftig, A. L. (1961), J. Chem. Phys. *34*, 23.

Eyring, H., Hirschfelder, J. O., and Taylor, H. S. (1936), J Chem. Phys. *4*, 479.

Firestone, R. F. (1957), J. Am. Chem Soc. *79*, 5593.

Firestone, R. F., and Dorfman, L. M. (1971), in *Actions Chimiques et Biologiques des Radiations*, v. 15 (Haissinsky, M., ed.), p. 7, Mason, Paris.

Franklin, J. L., Field, F. H., and Lampe, F. W. (1956), J. Am. Chem. Soc. *78*, 5697.

Freeman, G. R. (1968), Radiat. Res. Revs. *1*, 1.

Glasstone, S., Laidler, K. J., and Eyring, H. (1941), *Theory of Rate Processes*, McGraw Hill, New York.

Langevin, M. P. (1905), J. Chim. Phys. *5*, 245. English translation in McDaniel, E. W. (1964), *Collision Phenomena in Ionized Gases*, app. 1, Wiley, New York.

Levin, R. D. (1966), J. Chem Phys *44*, 2046.

Lind, S. C. (1961), *Radiation Chemistry of Gases*, Reinhold, New York.

Magee, J. L., and Burton, M. (1951), J. Am. Chem. Soc. *73*, 523.

Magee, J. L., and Funabashi, K. (1959), Radiat. Res. *10*, 622.

Meisels, G. G. (1965), J. Chem. Phys. *42*, 3237.

Meisels, G. G. (1968), in *Fundamental Processes in Radiation Chemistry* (Ausloos, P., ed.), p. 347, Interscience, New York.

Mund, W. (1956), in *Actions Chimiques et Biologiques des Radiations*, v. 2 (Haissinsky, M., ed.), p. 3, Masson, Paris.

Noyes, W. A., Jr., and Leighton, P. A. (1941), *The Photochemistry of Gases*, Reinhold, New York.

Rosenstock, H. M., Wallenstein, M. B., Wahrhaftig, A. L., and Eyring, H. (1952), Proc. Nat. Acad. Sci. *38*, 667.

Spinks, J. W. T., and Woods, R. J. (1976), *An Introduction to Radiation Chemistry*, 2nd Ed., ch. 6, John Wiley, New York.

Swallow, A. J. (1973), *Radiation Chemistry, An Introduction*, 2nd Ed., ch. 6, John Wiley, New York.

Talroze, V. L., and Lyubimova, A. K. (1952), Dokl. Akad. Nauk SSSR *86*, 909.

Theard, L. P., and Hamill, W. (1962), J. Am. Chem. Soc. *84*, 1134.

Vestal, M. L. (1968), in *Fundamental Processes in Radiation Chemistry* (Ausloos, P., ed.), p. 59, Interscience, New York.

Vestal, M. L., Wahrhaftig, A. L., and Johnston, W. H. (1962), J. Chem. Phys. *37*, 1276.

Woodward, T. W., and Back, R. A. (1963), Can. J. Chem. *41*, 1463.

The Solvated Electron

6.1 BACKGROUND

Solvated electrons were first produced in liquid ammonia when Weyl (1864) dissolved sodium and potassium in it; the solution has an intense blue color. Cady (1897) found the solution conducts electricity, attributed by Kraus (1908) to an electron in a solvent atmosphere. Other workers discovered solvated electrons in such polar liquids as methylamine, alcohols, and ethers (Moissan, 1889; Scott *et al.*, 1936). Finally, Freed and Sugarman (1943) showed that in a dilute metal–ammonia solution, the magnetic susceptibility corresponds to one unpaired spin per dissolved metal atom.

Debierne (1914) was the first to suggest a radical reaction theory for water radiolysis (H and OH). In various forms, the idea has been regenerated by Risse (1929), Weiss (1944), Burton (1947, 1950), Allen (1948), and others. Platzman (1953), however, criticized the radical model on theoretical grounds and proposed the formation of the hydrated electron. Stein (1952a, b) meanwhile had suggested that both electrons and H atoms may coexist in radiolyzed water and proposed a model in which the electron digs its own hole. Later, Weiss (1953, 1960) also favored electron hydration with ideas similar to those of Stein and Platzman. In some respects, the theoretical basis of these ideas is attributable to the polaron (Landau, 1933; Platzman and

Franck, 1954). Platzman (1953) made several conjectures for a slow electron in water:

1. It thermalizes in ~10^{-13} s.
2. Following thermalization, it would hydrate rather than react chemically with H_2O.
3. It takes ~10^{-11} s, the normal dielectric relaxation time for water, to form the hydrated electron, and ~10^{-9} s for the electron to disappear by reacting with the water molecule (the former is an overestimate, the latter an underestimate).
4. The heat of solution of the hydrated electron, by analogy with ammonia, would be about 2 eV.

Notwithstanding Platzman's theory, most calculations of radiation-chemical yields in water and aqueous solutions were performed using the free-radical model (see Magee, 1953; Samuel and Magee, 1953; Ganguly and Magee, 1956). The hypothesis was that the recapture time of the electron would be shorter than the dielectric relaxation time. Therefore, recombination would outcompete solvation.

On the experimental side, evidence was accumulating that there is more than one kind of reducing species, based on the anomalies of rate constant ratios and yields of products (Hayon and Weiss, 1958; Baxendale and Hughes, 1958; Barr and Allen, 1959). The second reducing species, because of its uncertain nature, was sometimes denoted by H'. The definite chemical identification of H' with the hydrated electron was made by Czapski and Schwarz (1962) in an experiment concerning the kinetic salt effect on reaction rates. They considered four reactions of the reducing species, designated here as X, with H_2O_2, O_2, H_3O^+, and NO_2^-, respectively, as follows:

$$X + H_2O_2 \rightarrow OH + OHX, \tag{I}$$
$$X + O_2 \rightarrow O_2X, \tag{II}$$
$$X + H_3O^+ \rightarrow H_2O + XH^+, \tag{III}$$
$$X + NO_2^- \rightarrow NO_2^-X. \tag{IV}$$

The logarithm of the rate ratios is plotted versus $\mu^{1/2}/(1 + \mu^{1/2})$ in Figure 6.1. In terms of the Brønsted model of ionic reactions and application to the Debye theory of ionic solutions, we may write

$$\log \frac{k}{k_0} = \frac{1.02\, z_1 z_2 \mu^{1/2}}{1 + \mu^{1/2}}, \tag{6.1}$$

where k is the reaction rate at ionic strength μ between the species having charges z_1 and z_2, and k_0 is the reaction rate at zero ionic strength. Consistent with the assumption that X indeed is an electron, Czapski and Schwarz found that k_{II}/k_I was independent of ionic strength, whereas k_{III}/k_I decreased and k_{IV}/k_I increased with ionic strength as given by the Brønsted-Bjerrum

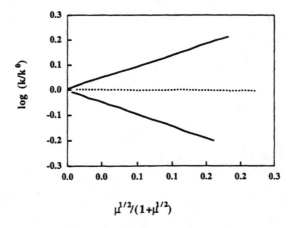

$$\mu^{1/2}/(1+\mu^{1/2})$$

FIGURE 6.1 Plot of the logarithm of rate constant ratios against $\mu^{1/2}/(1 + \mu^{1/2})$, where μ is the ionic strength. See text for explanation. Reprinted from Czapski and Schwarz (1962), with permission from Am. Chem. Soc.©

equation (Eq. 6.1). Their conclusion was verified by Collinson *et al.* (1962) and by Dainton and Watt (1963).

Keene (1960) apparently was among the first to report on the absorption spectrum of the hydrated electron. In 1962, Hart pulse-irradiated formic acid and carbon dioxide solutions and found not only a UV spectrum due to the CO_2^- ion, but also an intense absorption in the visible. By various chemical tests and using pure water, Hart and Boag (1962; Boag and Hart, 1963) demonstrated unequivocally that the spectrum belonged to the hydrated electron. This is the historical culmination of the discovery of the hydrated electron, the most important species generated in the radiolysis of water, the most fundamental reducing entity in chemistry, and a convenient species for dosimetry and monitoring of many other reactions. Figure 6.2 shows the spectrum of the hydrated electron (Keene, 1964; Gordon and Hart, 1964; Gottschall and Hart, 1967; Fielden and Hart, 1967).

6.2 THE HYDRATED ELECTRON

6.2.1 METHODS OF FORMATION OF THE HYDRATED ELECTRON

Following Walker (1968), we can distinguish three major methods of formation of e_h ; (1) chemical, (2) photochemical, and (3) radiation-chemical. The first of these is by the use of reagents without the help of external radiation.

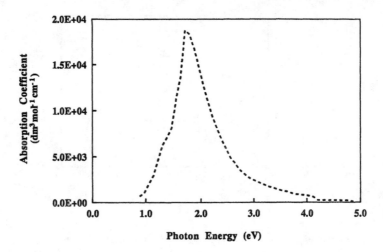

FIGURE 6.2   Absorption spectrum of the hydrated electron. The spectrum is structureless, broad (half-width ~ 0.84 eV), intense (oscillator strength ~ 0.75), and has a single peak at 1.725 eV. (See text for details.)

6.2.1a   Chemical Methods

Reaction of Metals with Water

The standard free energy for the formation of e_h is about 1.7 eV (Jortner and Noyes, 1966). Therefore, a reducing agent with a redox potential somewhat greater than 2.6 eV can donate an electron to water. This condition is satisfied for a number of metals. Taking sodium amalgam as an example, we can write the reactions leading to H_2 formation as follows (Hughes and Roach, 1965; Walker, 1966):

$$Na(Hg) + nH_2O \rightarrow Na_h^+ + e_h,$$
$$2e_h \rightarrow H_2 + 2OH^- \tag{6.1}$$

That e_h is the intermediate species and not the H atom has been verified by adding N_2O and methanol to water; then, N_2, not H_2, is the principal product. Alkali and alkaline earth metals above Na in the electrochemical series will also generate e_h on dissolution in water. Moreover the H/D isotope effect in water containing 50% D is consistent with the reaction $2e_h \rightarrow H_2 + 2OH^-$ (Anbar and Meyerstein, 1966; Hart and Anbar, 1970).

Although it is very hard to observe the absorption spectrum of e_h when metal is dissolved in water because of its high reactivity, some attempts were made in water and ice (Jortner and Stein, 1955; Benett *et al.*, 1964, 1967). Furthermore ESR (electron spin resonance) studies revealed that the trapped or solvated electron in ice interacts with six equivalent protons, thus ruling out H_2O^-.

Electrochemical Reduction of Water

Some experiments indicate that the cathode reaction may be due to e_h; at least, it is an open question (Walker, 1966, 1967; Hills and Kinnibrugh, 1966). Walker (1966), using platinum cathode, found that N_2O reduces the H_2O by about 65% but further addition of 0.1 M methanol did not change the N_2/H_2 ratio. The result may be explained by the cathodic reaction

$$e(\text{cathode}) \rightarrow e_h \qquad\qquad (6.II)$$

or by the assumption that N_2O reacts directly on the cathode with electrons and H atoms recombine catalytically.

When an aqueous solution containing an irreducible cation M^+ is electrolyzed, H_2 evolves at the cathode with the overall reaction $H_{aq}^+ + e(\text{cathode}) \rightarrow (1/2)H_2(\text{gas})$. The detailed mechanism of this reaction is somewhat ambiguous, as it could be attributed either to absorbed H atoms or absorbed H_2^+ ions. According to Walker (1966, 1967), the basic cathodic reaction is (6.II) followed by (6.I) to give H_2. There are several possibilities for reaction (6.II) (Walker, 1968): (1) direct electron donation by the cathodic metal to water, (2) electron liberation from the diffuse double layer, and (3) neutralization of the irreducible cation M^+ (e.g., Na^+) at the cathode, followed by the reaction of the neutral atom with water:

$$M^+ + e(\text{cathode}) \rightarrow M,$$
$$M + nH_2O \rightarrow M^+ + e_h.$$

Ionic Dissociation of H Atoms and Other Free Radicals

The basic requirement is that a free radical reacting with a negative ion in solution releases enough energy for the detachment of the electron from the negative ion and its eventual hydration. Consider the reaction

$$H + (OH^-)_{aq} \rightarrow H_2O + e_h, \qquad\qquad (6.III)$$

first confirmed by Jortner and Rabani (1962) by passing H atoms into an alkaline chloroacetate solution. Later, Matheson and Rabani (1965) verified the conversion of H into e_h in alkaline solution by the absorption spectrum. A similar reaction with the H atom can proceed with F^-:

$$H + (F^-)_{aq} \rightarrow HF + e_h.$$

Instead of H, one can also use CH_2OH or CF_3.

Reaction of Other Solvated Electrons with Water

Dewald et al. (1963) dissolved Cs in ethylenediamine (e_d) producing solvated electron e_s in e_d. When mixed with a solution of water in e_d in a fast-flowing system, evidence was obtained for rapid conversion of e_s to e_d. Dewald and Tsina (1967) also generated e_h by mixing water with Na–ammonia solution at $-34°$. There is considerable evidence that alkali halide crystals containing stable F-centers, produced for example by irradiation, when dissolved in water generates e_h (Westermark and Grapengiesser, 1960; Mittal, 1971; Arnikar et al., 1970).

6.2.1b   Photochemical Method

Direct one-photon ionization of liquid water is difficult to observe; however, hydrated electrons with a small yield have been found at photon energies somewhat less than the gas-phase ionization potential (Anbar et al., 1969; Asmus and Fendler, 1969). There has long been a widespread belief that ionization of liquid water requires much less energy than the gas-phase ionization potential, which is 12.6 eV. Indeed, Gutman and Lyons (1967) estimate the energy needed as 7 eV. It is likely that this ionization corresponds not to a prompt process, but to a multistep reorganization processes. The yield is expected to be very low, because this process has to compete with fast dissociation. Sokolov and Stein (1966) failed to observe the hydrated electron with 147-nm UV. Asmus and Fendler (1969) reported a quantum yield of ~0.05 at a somewhat longer wavelength using SF_6 as a monitor for e_h.

Hydrated electrons are easily produced when a solute ion is photolyzed at $\lambda > 200$ nm, where water is transparent, or when the solute ion has a charge-transfer-to-solvent (CTTS) absorption band. In the photolysis of halide ions (X^-; $X = $ Cl, Br, or I), there are three important observations (Stein, 1969): (1) the UV absorption has high oscillator strengths without any correspondence in the gas phase; (2) the band splittings for Br^- and I^-, respectively 0.44 and 0.94 eV, are nearly the same as the energy differences in the atomic states $^2P_{1/2}$ and $^2P_{3/2}$; and (3) in acidic solutions, light absorption by X^- results in hydrogen production via

$$2X^- + 2H^+ \rightarrow X_2 + H_2 .$$

Franck and Scheibe (1928) interpreted the absorption band of the halide ions in solution as a CTTS spectrum. This spectrum was further elaborated by Platzman and Franck (1954) and by Stein and Treinin (1959, 1960).

The formation of hydrated electrons by the photolysis of halide ions in solution may be envisaged in two steps. The first step is the CTTS absorption leading to $(X_{aq}^-)^*$. The second step is a slow, thermal process releasing the electron in competition with degradation and recapture. In the presence of acid and alcohol, photolysis of halide solutions generates H_2 with a yield that increases both with acid and alcohol concentrations (see Jortner et al., 1962, 1963, 1964). At $25°$, the limiting quantum yields are 0.98 for Cl^- at 185 nm, 0.6 and 0.5 for Br^- at 185 and 229 nm, respectively, and 0.3 and 0.25 for I^- at 254 and 229 nm, respectively. Since most of these yields are less than 1, the direct reaction of H_3O and $(X_{aq}^-)^*$ is ruled out. Instead, it is proposed that e_h is produced from the excited halide ion in competition with geminate recombination. The acid reacts with e_h to give H atoms, which then undergo abstraction reaction with the alcohol as follows (Stein, 1969):

$$X_{aq}^- \xrightarrow{h\nu} (X_{aq}^-)^* \to X + e_h \xrightarrow{H_3O^+} X + H \to X + R + H_2 .$$

Many other ions in solution can be photolyzed to give e_h. In particular, Schmidt and Hart (1968) emphasized alkaline H_2-saturated solution under flash photolysis. Here, one gets two e_h per photolysis, one by detachment from OH⁻ and another by fast successive reaction as follows:

$$OH^- + h\nu \to OH + e_h ,$$
$$H_2 + OH \to H_2O + H,$$
$$H + OH^- \to H_2O + e_h .$$

The identity of e_h in the photolysis of negative ions in solution is demonstrated by competitive scavenging (Jortner et al., 1962) or by its absorption spectrum (Ottolenghi et al.,1967). Alternatively, Grossweiner (1968) has used organic ions, and Delahay and Srinivasan (1966) have used photoionization from metallic surfaces immersed in water. In the latter case, using immersed Hg electrodes one gets photoemission at ~3 eV instead of 4.5 eV, which is the true work function of Hg, suggesting strongly that the binding energy of the electron in water, ~1.6 eV, helps to stabilize the ejected electron from Hg into water.

6.2.1c Radiation-Chemical Method

The most common method for generating hydrated or other solvated electrons is by the use of ionizing radiation. Electrons injected into water or produced by ionization will hydrate (see Sect. 6.1). *The rest of section 6.2 is devoted to the behavior of e_h produced by irradiation.* Note that the generation of hydrated electrons by ionizing radiation is a prompt process, different from UV photolysis of halide ions in solution. For radiation-chemical studies, pulse radiolysis is more convenient and is a superior method, when used in conjunction with kinetic spectrophotometry, for studying electron reactions or reactions of other transient species that have visible absorption. Various technical improvements of the basic pulsed electron irradiation setup and detection equipment have been made; these now allow the investigator to go down progressively into the time scales of nanoseconds and on the order of picoseconds (Wolff et al., 1973; Hunt et al., 1973; Jonah et al., 1973, 1976). When pulsed X-ray units are used for analytical purposes, the dose requirement is about 1 rad per pulse (Hart, 1966). Finally ^{60}Co-γ radiation can be used if the required steady state concentration of e_h is less than ~10 nM. At these concentrations in a H_2-saturated solution, the main mode of decay of e_h is by reaction (6.1) (Hart, 1966). The estimated dose rate requirement is $\sim 10^3$ rads/s, implying, in a typical setup, a source $\sim 10^4$ Ci strength.

6.2.2   YIELD

We must remember that measured yields may depend on a number of factors, such as pH, LET, scavenger concentration, and time. Yields depend on the sequence of reactions, which should be called the *kinetic scheme* but often the phrase *reaction mechanism* is loosely applied to it. Fortunately for the hydrated electron, which has a strong absorption signal, the directly measured yields have now been progressively pushed to the time scale of picoseconds or a little less (Wolff *et al.*, 1973; Jonah *et al.*, 1973, 1976). In most cases, however, a fundamental question is how to relate the measured yield at a given time to what is *originally* produced by the absorption of radiation. The terminology and the underlying concepts used by the practicing radiation chemists have been confusing in this respect. Following Hart and Platzman (1961), we will use the symbol $g(X)$ for the 100-eV primary yield, and the symbol $G(X)$ for the 100-eV observed yield for the species X.

First, we will review the stationary primary yield of the hydrated electron at neutral pH for low-LET radiation at a small dose. The primary species are e_h, H_3O^+, H, OH, H_2, and H_2O_2. Material balance gives

$$g(-H_2O) = 2g(H_2) + g(H) + g(e_h) = 2g(H_2O_2) + g(OH). \qquad (6.2)$$

Since charge conservation requires $g(e_h) = g(H_3O^+)$, the latter yield will not be considered further. The chemical measurement of $g(e_h)$ uses Eq. (6.2) and the measurements of primary yields of H, H_2, OH, and H_2O_2 in a suitable system. Various systems may be used for this purpose (see Draganic' and Draganic', 1971). For example, in methanol solution radiolysis, H_2 is produced by the reaction $H + CH_3OH \rightarrow H_2 + CH_2OH$. Therefore, in this system, $G(H_2) = g(H_2) + g(H)$. If, in addition, there is excess oxygen, the H atoms would be removed by the reaction $H + O_2 \rightarrow HO_2$. Therefore, from these two measurements, both $g(H)$ and $g(H_2)$ may be obtained.

Another way of measuring $g(H_2)$ directly is to design a system in which (1) further reactions of H and e_h do not generate H_2; and (2) OH reacts efficiently with the scavenger, preventing loss of H_2 by the reaction $H_2 + OH \rightarrow H_2O + H$. Various scavengers including the bromide will satisfy these requirements. It has been customary to plot $G(H_2)$ in this system as a function of the cube root of scavenger concentration (see Mahlman and Swarski, 1967). Although the so-called cube root rule has no sound theoretical basis (Kuppermann, 1961; Mozumder and Magee, 1975), it has been useful in extrapolating $G(H_2) = 0.45$ to the zero scavenger limit. Thus, in this system,

$$g(H_2) = G(H_2) = 0.45. \qquad (6.3)$$

The primary yield of H_2O_2 may be obtained by measuring the H_2O_2 yield in a system containing excess oxygen. In this case, hydrated electrons and H atoms are converted into O_2^- and HO_2, respectively, with an equilibrium between these species:

$$e_h + O_2 \rightarrow O_2^-, \tag{6.IV}$$

$$H + O_2 \rightarrow HO_2, \tag{6.V}$$

$$HO_2 \longleftrightarrow H^+ + O_2^-.$$

The HO_2 radicals react with themselves, giving H_2O_2 and O_2, and the only reaction of the OH radical is with H_2O_2, partially regenerating the HO_2 radical:

$$2HO_2 \rightarrow H_2O_2 + O_2,$$

$$OH + H_2O_2 \rightarrow H_2O + HO_2.$$

Considering the foregoing reaction one has, in this system,

$$G(H_2O_2) = g(H_2O_2) + \frac{1}{2}[g(e_h) + g(H) - g(OH)]. \tag{6.4}$$

From Eqs. (6.2) and (6.4), one gets

$$g(H_2O_2) = \frac{1}{2}[G(H_2O_2) + g(H_2)].$$

In this way, $g(H_2O_2)$ has been determined to be about 0.71. To find $g(OH)$, one uses a solution of sodium formate, a mild reducing agent, and oxygen. In this system, all radicals react to give HO_2 or O_2, the electron, and H atom by reactions (6.IV) and (6.V), and the OH radical by the following reactions:

$$OH + HCOO^- \rightarrow H_2O + CO_2^-,$$

$$O_2 + CO_2^- \rightarrow CO_2 + O_2^-.$$

In this system, then,

$$G(H_2O_2) = g(H_2O_2) + \frac{1}{2}[g(e_h) + g(H) + g(OH)]. \tag{6.5}$$

From (6.2) and (6.5), one obtains

$$g(e_h) = G(H_2O_2) - g(H) - g(H_2) \tag{6.6}$$

and

$$g(OH) = G(H_2O_2) + g(H_2) - 2g(H_2O_2). \tag{6.7}$$

In this manner, one obtains $g(e_h) = 2.7$ at neutral pH. In the past, considerable controversy existed about these "primary yields" in view of work at different laboratories using slightly different techniques, under somewhat different conditions, and so forth (see Farhataziz et al., 1966). Table 6.1 illustrates the results and possible reconciliation. The controversies have been basically resolved, and most workers now agree that $g(e_h) = g(OH) = 2.7 \pm 0.1$, $g(H) = 0.56 \pm 0.05$,

TABLE 6.1   Observed Yields of Radical and Molecular Species in Water Under High-Energy Radiation Collected from Various Laboratories

System	$G(eh)$	$G(H)$	$G(OH)$	$G(H_2)$	$G(H_2O_2)$	Laboratory[a]
$O_2 + H_2O_2$	2.8	—	—	0.40	0.70	1
TNM	←3.4→		2.7	0.34	0.75	1
$O_2 + H_2O_2 + EtOH$	2.3	0.6	2.3	0.46	0.7	1
$H_2 + OH + EtOH$ (pH 13)	3.1	0.5	2.9	—	—	2
$O + HCOOH$ (pH 3–11.8)	←3.0→		3.0	0.45	0.45	2
$O_2 + CO$	2.58	0.55	2.6	0.45	0.71	3
2mM NO (pH 4–12)	←3.1→		2.9	0.40	0.54	4
O + oxalate (pH 7–14)	←2.75→		2.0	0.45	0.75	5
OH scavenger + N_2O in low conc.	2.3–2.8	0.6	—	0.4	—	6
Ferrous sulfate or ceric sulphate + 0.8N H_2SO_4	←3.65→		2.9	0.45	0.8	6

[a]Laboratory key: 1, Brookhaven; 2, Argonne; 3, Oak Ridge; 4, Edinburgh; 5, Vinca; 6, Miscellaneous.

Adapted from Farhataziz et al. (1966). Note that these are observed yields from which the primary yields are calculated and referred to in the text.

$g(H_2) = 0.44 \pm 0.02$, and $g(H_2O_2) = 0.72 \pm 0.02$ under ideal conditions. These yields, established by purely chemical methods, nominally satisfy the material balance equation (6.2) and give an *observable* water decomposition yield of $g(-H_2O) = 4.14$, all referred to a ~0.1-μs time scale.

The pulse radiolysis technique gives a direct way for measuring the hydrated electron yield. To get the stationary yield, one can simply follow the electron absorption signal as a function of time and, from the known value of the extinction coefficient (Table 6.2), evaluate $g(e_h)$. Alternatively, the electron can be converted into a stable anion with a known extinction coefficient. An example of such an ion is the nitroform anion produced by reaction of e_h with tetranitromethane (TNM) in aqueous solution:

$$e_h + C(NO_2)_4 \rightarrow C(NO_2)_3^- + NO_2 .$$

Initially, TNM solutions tended to give a somewhat larger yield, but now it is believed that the yields obtained by absorption spectroscopy and by chemical scavenging studies give identical values under ideal conditions.

6.2.2a   The Effect of pH

In the pH range 3 to 13, all primary yields, corrected for solute reactivity, are pH independent, and $g(e_h) = g(OH) \approx 2.7$ (Fielden and Hart, 1967, Draganic' et al., 1969). In the pH range 3 to 1.3, Draganic' et al. (1969) find ~7% increase in the

TABLE 6.2 Summary of Physical Data for the Hydrated
Electron at 23°

Radiation yield	2.7 per 100 eV at ~μs.
	4.6 per 100 eV at ~10 ps.
Hydration energy	−38 Kcal/mole
Wavelength at maximum absorption	720 nm
Maximum molar extinction coefficient	$1.85 \times 10^4 M^{-1}cm^{-1}$
Half-width of absorption spectrum	0.9–1.0 eV
Oscillator strength	>0.7
Temperature coefficent of absorption maximum	−0.003 eV/°
g factor (esr)[a]	2.0002 ± 0.0002
Line width of esr spectrum[a]	~0.5 gauss
Difussion coefficient	$4.9 \times 10^{-5} cm^2 s^{-1}$
Effective ground state radius[b]	2–3 Å
Natural lifetime	~1 ms
Molar volume[c]	−3 ml/M
Solvation time	<1 ps

[a]At 5° using CH_3OH as OH radical scavenger (Avery et al., 1968).
[b]Hart and Anbar (1970, Tables III.3 and III.4).
[c]Calculated.

water decomposition yield and ~10% increase in the total reducing yield. Fielden and Hart (1967) report an increase of $G(e_h)$ from 2.7 in neutral solution to about 3.1 in strong alkaline solutions; they conclude that at pH > 12, total reducing yield remains constant. These findings are consistent with the hypothesis that the various original yields are pH independent (Haissinsky, 1967), and small variations at extreme pH values may be expected on account of secondary reactions.

The reaction $H + OH^- \rightarrow e_h$ is undoubtedly responsible for the increase of $G(e_h)$ at high pH. Similarly, the reaction $e_h + H^+ \rightarrow H$ must be responsible for the reduction of the hydrated electron yield in acid solution. The increase of total reducing yield and water decomposition yield at pH = 1.3 is not clearly understood, but it may also be due to secondary reactions.

6.2.2b The Effect of Temperature

Early experiments by Gottschall and Hart (1967) and by Michael et al. (1971) covering the range −4 to 390° showed that the product $G\varepsilon_{max}$ decreases slightly with temperature, where ε_{max} is the transition energy at the peak of its absorption curve. Since other primary yields were believed to be relatively independent of temperature (Hochanadel and Ghormley, 1962), these authors concluded that ε_{max} decreases with temperature and the spectrum becomes more asymmetric, favoring higher-energy transitions. Hart et al. found $d(\varepsilon_{max})/dT = -3 \times 10^{-3}$ eV/° below 100° and $= -1.5 \times 10^{-3}$ eV/° at 350°. The half-width of the spectrum has no significant temperature dependence, being about 0.8–0.9 eV. The variation of ε_{max} with temperature is said to be consistent with Jortner's (1964) theory, although the invariance of the half-width remained to be explained.

6.2.2c The Effect of Pressure

There is experimental evidence that the primary yield of the hydrated electron is independent of pressure up to about 9 Kbar and the different observed variations are attributable to the change of rate constants of secondary reactions with pressure, following the usual thermodynamics. Schindewolf et al. (1969) observed a blue shift of 20 nm in the spectrum of e_h at 27° under a pressure of 1000 atm. Although the authors interpreted their result in terms of the cavity model, the high pressure diminishing the cavity volume, later extensive experiments by Hentz et al. (1967a–c; Hentz and Knight, 1970) showed that the hydrated electron occupies negligible volume. Accordingly, pressure affects the ratios of electron reaction rates but not the primary yields.

6.2.2d The LET Effect

Much work has been done with protons and α-particles absorbed in the Fricke dosimeter system (see Draganic' and Draganic', 1971). Although the experimental results are not always clear-cut, it may be said that the hydrated electron yield, together with the total reducing yield, decreases with LET (Allen, 1961). For example, using ^{210}Po α-particles (5.3 MeV) Lefort and Tarrago (1959) obtained $g(H) + g(e_h) = 0.6$. Kuppermann's (1967) calculation using diffusion kinetics shows rather fast decrease of the yield with LET in the region 10^{-1} to 10^1 eV/Å.

6.2.2e The Time Dependence of the Hydrated Electron Yield

The first experimental measurements of the time dependence of the hydrated electron yield were due to Wolff et al. (1973) and Hunt et al. (1973). They used the stroboscopic pulse radiolysis (SPR) technique, which allowed them to interpret the yield during the interval (30–350 ps) between fine structures of the microwave pulse envelope (1–10 ns). These observations were quickly supported by the work of Jonah et al. (1973), who used the subharmonic prebuncher technique to generate very short pulses of ~50-ps duration. Allowing for the photodiode rise time, the authors reported time-dependent e_h yield beyond 200 ps. Both sets of experiments showed no e_h decay from 30 ps to 1 ns, and a fast decay from 1 to 10 ns (Figure 6.3). This finding was inconsistent with the Schwarz (1969) form of the diffusion model, which was highly satisfactory in the analysis of the variation of radical and molecular yields with LET and scavenger concentration (see Sect. 7.2.2). Diffusion theory predicted more decay in the time scale less than 1 ns and less in longer time scales.

Kuppermann (1974) attempted a rationalization invoking bigger spurs, which seems physically implausible. In any case, these larger spurs are not consistent with LET and scavenger effects (see Jonah *et al.*, 1976). Rzad and Schuler (1973) attempted another rationalization based on scavenging work and the (inverse) Laplace transformation. Although it rationalizes the direct and indirect time dependence of the e_h yield, it does not base either on a physical model (Jonah, 1974). Jonah *et al.* (1976) later reexamined the evolution of the e_h yield form 100 ps to 3 ns and found 17% decay, about half of which occurred before 700 ps. The situation, also shown in Figure 6.3, is nearer the Schwarz calculation, although there are still significant discrepancies. Jonah *et al.* (1976) concluded that the 100-ps yield is 4.6±0.2 and that the absorption spectrum is fully developed at the smallest time scale experimentally accessible (~30 ps). These indicate (1) the macroscopic dielectric relaxation time (~10 ps) can only be an upper bound to the hydration time, and (2) there must be additional sources of ionization in liquid water. The primary e_h yield of 4.6 and a "dry" electron yield of 0.8 (Wolff *et al.*, 1973) would put the ionization yield in liquid water to 5.4, which is hard to explain theoretically.

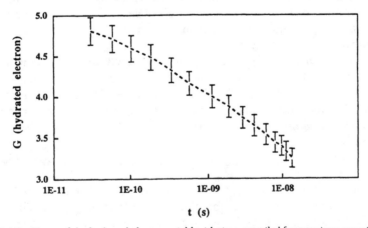

FIGURE 6.3 Decay of the hydrated electron yield with time compiled from various experiments. There is relatively little decay from ~30 ps to ~1 ns and a fast decay from 1 to 10 ns. These results were found difficult to reconcile with diffusion theory. The error bars indicate experimental uncertainties.

Following Platzman (1967), Magee and Mozumder (1973) estimate the total ionization yield in water vapor as 3.48. The yield of superexcited states that do not autoionize in the gas phase is 0.92. Assuming that all of these did autoionize in the liquid, we would get 4.4 as the total ionization yield. This figure is within the experimental limits of e_h yield at 100 ps, but it is less than the total experimental ionization yield by about 1. The assumption of lower ionization potential in the liquid does not remove this difficulty, as the total yield of excited states in the gas phase below the ionization limit is only 0.54.

6.2.3   THE SPECTRUM AND OTHER PHYSICAL PROPERTIES

The absorption spectrum of e_h is intense, has a broad peak at about 720 nm (half-width ~1 eV), and is structureless (see Sect. 6.1 and Figure 6.2). It covers at least 220 to 1000 nm and possibly extends on either side. There is some evidence that the absorption rises somewhat in the UV, which has been interpreted as the water absorption perturbed by the hydrated electron (Nielsen et al., 1969, 1976).

The intensity of absorption gives the product $G\varepsilon$, where G is the observed yield and ε is the molar extinction coefficient. The absolute value of ε was determined by Fielden and Hart (1967) using an H_2-saturated alkaline solution and an alkaline permanganate–formate solution, where all radicals are converted into MnO_4^{2-}. They thus obtained $\varepsilon = 1.09 \times 10^4$ $M^{-1}cm^{-1}$ at 578 nm, which is almost identical with that obtained by Rabani et al. (1965), who converted the hydrated electron into the nitroform anion in a neutral solution of tetranitromethane. From the shape of the absorption spectrum and the absolute value of ε at 578 nm, one can then find the absolute extinction coefficient at all wavelengths. In particular, at the peak of absorption, $\varepsilon(720)/\varepsilon(578) = 1.7$ gives ε at 720 nm as 1.85×10^4 $M^{-1}cm^{-1}$.

The oscillator strength for absorption is a very important quantity signifying the nature of the transition. If the absorption spectrum is known, the oscillator strength can be calculated using Eq. (4.20). Instead of numerical integration, one often assumes that the spectrum is approximately gaussian with the same half-width Δv (cm^{-1}) as experimentally observed. One then obtains f, the oscillator strength, as

$$f = 4.6 \times 10^{-9}\varepsilon_{max} \, \Delta v.$$

For the hydrated electron, $\varepsilon_{max} = 1.85 \times 10^4$ $M^{-1}cm^{-1}$. Taking the half-width as 0.93 eV or $\Delta v = 8200$ cm^{-1}, the foregoing equation yields $f = 0.7$. This is almost exactly the same as for the solvated electron in ammonia, where $\varepsilon_{max} = 5.0 \times 10^4$ $M^{-1}cm^{-1}$ and $\Delta v = 3000$ cm^{-1}, giving $f = 0.69$. Both absorptions are therefore allowed. The actual oscillator strength for e_h is probably greater than 0.7, as at the extremities the curve lies above the gaussian. Some values of f for e_h and e_{am} are quoted as high as 0.8 and 0.9, respectively (see Matheson, 1975, Table VIII).

The mobility of e_h was determined by measuring the equivalent conductance following pulse irradiation (Schmidt and Buck, 1966; Schmidt and Anbar, 1969). After correcting for the contribution of H_3O^+ and OH^- ions, they found the equivalent conductance of $e_h = 190\pm10$ mho cm^2. From this, these authors obtained the mobility $\mu(e_h) = 1.98 \times 10^{-3}$ cm^2/v.s. and the diffusion coefficient $D(e_h) = 4.9 \times 10^{-5}$ cm^2/s using the Nernst–Einstein relation, with about 5% uncertainty. The equivalent conductance of e_h is the same as that for the OH_{aq}^- ion within experimental uncertainty. It is greater than that of the halide ion and smaller than that of e_{am}.

The mechanism of diffusion is believed to be either rapid transition from trap to trap followed by relaxation, or smooth slithering from one site to next. Hart and Anbar (1970) favor a trap-to-trap tunneling of the electron in analogy with many electron transfer reactions in water, which have activation energies of only 3–4 Kcal/mole. Considering that the hydration energy alone is about 40 Kcal/mole, one would expect an activation energy an order of magnitude more if the mechanism involved jumping over the potential barrier. In the case of electron diffusion, the traps either preexist or are self-dug.

Tachiya (1974) has applied the configuration coordinate model to the diffusive motion of the hydrated electron. In this model, one considers the orientational polarization as a kind of configurational coordinate. For diffusion, the electron is considered to slither from one point to next separated by 3 Å, which is somewhat greater than the diameter of the water molecule. In between these points is the activated complex, the energy of which is evaluated by continuously varying the orientational polarization parametrically and linearly. A plot of total energy gives the potential energy diagram for diffusive motion, the wavefunction for intermediate positions being determined by a variational procedure involving a linear combination of wavefunctions at the initial and final positions. The computed activation energy for diffusion, given by the difference of total energy at the activated complex and that of e_h in normal state, is 3.7 Kcal/mole. This is remarkably close to the experimental value, although D itself is not calculated since time does not appear explicitly in this theory.

Table 6.2 lists some of the physical data for the hydrated electron. Most of these data are experimental. The molar volume is calculated, as experimental measurements are not reliable. The oscillator strength and the natural lifetime against reaction with water molecules are lower bounds, whereas the salvation time is possibly an upper bound.

6.3 THE SOLVATED ELECTRON IN ALCOHOLS AND OTHER POLAR LIQUIDS

The absorption spectrum of radiation-produced e_{am} is identical to that in dilute metal–ammonia solutions. It has a broad, structureless absorption in the red and IR, with a peak at about 1.88 μm and a half-width of 0.2 eV on the high-energy side. The absorption is intense with $\varepsilon_{max} = 4.8 \times 10^4$ M^{-1}cm^{-1}, giving an oscillator strength greater than 0.69, which is comparable to e_h. ESR studies in dilute metal-ammonia solutions indicate a g factor of 2.002319 independent of the metal and temperature (Catterall, 1976), indicating the isolated nature of the electron. The radiation yield of e_{am} was earlier considered to be low (Dainton et al., 1964; Renaudiere and Belloni, 1973), possibly due to the reaction of e_{am} with NH_4^+ on the longer time scale. Later, Farhataziz et al. (1974a) obtained the radiation yield as 3.2 at 0.2 μs, which is insensitive to temperature and is consistent with the measurements of Ward (1968), Seddon et al. (1974), and Belloni et al. (1974).

Farhataziz et al. (1974a, b) studied the effect of pressure on e_{am} and found that as the pressure is increased from 9 bar to 6.7 Kbar at 23°: (1) the primary yield of e_{am} decreases from 3.2 to 2.0; (2) hv_{max} increases from 0.67 to 0.91 eV; (3) the half-width of the absorption spectrum on the high-energy side increases by 35%; and (4) the extinction coefficient decreases by 19%, which is similar to e_h. The pressure effects are consistent with the large volume of e_{an} (98 ml/M), whereas the reduction in the observed primary yield at 0.1 μs is attributable to the reaction $e_{am} + NH_4^+$. Some of the properties of e_{am} have been discussed by several authors in Solvated Electron (Hart, 1965).

Solvated electron in aliphatic alcohols was discovered using pulse radiolysis soon after the hydrated electron (Adams et al., 1964; Sauer et al., 1964; Taub et al., 1963, 1964). Dorfman (1965) reviewed the early situation. The absorption spectrum covers the entire visible and goes into the UV and IR regions. As in water, the absorption is strong but structureless; the peak extinction coefficients in the various alcohols are about the same as for e_h, being $\sim 10^4$ $M^{-1}cm^{-1}$ (Busi and Ward, 1973; Jha et al., 1972). A broad maximum absorption is seen either in the red or IR. Based on $G(e_h) = 2.6$ in the microsecond time scale, the yields of e_s in various alcohols were determined to be about 1.1±0.1. However, it is fairly certain that the yields in the shorter time scales are much greater (Hentz and Kenney-Wallace, 1974; Lam and Hunt, 1974; Dixon and Lopata, 1975). Since the experimentally measured oscillator strength is in the range 0.6 to 0.9, Dorfman (1965) concludes that there is not likely to be an additional absorption band in the UV. He describes various neutralization and attachment reactions of e_s in alcohols with rate constants in the range 10^9–10^{10} $M^{-1}s^{-1}$ at room temperature, but draws attention to the fact that the analog of the reaction in water with itself—namely, $e_s + e_s \rightarrow H_2 + 2RO^-$—has not been found. Grossweiner et al.'s (1963) flash-photolytic studies generally corroborate the spectra found by pulse radiolysis.

Hentz and Kenney-Wallace (1972, 1974) made a detailed study of e_s in 25 neat alcohols and three alkane solutions in 1-hexadecanol at 30° using a 5-ns electron pulse. Most data were new, but in some cases they confirmed earlier observations (Dorfman, 1965; Baxendale and Wardman, 1971). The authors found the spectrum fully developed at the end of the pulse, with no spectral change thereafter. The spectra are all broad, asymmetric, and structureless, having peaks in the visible or the IR. The half-widths for most normal alcohols are ~1.5 eV in accordance with earlier observation (Dorfman, 1965). Table 6.3 shows the static dielectric constant ε_s, transition energy at maximum absorption, and half-width of the spectrum.

Contrary to earlier expectations (see Dorfman, 1965), Hentz and Kenney-Wallace (1972, 1974) failed to find any correlation between ε_s and E_{max}. Actually, there is a better correlation of matrix polarity with the spectral *shift* from e_t to e_s upon solvation and the time required to reach the equilibrium spectrum (Kevan, 1974). Furthermore, Hentz and Kenney-Wallace point out that e_{max} is smaller for alcohols with branched alkyl groups, the spectrum being sensitive to the number, structure, and position of these groups relative to OH. Clearly, a steric effect is called for, and the authors claim that a successful theory must not rely too heavily on continuum interaction as appeared in the earlier theories of Jortner (1959, 1964). Instead, the dominant interaction must be of short range, and probably the spectrum is determined by optimum configuration of dipoles within the first solvation shell.

Although various structural models (Raff and Pohl, 1965; Natori and Watanabe, 1966; Newton, 1973) and semicontinuum models (Copeland et al., 1970; Kestner and Jortner, 1973; Fueki et al., 1973) have been proposed for the solvated electron, the basis of the agreement or disagreement between theory and experiment is not well established. Another complication with the continuum or the semicontinuum models is the fact that in a number of polar systems the spectrum is fully developed in a time *far* shorter than the dielectric relaxation times (see, e.g., Bronskill et al., 1970; Baxendale and Wardman, 1973; Rentzepis et al., 1973).

Hentz and Kenney-Wallace (1974) obtained the evolution of e_s yield in some common alcohols by comparison with the corresponding yield of e_h and extrapolated the results to 30 ps. The picosecond data for the alcohols were obtained from the work of Wolff et al. (1973) and Wallace and Walker (1972); the nanosecond work was in substantial agreement with Baxendale and Wardman (1971). The evolution of the e_s yields in the common alcohols shows considerable decay from the picosecond to nanosecond regime and a comparable decay from the nanosecond to microsecond time scales. However, the microsecond yields are also probably somewhat larger than previously reported, especially for methanol and ethanol (see Dorfman, 1965). In agreement with this, Lam and Hunt (1974) report e_s yields in aliphatic alcohols at ~100 ps to be greater than 3. Nevertheless, there is room for neutralization of the "dry" electron in the presolvated state.

Solvated electrons are known to be formed in amines, amides, dimethyl sulfoxide, and many other liquids that will not be discussed here. Note that, except for the yield and time scale of observation, the production of e_s itself is not related to polarity. Thus, the e_s absorption spectrum has indeed been observed in nonpolar liquids both at low temperatures and room temperature (Taub and

TABLE 6.3   Transition Energy at Maximum Absorption (E_{max}) and Half Width ($W_{1/2}$) for e_s in Various Alcohols

	$\varepsilon_s{}^a$	E_{max} (eV)	$W_{1/2}$ (eV)
Neat Liquids			
Methanol	32.6	1.93	1.4
Ethanol	24.3	1.70	1.3
1,2-Ethanediol	37.7	2.13	1.4
2-Methoxyethanol		1.67	1.5
2-Ethoxyethanol		1.67	1.5
1-Propanol	20.1	1.67	
2-Propanol	18.3	1.49	
1-Butanol	17.1	1.82	1.5
2-Butanol	17.9	1.67	1.4
2-Methyl-2-propanol	10.9	0.97	0.8
1-Pentanol	13.9	1.90	1.4
3-Methyl-1-butanol	14.7	1.79	
3-Methyl-2-butanol	5.8	0.99	0.8
Cyclopentanol	15	1.50	1.4
3-Methyl-3-pentanol	5	<0.82	>0.9
3-Ethyl-3-pentanol	5	<0.82	>0.9
1-Hexanol	13.3	1.84	1.5
Cyclohexanol	15.0	1.65	1.5
4-Methylcyclohexanol		1.54	1.6
4-Heptanol	5.9	1.34	1.5
1-Octanol	10.3	1.90	1.5
2-Octanol	7.8	1.44	1.6
1-Nonanol	9.1	1.85	1.5
1-Decanol	7.8	1.90	1.5
1-Undecanol	5.9	1.84	1.5
Solutions[b]			
Cyclohexane (5 mole %)	2	1.54	1.4
2,2,4-Trimethyl-pentane (5 mole %)	2	1.24	
Hexadecane (10 mole %)	3	1.65	

[a]Static dielectric constant.

[b]Solution of 1-hexadecanol in alkanes.

Source: After Hentz and Kenney-Wallace (1974).

Gillis, 1969; Baxendale and Rasburn, 1974; Baxendale and Wardman, 1971; Baxendale *et al.*, 1971, 1973). However, it may be best to consider these cases as extensions of the trapped electron in low-temperature matrices.

6.4   TRAPPED AND SOLVATED ELECTRONS AT LOW TEMPERATURES

Electron trapping in condensed media is ubiquitous in liquids, glasses, and (imperfect) crystals. On the other hand, there is no evidence for trapped electrons in liquid methane, Ar, Kr, or Xe, although the question of trapped electrons with very small binding energy may be a semantic one. Trapped electrons (e_t) are formed at earliest times or at very low temperatures (4 K) by various ionization processes. They exhibit IR or far-IR absorption. On warming or with the progress of time, they relax to form solvated electrons while the spectrum shifts toward the visible. In this sense, trapped electrons in nonpolar media eventually solvate, although the distinction between these two species is not as great as in the polar media. According to Higashimura *et al.* (1970), trapped electrons in all glassy matrices are fully solvated at 77 K in the long-time limit.

Lifetimes of e_t increase upon cooling, being on the order of hours at 77 K. The yield of e_t is quite good at 77 K, the saturation density being ~10^{17} g^{-1} in nonpolar glasses and ~10^{19} g^{-1} in polar glasses. Crystalline ice traps electrons with a small yield, which decreases drastically on cooling. However, the yield in glassy alkaline ices is comparable to the nanosecond yield of hydrated electrons (Willard, 1968; Hamill, 1968; Eckstrom, 1970; Kevan, 1974; Funabashi, 1974). Figure 6.4 shows a typical spectrum of e_t at 4 K and that of the corresponding e_s at 77 K.

Table 6.4 gives the radiation yields of stabilized electrons in some selected glasses at 77 K with about 15% uncertainty. A detailed discussion of these yields has been given by Kevan (1974). Hase *et al.* (1972a) observed e_t spectrum in ethanol at 4 K with a peak at 1500 nm; on quick warming to 77 K, the spectrum relaxed with the same peak in the visible as that of e_s in liquid ethanol. Hase *et al.* (1972b) also observed e_t spectrum at 4 K in 3MP with no clear maximum. On quick warming to 77 K, the spectrum relaxed with a clear maximum at ~1700 nm.

It is not surprising that addition of hole traps will increase the metastable yield of e_t (Gallivan and Hamill, 1966; Bonin *et al.*, 1968). The yield of e_t becomes sublinear at doses ~10^{20} eV g^{-1}, reaches a peak, and eventually decreases at very high doses. There are three possible explanations: (1) reaction with radiation products (Eckstrom *et al.*, 1970); (2) electron tunneling to radiation-produced scavengers (Miller, 1972); and (3) dielectron formation (Feng *et al.*, 1973).

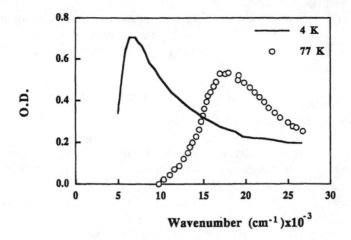

FIGURE 6.4   A typical trapped electron absorption spectrum in ethanol at 4 K and the corresponding solvated electron spectrum at 77 K. The irradiation is at 4 K in both cases. Reproduced from Hase *et al.* (1972a), with permission from Am. Inst. Phys.©

Although there are several theories for the solvated electron (*vide infra*), there does not seem to be enough theoretical work for the trapped electron. Tachiya and Mozumder (1974a, b) considered preexisting traps in *polar* media. In this model, the trapping potential is provided by fortuitous arrangement of molecular dipoles in a cell. After variationally solving the Schrödinger equation for the excess electron with a classical random orientation of the dipoles, they concluded the following: (1) a minimum value (~3.8 D) of the sum of the central components of the dipole moments is required to trap an electron; (2) at any temperature, shallow traps are more likely than the deeper ones; and (3) the relative probability of occurrence of shallow versus deep traps increases rapidly upon cooling.

Kevan (1974) has exhaustively reviewed e_t in organic glasses, to which the reader's attention is drawn. He points out that the effective spur radius r for trapped electrons may be operationally given in angstroms as

$$r = \left(\frac{4.8 \times 10^6}{4\pi D\rho} \right)^{1/3}, \tag{6.8}$$

where ρ is the medium density and D is the dose in Mrad at which the effective relaxation time $(T_1 T_2)^{1/2}$ begins to decrease. In most media, D is between 0.9 and 10. Equation (6.8) then gives r between about 40 Å for alkaline ice, and greater than 130 Å for 3MP and 3MHX. These large values are interpreted as the trapping distance for the ionized electrons rather than the size of the sphere of

TABLE 6.4   Radiation Yields (G Values) of Stabilized
Electrons in Organic Glasses Under γ-Irradiation at 77 K

Glass	Method	Yield
Alkanes		
3-Methylpentane	Scavenging	1.2
3-Methylpentane	esr	0.6
Methylcyclohexane	Scavenging	~2
Methylcyclohexane	esr	0.3
3-Methylhexane	esr	0.9
Polar		
Ethanol	Scavenging	2.5
Ethanol	esr	2.3
Methanol (5% H_2O)	esr & Scavenging	2.6
1-Propanol	Scavenging	2.0
1-Propanol	esr	1.5
2-Propanol	Scavenging	2.0
Ethylene glycol	esr	1.5
Miscellaneous		
Cumene	Scavenging	0.8
Benzene	Scavenging	1.4
Toluene	Scavenging	1.4
Methylcyclohexane	Scavenging	1.8
Amines and ethers	Scavenging	~2

influence of the trapped electron itself. It is remarkable that this spur size decreases with matrix polarity, which is consistent with other experimental findings. The discussion of theoretical models presented by Kevan (1974) actually applies to the solvated rather than to the trapped electron in the sense we have used these terms here.

Kevan and co-workers extended the semicontinuum model of liquids to low-temperature glasses (Fueki *et al.*, 1970, 1971, 1973). The authors claim good agreement between the semicontinuum theory and experiment in various polar organic glasses such as alcohols, ethers, and amines. The extension of the semicontinuum model to nonpolar glasses is not straightforward. Nevertheless, Feng *et al.* (1974) attempt to do that, considering the microdipoles of the nonpolar molecules in the C—H bonds. The experiments of Narayana and Kevan (1976) indicate that this extension of the semicontinuum model is not consistent.

6.5   THEORETICAL MODELS
OF THE SOLVATED ELECTRON

The basic requirement of a satisfactory theoretical model is the ability to explain the following: (1) electron binding, with a reasonable estimate of the heat of solution; (2) the optical absorption spectrum, with the peak and the oscillator strength approximately as found in experiment; and (3) a distribution of unpaired electron spin over neighboring molecules, as determined by ESR experiments. In addition, it is desirable that the model should give correctly the photoionization threshold, the line shape, or the half-width of the absorption spectrum and the time of spectral relaxation, and it should be consistent with respect to (energetic and configurational) stability. No theory at present satisfies the basic requirements and has the desirable additional features. All too often, a model will assume one key experimental result such as the peak of absorption and attempt to explain other features. Also the theories, with few exceptions, seem to be wanting in predictive powers.

Theories of solvated electrons may be divided as follows (Jortner, 1970; Webster and Howat, 1972; Kevan, 1974; Kestner, 1976): (1) molecular orbital models, (2) structural models, (3) continuum models, and (4) semicontinuum models. We will consider these models a little in detail.

A *molecular orbital model* (MO) treats all electrons belonging to a fixed number of solvent molecules plus an excess electron in the resultant field of the nuclei of the molecules as being in a fixed configuration. The nuclei belonging to a particular molecule normally keep the ground state structure of that molecule. The relative distances and orientations of these molecules are varied until energetic, and if possible configurational, stability is obtained. In some cases, molecular distortions have been considered.

Numerical solutions of the Schrodinger equation are obtained for the ground and excited states via the variational principle using a suitable basis set. However, it is not possible to determine the stability of the localized electron relative to the quasi-free electron by the MO method. One has to be content in calculating the energy difference of the complex M_n^- and M_n in a suitable arrangement for n solvent molecules. Even this requires extreme precision in calculation when one realizes that the energy differential is ~1 eV, whereas the molecular electronic energy is ~10^4 eV. Furthermore, Kestner (1976) points out that usually one aims for energetic stability, whereas one should look for thermodynamic stability. The distinction is unimportant at low temperatures (e.g., liquid helium) but cannot be dismissed at room temperature.

The various MO calculations use different basis sets and have different ways of calculating multicenter coulomb and exchange integrals. The current trend in MO is to expand as a linear combination of atomic orbitals (LCAO). The atomic orbitals are represented by Slater functions with expansion in gaussian functions, taking advantage of the additive rule. When the calculation is performed in this way, treating the overlap fully, the method is designated as *ab initio*. For larger systems or for greater configurational flexibility, approximations are often used. These are called complete neglect of differential overlap (CNDO) and intermediate neglect of differential overlap (INDO), while even more approximate results are obtained using the Huckel or extended Hückel approximations.

Ab initio calculations for the solvated electron have been made by Naleway and Schwartz (1972) and by Newton (1973, 1975), CNDO calculations by Weissmann and Cohan (1973), and INDO calculations by Howat and Webster (1972) and by Ishimaru *et al.* (1973). Many of the conclusions concerning stability reached by the different investigators are similar, whereas differences exist in relation to excitation energy, oscillator strength, and void volume when cavity formation is also considered in the model. No electron binding is obtained for the dimer or trimer ion of water. Even for $(H_2O)_4^-$, Newton (1973, 1975) finds the binding too weak to survive thermal disruption. This perhaps should indicate energetic stability brought about by long-range interaction. Indeed, Newton calculates a heat of solution ~1 eV and a distance of the center of molecules in the first solvation layer from the origin as 2.45 Å when the tetramer ion was put in the background of the continuum. Howat and Webster (1972) have also demonstrated relative configurational stability using MO and by varying one distance parameter.

In the structural model, the solvated electron, treated separately from the rest of the electrons, moves in the field of the *few* adjacent (polar) molecules. Natori and Watanabe (1966) and Natori (1968, 1969) applied this model to e_h. As shown in Figure 6.5, e_h is considered trapped at the center of a tetrahedron (vacant site) with vertices occupied by the O atoms of the surrounding water molecules. In the completely relaxed configuration used by these authors, the four H atoms inside the tetrahedron compose the innermost shell. In the next shell are the O atoms, and in the outermost shell are the H atoms outside the tetrahedron. The potential experienced by the excess electron is due to the nuclei and the averaged motion of the electrons in the molecules, which is treated approximately. The ground state energy was calculated to be about -1.5 eV, with fair agreement for the heat of solvation (1.7 eV) when the excess electron wavefunction was required to be orthogonal to those of the molecular electrons.

The calculated stabilization energy in this model refers simply to the difference between the energy of the entity in the ground state and the energy of the oriented cluster of water molecules plus an electron at infinity. Because of the neglect of quasi-free electron energy, this is *not* the true stabilization energy. Nevertheless, the structural model helps to illustrate the point that electron binding is possible with short-range forces only.

The excited state is formed out of a combination of $2p$ orbitals, and the absorption spectrum is seen as a $1s \rightarrow 2p$ transition. The excited state is weakly bound in this model, by ~0.9 eV. The calculated oscillator strength is too low, which seems to be a feature of all structural models. There is no configurational

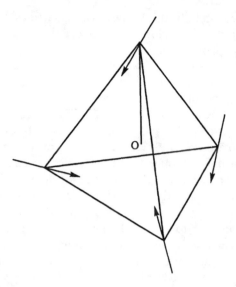

FIGURE 6.5 Schematic of the structural model of the solvated electron. The electron is considered trapped at the center of the tetrahedron, whereas for the hydrated electron, the vertices are occupied by O atoms. Arrows indicate the direction of molecular dipoles that may differ from cell to cell.

stability in this model; thus, either the tetrahedron size is to be fitted to give the right peak in the absorption spectrum, or a calculation should be performed for the tetrahedron size in ice. According to the structural model of e_h, there should be a correlation of hydrated electron yield with vacancy concentration in the ice structure. There is no such evidence.

In the continuum and semicontinuum models of e_s, long-range forces due to distant solvent molecules are usually represented by the optical and static dielectric constants. In a true continuum model, the continuity is extended to the origin or to the surface of the cavity. In some sense, the continuum and semicontinuum models both contain both short- and long-ranged interactions. The main difference is that in the semicontinuum model, the molecules in the first shell(s) are structured.

An electron in a condensed medium is considered localized if the lowest energy in that state is less than V_0, the ground state energy of the quasi-free electron. According to Springett et al. (1968), the condition for localization is expressed as

$$\frac{2\pi\hbar^2\gamma}{mv_0^2} \le 0.047,$$

where m is electron mass and γ is the surface tension of the liquid. Accordingly, Jortner (1970) reasons that electrons in hydrocarbons should be quasi-free. However, it is well known that only in very high mobility liquids such as methane or tetramethyleilane may the electron be considered quasi-free (Schmidt and Bakale, 1972). In most liquid hydrocarbons, the electron is largely localized, although a dynamic equilibrium is possible between localized and quasi-free states (Minday *et al.*, 1971).

Landau (1933) originally conceived the polaron as an electron trapped by the polarization field set up by itself in the dielectric. Davydov (1948) first applied the polaron model to e_{am}, later pursued by Diegen (1954) and by Platzman and Franck (1954) for aqueous solutions. However, these authors neglected volume expansion in metal-ammonia solutions. They also used the electronic adiabatic approximation (*vide infra*), which is consistent only if the excess electron binding energy is *very low* compared with the binding energy of electrons in the solvent molecules. Jortner (1962, 1964) criticized these approximations and constructed a self-consistent field (SCF) method for the solvated electron considering, in the case of liquid ammonia, the appropriate cavity volume.

The electronic energy E of e_s is given by the Schrodinger equation

$$\left[-\frac{\hbar^2}{2m} \nabla^2 + V \right] \psi = E\psi,$$ (6.9)

where ι is the stationary wavefunction in the ground or excited state and the potential V due to orientational polarization $P(r)$ is given in the spherically symmetric case by

$$V(r) = 4\pi e \int_r^\infty P(x)\,dx.$$ (6.10)

For $P(r)$, one usually has two choices: (1) the electronic adiabatic approximation, or (2) the SCF method. In the adiabatic approximation, the velocity of the excess electron is assumed to be small compared with that of molecular electrons. Then, electronic polarization of the medium does not contribute to binding. Jortner (1962, 1964) questioned the validity of this approximation for e_h or e_{am}, since the binding energy of the excess electron ($\sim$1–2 eV) is not insignificant compared with that of the medium electrons. He used the SCF method, in which all electrons are treated on equal footing. The resultant potential $V(r)$ is now given by (see Eq. 6.10)

$$V(\mathrm{r}) = ef(1 - \varepsilon_s^{-1}),$$ (6.11)

where the electrostatic potential f is given through the Poisson equation

$$\nabla^2 f = 4\pi e |\psi|^2.$$ (6.12)

In terms of $V(r)$, the electronic energy is given by

$$E' = \left\langle \psi \left| -\frac{\hbar^2}{2m} \nabla^2 + V \right| \psi \right\rangle,$$ (6.13a)

and the total energy (electron plus medium) by (Jortner, 1962)

$$E = \left\langle \psi \left| -\frac{\hbar^2}{2m} \nabla^2 + \frac{V}{2} \right| \psi \right\rangle. \tag{6.13b}$$

If the ground state wavefunction is used for ψ, then $-E$ is the heat of solution of e_s. The difference $E - E'$ is the work done Π on the medium, called the medium reorganization energy, to create the potential $V(r)$. From (6.11)–(6.13), Π can be given as

$$\Pi = -\frac{e}{2}\left(1 - \varepsilon_s^{-1}\right)\left\langle \psi \left| f \right| \psi \right\rangle.$$

The procedure for calculating the energy of the solvated electron is to assume a particular form for ψ containing one or more adjustable parameters and then calculate f, V, and E self-consistently through Eqs. (6.11)–(6.13). When a finite cavity is considered, the potentials f and V are taken constant inside the cavity with continuity at the surface. Then, E is varied with respect to the adjustable parameter(s) until a minimum is obtained. For the ground state of e_s, one uses a 1s wavefunction

$$\psi_{1s} = \left(\frac{\lambda^3}{\pi}\right)^{1/2} \exp(-\lambda r), \tag{6.14a}$$

and for the excited state,

$$\psi_{2p} = \left(\frac{\mu^5}{\pi}\right)^{1/2} r \cos\theta \exp(-\mu r), \tag{6.14b}$$

where λ and μ are the variational parameters. Imposing the variational condition $\partial E_{1s}/\partial\lambda = 0$, one obtains, for zero cavity radius,

$$\lambda = \frac{5}{16a_0}\left(1 - \varepsilon_s^{-1}\right) \tag{6.15a}$$

and

$$E_{1s} = -\frac{25}{256}\left(1 - \varepsilon_s^{-1}\right)^2 R, \tag{6.15b}$$

where a_0 is the Bohr radius (0.529 Å) and R is the Rydberg energy (13.6 eV).

The optical absorption of the solvated electron, in the continuum and semicontinuum models, is interpreted as a $1s \rightarrow 2p$ transition. Because of the Franck–Condon principle, the orientational polarization in the $2p$ state is given by what is consistent with the ls charge distribution, although the electronic polarization is given by the $2p$ charge distribution. Designating the thusly obtained total energy as $E_{2p(1s)}$, Jortner (1962) obtains formally

$$E_{2p(1s)} = \left\langle 2p \left| -\frac{\hbar^2}{2m} \nabla^2 \right| 2p \right\rangle + \frac{e}{2}(1 - \varepsilon_{op}^{-1}) <2p|f_{2p}|2p>$$

$$- \frac{e}{2}(\varepsilon_{op}^{-1} - \varepsilon_s^{-1}) \left\langle 1s|f_{1s}|1s \right\rangle + e(\varepsilon_{op}^{-1} - \varepsilon_s^{-1}) \left\langle 2p|f_{1s}|2p \right\rangle.$$

Again the energy $E_{2p(1s)}$ is minimized by varying μ and the optical transition frequency is given by

$$h\nu = E_{2p(1s)} - E_{1s}. \tag{6.16}$$

The preceding calculations can also be performed for finite cavity sizes. For this case, there are some additional sources of small amounts of energy associated with cavity formation arising from surface tension, pressure-volume work, and electrostriction. Because of the Franck-Condon principle these do not affect the transition energy, but they have some influence on the heat of solvation. Jortner's (1964) results are summarized as follows:

1. *For zero cavity radius,* e_h has a mean radius of charge distribution in the ground state equal to 2.54 Å, and $E_{1s} = -1.32$ eV, which is numerically somewhat less than the experimental heat of solution (1.7 eV). For the excited state, the mean radius of charge distribution is 4.9 Å, with $h\nu$ (see Eq. 6.16) = 1.35 eV. Note that $h\nu > -E_{1s}$, implying that the $2p(1s)$ is actually in the continuum.

2. *For a cavity radius of 3.3 Å,* appropriate to volume expansion in liquid ammonia, the experimentally determined optical absorption peak is at ~0.8 eV. However, the continuum model of Jortner (1964) does not give configurational stability, although it predicts energetic stability. It was stressed by Jortner that cavity formation does not lead to stability, which is due to long-range interactions. As a result, a popular procedure has evolved in which energetic stability is determined for a given cavity size. The latter is varied to give agreement with the experimental peak in the absorption spectrum.

3. The oscillator strength of the transition in e_h, obtained by the dipole length formulation, is 1.1; the experimental value is >0.7. This overestimation seems to be common among all continuum models. No information on line shape is obtained in this simple model.

4. The model predicts a red shift of the absorption spectrum with temperature for both e_h and e_{am}. However, in ammonia the effect is about four times bigger, because the volume expansion of the cavity is the main contributing factor. In water, the temperature variation of the

dielectric constants are the only contributing factors, giving $d(h\nu)/dT \sim -2.8 \times 10^{-4}$ eV/°, which has been verified approximately (Gottschall and Hart, 1967).

5. The e_2-center is considered qualitatively as an intermediate in H_2 production. It is predicted that if that center has a reasonable lifetime, then its absorption spectrum should be red-shifted. There is no compelling experimental evidence for the e_2-center. Various theoretical speculations exist as to its spectrum and binding energy, but they are discordant (see Kestner, 1976, and Kevan, 1974).

There are basically two semicontinuum models: one owing to Copeland, Kestner, and Jortner (1970) (CKJ) and another to Fueki, Feng, and Kevan (1970, 1973; Fueki et al., 1971) (FFK). The calculations were designed for e_h and e_{am}, but have been extended to other polar media (Fueki et al., 1973; Jou and Dorfman, 1973). In these four or six solvent molecules form the first solvation layer in definite arrangement. Beyond that, the medium is taken as a continuum with two dielectric constants and a value of V_0, the lowest electron energy in the conduction state.

The localizability criterion may be expressed as $E_t < V_0$, where E_t is given by the sum of the electron self-energy E_e (−ve) and the medium rearrangement energy E_M (+ve). E_e is the sum of the *electronic* polarization energy and ε_{nl}, the characteristic energy obtained by solving the one-electron Schrodinger equation with the appropriate trapping potential $V(r)$ in the state with quantum numbers n and l. It comprises the electron kinetic energy, the polaron potential in the continuum, and a potential acting on the electron in the first solvation layer and in the cavity. The CKJ model considers effective radii for the cavity, the solvent, and for the hard core of the molecule. It prescribes $V(r)$ in different regions of space on a physical basis. *The interior potentials are temperature-dependent,* since thermally averaged values (the Langevin result) are used for the radial component of molecular dipole moments. (For details, see the original references.)

The medium rearrangement energy is given by

$$E_M = E_v + E_{dd} + \Pi + E_{HH}. \tag{6.17}$$

Here E_v is the energy necessary to create a void of radius R given by $4\pi R^2\gamma + (4/3)\pi R^3\Gamma$, where γ and Γ are respectively the surface tension and pressure. E_{dd} is the dipole-dipole repulsive energy in the first solvation shell, Π is the (long-range) polarization energy of the medium, and E_{HH} is the molecular repulsive energy within the first shell, determined mostly by hydrogen-hydrogen repulsion.

The actual evaluations of E_{dd}, Π, and E_{HH} are complex. Note that for E_{dd}, CKJ uses Buckingham's (1957) prescription for the number of dipoles in the first solvation shell and considers both the thermally averaged dipole moment and the induced moment. The polarization energy is obtained from Land and O'Reilly (1967).

The molecular repulsive energy E_{HH} is very important in the sense that it prevents the cavity from collapsing. Without it, configurational stability could not be obtained. CKJ used an empirical, exponential form for ammonia. This was later modified by Gaathon and Jortner (1973) [GJ] for water, with E_{HH} varying inversely as the ninth power of the distance of the first shell molecular center from the center of the cavity. Apparently, Newton's (1973) MO calculation supports this form of E_{HH} in water. GJ further split the intermediate region of the trapping potential $V(r)$, implying that the first solvation shell has its own specific value of V_0, called V_{0s}. All electron-molecule interaction except the charge dipole part is lumped in V_{0s}; it is calculated with a molecular density in the first shell, which is different from that of the bulk.

The method of calculating the total energy proceeds with a suitable one-electron wavefunction (1s or 2p) containing a variational parameter. Then, the electronic energy is determined for the potential $V(r)$. To this, the electronic polarization energy is added to get the self-energy E_e of the electron. Finally, the medium rearrangement energy is calculated as described before, and the total energy computed as the grand sum. For the excited state wavefunction, due attention is paid to the Franck-Condon principle, as discussed previously. In this sense, this excited state is also of $2p(1s)$ type; it is called the unrelaxed excited state (Kevan, 1974). By parametric variation, the energetic and configurational stability are obtained from the ground state calculation, and the transition energy, oscillator strength, and so forth, are obtained from the ground and excited state wavefunctions.

Kestner (1976) has analyzed the results of various semicontinuum models by comparing their features and reviewing experimental results. Some of his points are summarized in the following subsections.

6.5.1 ENERGETIC AND CONFIGURATIONAL STABILITY

The real advancement in the semicontinuum model is that, in addition to energetic stability, a configurational stability is also obtained. A minimum total energy is found for a finite cavity volume by balancing the electronic self-energy E_e against the medium rearrangement energy E_M. In ammonia, CKJ obtained a cavity radius consistent with volume expansion data. In water, the situation is not clear. CKJ obtained a small cavity, about the size of a water molecule, which is consistent with the pressure work of Hentz *et al.* (1967b, c; Hentz and Knight, 1970). However, Gaathon and Jortner (1973) found it necessary to use a bigger cavity radius, 2.3 Å, for configurational stability; apparently, the MO calculation of Newton (1973) also supports a similar radius.

6.5.2   COORDINATION NUMBER

N = 4 or 6 seems to give most acceptable results, although somewhat larger values of N are within the error of calculation (see Yoshida et al., 1973). Theoretical calculation of the coordination number depends on the value of V_0, which is uncertain in water and ammonia. CKJ conclude that for a reasonable value of V_0, between 0.5 and −0.5 eV, N = 4 or 6 is most appropriate for both water and ammonia.

6.5.3   HEAT OF SOLUTION

Defining $\Delta H = -E_t(1s)$, the calculated value (GJ) in water, 1.5 eV, compares well with experiment, 1.7 eV. In ammonia, GJ calculate 1.0 eV, which is within the large uncertainty of the experimental value, 1.7±0.7 eV. The agreement may not be much better than Jortner's (1964) earlier calculation via a continuum model, especially if the contribution of H bonds is explicitly considered.

6.5.4   OPTICAL ABSORPTION

In general, semicontinuum models explain the band maximum position rather well. GJ calculate the energy at maximum absorption as 0.9 and 2.0 eV respectively for e_{am} and e_h, which compare reasonably well with the experimental values, 0.8 and 1.7 eV respectively.

The oscillator strength for e_h was calculated earlier by Jortner (1964) as 1.1 in the continuum model. The semicontinuum models give better agreement with experiment, the latter being >0.7 for both e_{am} and e_h. Using the velocity form, Kestner et al. (1973) calculate f = 0.9 and 0.8 respectively for e_{am} and e_h. The corresponding values for the length expression are calculated as 0.6 and 0.7, respectively. Apparently, the length expression gives better agreement with experiment; the reason for this is not clear.

Much worse than the oscillator strength is the line shape. The calculated absorption spectra has no similarity with what is experimentally seen. The calculated half-width is always smaller, typically by a factor of 2; the exact reasons for this are only speculated. It is common knowledge that a photodetachment process is capable of giving a very broad absorption spectrum, but a satisfactory method has not been developed to adopt this with the bound-bound transition of the semicontinuum models. Higher excited states (3p, 4p, etc.) have been proposed for the solvated electron, but they have never been identified in the absorption spectrum.

The position of absorption maximum shifts to lower energy with temperature, attributed by Jortner (1959, 1964) to the temperature dependence of dielectric constants and to cavity expansion in the continuum model. Temperature variation of all parameters can be included in the semicontinuum models. However, a correction is needed to convert the constant-pressure experimental values to the constant-density theoretical calculations (Kestner *et al.*, 1973; Kestner, 1976). With this done, the agreement is fair but not excellent. In ammonia, the calculated temperature coefficient (in eV/°) is -3×10^{-4} versus the experimental value of -2×10^{-4}. In water, calculation gives -3×10^{-4} versus the experimental value of -1×10^{-3}.

The effect of pressure on the absorption spectrum arises from V_0 and the cavity volume. Surprisingly good agreement between theory and experiment is obtained for e_{am}, where the decrease of the cavity size under pressure contributes about half of the spectral shift. The pressure effect in the e_h absorption spectrum can be explained ignoring the cavity variation; including a cavity of 2.5Å radius improves the agreement between 2 and 6 Kbars by about 0.02 eV. A cavity of this radius has been used by Newton (1973) and by Gaathon and Jortner (1973). However, Hentz *et al.* (1972) found by measuring the volume of activation of non-diffusion-controlled reaction an effective cavity radius 1.6 Å for e_h, which includes about 3-ml/mole contribution from electrostiction. Thus, the cavity volume in water is still a debatable issue.

6.5.5 MOBILITY

Experimental mobility values, 1.2×10^{-2} cm²/v.s. for e_{am} and 1.9×10^{-3} cm²/v.s. for e_h, indicate a localized electron with a low-density first solvation layer. This, together with the temperature coefficient, is consistent with the semicontinuum models. Considering an effective radius given by the ground state wavefunction, the absolute mobility calculated in a brownian motion model comes close to the experimental value. The activation energy for mobility, attributed to that of viscosity in this model, also is in fair agreement with experiment, although a little lower.

6.5.6 THRESHOLD OF PHOTOIONIZATION INTO A VACUUM

If $V_0 > 0$, then the threshold for photoionization into a vacuum $E_{vac} = -E_e(1s)$. If $V_0 < 0$, then that additional energy has to be supplied. Delahay (1976) has summarized this process, but the actual mechanism is still debatable. If ΔH and E_{vac} are experimentally known, one can get an experimental measure of the medium rearrangement energy in the ground state, $E_M(1s) = E_{vac} - \Delta H$. On that basis, since E_M is always positive, Kestner (1976) prefers the lower experimental value of ΔH for e_{am}.

In many respects, the FFK formulation is similar to that of CKJ. The differences are the following:

1. The hard-core radius of the solvent molecules is absent in FFK.
2. FFK use a fully SCF method to calculate the total energy, composed of the electron self-energy and the medium rearrangement energy. The former consists of kinetic energy, short- and long-range interactions of the electron with the medium, and a special short-range repulsion due to medium electrons given by $E_q(i) = V_0(1 - C_i)$, where C_i is the charge enclosed within the first shell. The medium rearrangement energy is taken to be the sum of the surface tension energy for creating the void (usually negligible), the short-range dipole-dipole repulsion, and a long-range polarization contribution.
3. E_{HH}, which played an important role in CKJ (see Eq. 6.17) is absent in FFK except for ammonia (Feng *et al.*, 1973). FFK do not continue the trapping potential interior to the first shell with respect to V_0, which is done systematically by CKJ and GJ. This is partly compensated for by the additional term $E_q(i)$. FFK calculate the ground state energy variationally using a 1s wavefunction. In most cases, including polar glasses at 77 K, configurational stability is also obtained, giving an absolute minimum ground state energy. The unrelaxed 2p state energy is obtained by a variational procedure giving minimum energy with orientational polarization consistent with the 1s charge distribution. Finally, V_0, which in most cases is not well known, is varied until the transition energy $h\nu$ comes close to experimentally observed absorption maximum. This should be considered a provisional measure until V_0 values are reliably measured. The entire set of calculations may be performed for different values of the coordination number N (usually 4 or 6), and the net result compared with experiment. Kevan (1974) has discussed the details of the FFK model, especially those related to organic glasses at low temperatures; he has also given a critical comparison of this model with experiment and with the CKJ and other models.

FFK envisage a relaxed excited state $(2p')$ in which everything, including the orientational polarization, is consistent with its own $(2p')$ charge distribution. In practice, the lifetime of this state may be too short for observation due to fast resolution, except in rigid media. FFK also consider the lowest conduction state as unrelaxed and completely delocalized $(C = 0)$ and, therefore, take the electronic polarization and polarizability as zero. The total energy in this state is just the sum of V_0 and the medium rearrangement energy: $E_t(c) = V_0 + E_M$. Noting that E_M (unrelaxed) is given by the 1s charge distribution, no additional wavefunction is required to calculate $E_t(c)$. The photoconductivity threshold is

TABLE 6.5   Results of Calculations Using the FFK Semicontinuum Model[a]

Medium	T (K)	$V_o{}^b$ (eV)	$r_d{}^c$ (A)	ΔH (eV)	I (eV) Theory	I (eV) Experiment
H_2O^c	298	−1.0	1.9	2.75	3.6	—
Ice^c	77	−1.0	1.9	2.08	2.4	2.3 ± 0.1
$NH_3{}^f$	203	−1.0	2.66	1.94	2.07	—
Methanol	77	0.5	2.3	1.3	2.7	2.3
Ethanol	77	1.0	2.5	1.0	2.7	2.3
2MTHF	77	−0.5	2.9	1.4	1.4	1.6
Triethylamine	77	0.3	3.2	0.5	1.1	0.9

Medium	$h\nu$ (eV) Theory	$h\nu$ (eV) Experiment	Oscillation Strength Theory[d]	Oscillation Strength Experiment
H_2O^c	2.15	1.72	0.7	>0.7
Ice^c	1.84	1.9	0.4	>0.3
$NH_3{}^f$	0.96	0.8	—	>0.7
Methanol	2.1	2.4	0.5	0.7
Ethanol	2.2	2.3	0.6	0.7
2MTHF	1.0	1.0	0.5	0.6
Triethylamine	0.8	0.7	0.4	—

[a] Results for only $N = 4$; see Kevan (1974) for additional calculations for $N = 6$.

[b] Considerable uncertainties exist for the V_0 value. Values used in *certain* FFK model calculations are given.

[c] At the configurationally stable position.

[d] Calculated by the velocity formula.

[e] From Fueki *et al.* (1973).

[f] From Feng *et al.* (1973).

then given by $I = E_t(c) - E_t(1s)$. Table 6.5 gives a summary of some of the results obtained with the FFK model.

It is clear that in most cases better agreement between theory and experiment is obtained with $N = 4$. Unlike in the CKJ model, $N = 6$ gives substantially different results for transition energy, oscillator strength, and so forth. The agreement between calculation and experiment for oscillator strength, photoionization threshold, and heat of solvation is quite good. Values of V_0 needed for agreement between calculated and observed absorption maxima are also reasonable. However, as with the CKJ model, the line shape of absorption spectrum is not predicted well.

Kevan (1974) and Tachiya (1972) point out that CKJ use an SCF approximation to calculate the medium polarization energy, but in everything else they use the adiabatic approximation. This somewhat inconsistent procedure, which may be called the modified adiabatic approximation, gives results similar to those obtained by FFK. Varying the dipole moment and the polarizability in the semicontinuum models varies the result qualitatively in the same direction. It increases the electron-solvent attraction in the first shell and also increases the dipole-dipole repulsion. Both $h\nu$ and I increase with the dipole moment, but not proportionately.

6.6   REACTIONS OF THE SOLVATED ELECTRON

Of all the solvated electrons, e_h is the most reactive; several thousand of its reactions have been measured and, in many cases, activation energies, the effects of pH, and so forth, determined (Anbar and Neta, 1965; Anbar et al., 1973; Ross, 1975; CRC Handbook, 1991). Relatively few reactions of e_{am} have been studied because of its low reactivity. Rates of reactions of the solvated electron with certain scavengers are also available in alcohols, amines, and ethers.

Table 6.6 lists some reactions of the electron in water, ammonia, and alcohols. These are not exhaustive, but have been chosen for the sake of analyzing reaction mechanisms. Only three alcohols—methanol, ethanol, and 2-propanol—are included where intercomparison can be effected. On the theoretical side, Marcus (1965a, b) applied his electron transfer concept (Marcus, 1964) to reactions of e_s. The Russian school simultaneously pursued the topic vigorously (Levich, 1966; Dogonadze et al., 1969; Dogonadze, 1971; Vorotyntsev et al., 1970; see also Schmidt, 1973). Kestner and Logan (1972) pointed out the similarity between the Marcus theory and the theories of the Russian school. The experimental features of e_h reactions have been detailed by Hart and Anbar (1970), and a review of various e_s reactions has been presented by Matheson (1975). Bolton and Freeman (1976) have discussed solvent effects on e_s reaction rates in water and in alcohols.

6.6.1   REACTIONS OF THE HYDRATED ELECTRON

6.6.1a   Reactions with Water and Products of Water Radiolysis

e_h reacts very slowly with water, producing an H atom and an hydroxide ion [see reaction (6.III)]:

$$e_h + H_2O \longleftrightarrow (H)_{aq} + (OH^-)_{aq} ; \qquad \Delta G^0 = 8.44 \text{ Kcal/mole.} \qquad (6.III')$$

TABLE 6.6   Rates of Reaction of the Solvated Electron

The Hydrated Electron (~300K)

Reactant	pH	k $(M^{-1}s^{-1})$	Product(s)
H_3O^+	4.3	2.1×10^{10}	H
H	10.5	2.5×10^{10}	$H_2 + OH^-$
OH	10.5	3.0×10^{10}	OH^-
e_h	11–13	0.6×10^{10a}	$H_2 + 2OH^-$
H_2O_2	7	1.2×10^{10}	$OH + OH^-$
H_2O	8.4	16	$H + OH^-$
O_2	2	2×10^{10}	O_2^-
O^-	13	2.2×10^{10}	$2OH^-$
HO_2^-	13	3.5×10^9	$2OH^-$
NO	7	3×10^{10}	NO^-
N_2O	7	8.7×10^9	N_2O^-
NO_3^-	7	1.1×10^{10}	NO_3^-
NO_2^-	7	4.3×10^9	
CO		1.0×10^9	CO^-
CO_2	7	7.7×10^9	CO_2^-
I_2	7	5.1×10^{10}	I_2^-
I_3^-	7	2×10^{10}	$I^- + I_2^-$
Ag^+	7	3.6×10^{10}	Ag^0
Al_{aq}^{3+}	6.8	2×10^9	
Cd^{2+}	7	5.8×10^{10}	
Co^{2+}		1.4×10^9	
Cr^{2+}	6.9	4.2×10^{10}	
CrO_4^{2-}		1.8×10^{10}	
Cu^{2+}	6	4×10^{10}	Cu^+
$Fe(CN)_6^{3-}$	7–10	3×10^9	
Benzene	7	$<7 \times 10^6$	
Ethylene	7	$<2.5 \times 10^6$	
Acetaldehyde	6.5–11	3.5×10^9	
Acetone	7–11	5.9×10^9	
Acetylene	3.3	3.5×10^{10}	
Acrylamide	7	$2–3 \times 10^{10}$	

(Continued)

TABLE 6.7 Rates of Reaction of the Solvated Electron (*Continued*)

Reactant	pH	k (M^{-1}s^{-1})	Product(s)
CCl$_4$	7	3×10^{10}	
Chloroform	7	3×10^{10}	
Methylene Blue	7.8	2.5×10^{10}	
Methyl iodide		1.7×10^{10}	
Ethyl bromide		1.2×10^{10}	$C_2H_5 + Br^-$
Nitrobenzene	7	3×10^{10}	$C_2H_5NO_2^-$
Nitroethane	0–6	2.7×10^{10}	
Nitromethane	0–6	2.9×10^{10}	
Styrene		1.5×10^{10}	
Butadiene	7	8×10^9	
Methacrylate ion	10.1	8.4×10^9	
1,3-cyclo-hexadiene	11	1×10^9	
Uracil	7	9.3×10^9	
Adenine	6	3×10^{10}	
Cytosine	6	$(7–10) \times 10^9$	
Guanidine	6.1	2.5×10^8	
Thymine		1.7×10^{10}	

The Ammoniated Electron

Reactant	T (K)	k (M^{-1}s^{-1})	E_a (Kcal/mole)
NH$_4^+$	238–248	~10^6	
NH$_2$		2.5×10^{10}	
Imidazole	238	5×10^5	
Pyridine	238	~10^4	
Dimethyl sulfide	238	5.5	3.8
Diethyl sulfide	228	4.1×10^{-2}	6.3
Dimethyl sulphoxide	238	5.0	5.5
Pyrrole	228	2.8×10^{-1}	2.4
Thiophene	228	1.0	5.6

(Continued)

This reaction is important, because it gives the natural lifetime of e_h, has an isotope effect (the rate in D_2O being a factor ~13 times less than that in H_2O), and it can give us the reduction potential of e_h. The forward reaction is hard to observe because of the low rate, and special precautions are necessary (Hart and Anbar, 1970). Using the utmost care, Hart *et al.* (1966) established that rate as $k_f = 16\pm1$

TABLE 6.7   Rates of Reaction of the Solvated Electron (Continued)

The Electron in Alcohols (~300K)

Reactant	k in Alcohols (M^{-1}s^{-1})		
	Methanol	Ethanol	Propanol[b]
H$^+$	7×10^{10}	4.5×10^{10}	1.7×10^{10c}
O$_2$	2×10^{10}	2×10^{10}	
N$_2$O		7×10^9	
SF$_6$		1.3×10^{10}	
CCl$^+$	5×10^{10}		2×10^{10}
Nitrobenzene	3.5×10^{10}		2×10^{10}
Methylbromide	1.3×10^{10}	9.1×10^9	
Methylchloride		2.3×10^9	
Acetone	5×10^9	3×10^9	5×10^9
Acetaldehyde		4×10^9	
Benzyl chloride	5×10^9	5×10^9	
Biphenyl		4×10^9	
Benzophenone			1.1×10^{10}
Tetranitromethane		3.5×10^{10}	2.5×10^{10}
Napthalene	2×10^9	5.5×10^9	
p-Terphenyl		7×10^9	
Phenol		4.5×10^7	
Benzene	2×10^7		1.8×10^8
Chlorobenzene	8×10^8		4.7×10^9

[a]With the rate law written as $d[e_h]/dt = -2k[e_h]^2$.
[b]2-Propanol. unless otherwise mentioned.
[c]k in 1-propanol, 2.6×10^{10} M^{-1}s^{-1}.

M^{-1}s^{-1}. The rate of the reverse reaction is known to be 2.3×10^7 M^{-1}s^{-1}. This gives the free energy of (6.III') as 8.44 Kcal/mole using the relation $-\Delta G = RT \ln(k_f/k_b)$. Consider the reactions

$$(H^+)_{aq} + (OH^-)_{aq} \longleftrightarrow H_2O; \quad \Delta G = -21.6 \text{ Kcal/mole,}$$
$$(H)_g \longrightarrow (1/2)(H_2)_{aq}; \quad \Delta G = -48.6 \text{ Kcal/mole,}$$
$$H_{aq} \longrightarrow H_g; \quad \Delta G = -4.5 \text{ Kcal/mole.}$$

Adding all these together with reaction (6.III'), one gets $e_h + (H^+)_{aq} \longrightarrow (1/2)(H_2)_g$; $\Delta G = -66.3$ Kcal/mole. Thus, the standard reduction potential of e_h, e + water $\longleftrightarrow e_h$, is given as -2.85 V. The hydrated electron is a powerful reducing agent; it can reduce the silver ion but not the potassium ion. On the other hand, the mechanism of the forward reaction (6.III') is not understood. Its low activation energy, ~4.5 Kcal/mole (Fielden and Hart, 1968), is puzzling (see Swallow, 1965). The large isotope effect may be consistent with the participation of a large number of water molecules.

Reactions of e_h with H and OH were once considered diffusion-controlled; see, however, Elliot et al. (1990). The rate constants, 2.5–3.0×10^{10} $M^{-1}s^{-1}$ (see Table 6.6), are high. In both cases, a vacancy exists in the partially filled orbitals of the reactants into which the electron can jump. Thus, hydrogen formation by the reaction e_h + H may be visualized in two steps (Hart and Anbar, 1970): e_h + H$\rightarrow$H$^-$, followed by H$^-$ + $H_2O\rightarrow OH_2^-$. This reaction has no isotope effect, which is consistent with the proposed mechanism. The rate of reaction with OH is obtained from the e_h decay curve at pH 10.5 in the absence of dissolved hydrogen or oxygen, where computer analysis is required to take into account some residual reactions. At higher pH (>13), OH exists as O$^-$ and the rate of e_h + O$^-\rightarrow$O^{2-} has been measured as ~2.2×10^{10} $M^{-1}s^{-1}$.

Matheson and Rabani (1965) measured the rate of the reaction e_h + $e_h\rightarrow$H$_2$ + 2OH$^-$ at pH 13 under 100 atmospheres H$_2$ pressure, where all radicals are converted to e_h. From a pure second-order decay, the rate constant was determined as 6×10^9 $M^{-1}s^{-1}$. There are contradictory views on this reaction. According to some, this rate is too low for a diffusion-controlled reaction between like charges, by a factor of ~4 (see Farhataziz, 1976). This factor of 4 can be accounted for by spin considerations, since each electron is a doublet but the end product H$_2$ is a singlet. To be consistent, then, one has to consider the rate of reaction e_h + O$^-\rightarrow$O^{2-} as normal for diffusion control.

According to another school (Hart and Anbar, 1970), the reaction of e_h with itself is normal and diffusion-controlled, whereas the rate of the reaction e_h + O$^-\rightarrow$O^{2-} is abnormally high. They view the immediate product of the e_h + e_h reaction as the dielectron, $(e_h)_2$, which decomposes slowly by reacting with water, $(e_h)_2$ + $2H_2O\rightarrow$H$_2$ + 2OH$^-$. Justification of this hypothesis comes from the following: (1) The H/D isotope effect (4.7) in a 1 : 1 H_2O/D_2O mixture, indicating a reorganization in the intermediate state with a lifetime exceeding 1 ns; (2) Kevan's (1968) observation of a broad absorption at 1000 nm in alkaline ice; and (3) Fueki's (1969) theoretical calculation using a continuum model. Despite all this, the dielectron cannot be said as firmly established in water, and the nature of the reaction e_h + $e_h\rightarrow$H$_2$ + 2OH$^-$ is open to question. There is, however, experimental evidence for the dielectron in concentrated metal-ammonia solutions.

e_h reacts with H_3O^+, giving an H atom. The rate is high (see Table 6.6), but somewhat under the diffusion-controlled limit. The activation energy measured by Thomas et al. (1964) is 3.2 Kcal/mole, and the H/D isotope factor is 3.7 (Anbar and Meyerstein, 1966). The electron transfer then involves an appreciable entropy of activation. Anbar (1965) points out that the electron accommodation to $(H_3O^+)_{aq}$ is diffuse. Magee (1964) proposed that the immediate product of the neutralization reaction is H_3O in equilibrium with the reactants:

$$e_h + (H_3O^+)_{aq} \longleftrightarrow H_3O \rightarrow H + H_2O.$$

The large isotope factor shows that the H atom is *not* formed by proton transfer. Sawai and Hamill (1969) have shown that the nonhydrated ("dry") electron in water is unreactive toward $(H_3O^+)_{aq}$, although it will react with cations such as Cd^{2+}.

The hydrated electron reacts with H_2O_2 with a diffusion-controlled rate (see Table 6.6), giving OH and OH^-. An intermediate product of this reaction, $H_2O_2^-$, may be responsible for prolonged conductivity in pulse-irradiated water. The rate of this reaction is consistent with rates of similar one-electron reduction reactions of H_2O_2.

6.6.1b Reactions with Inorganic Compounds and Ions

The reaction of e_h with H_2 is very slow (~10^3 $M^{-1}s^{-1}$), due to a positive free energy barrier. Nitrogen has negative electron affinity; it is unreactive toward e_h. O_2 has a high electron affinity, and it reacts with e_h with a diffusion-controlled rate (see Table 6.6). The immediate product of the reaction, O_2^- or its acidic form HO_2, reacts further with itself, giving H_2O_2 and O_2.

Both CO and CO_2 are reduced by e_h. The immediate product of the first reaction is CO^-, which reacts with water, giving OH and the formyl radical; the latter has been identified by pulse radiolysis. The product of carbon dioxide reduction, CO_2^-, is stable in the condensed phase with an absorption at 260 nm. It reacts with various organic radicals in addition reactions, giving carboxylates with rates that are competitive with ion-ion or radical-radical combination rates.

Oxides of nitrogen are popular electron scavengers. The final product of reaction of e_h with N_2O is N_2 and OH, but the mechanism is not well understood. The rate approaches diffusion control. A possible reactions scheme is

$$e_h + N_2O \rightarrow N_2O^-,$$
$$N_2O^- + H_2O \rightarrow N_2OH + OH^-,$$
$$N_2OH \rightarrow N_2 + OH.$$

The gas-phase lifetime of N_2O^- is ~10^{-3} s; in alkaline solutions, it is still >10^{-8} s. Under suitable conditions, N_2O^- may react with solutes, including N_2O. The hydrated electron reacts very quickly with NO (see Table 6.6). The rate is about three times that of diffusion control, suggesting some faster process such as tunneling. NO has an electron affinity in the gas phase enhanced upon solvation. The free energy change of the reaction $NO + e_h \rightarrow (NO^-)_{aq}$ is estimated to be ~ -50 Kcal/mole. Both NO_2^- and NO_3^- react with e_h at a nearly diffusion-controlled rate. The intermediate product in the first reaction, NO_2^-, generates NO and

$N_2O_3^-$. The chemistry of the immediate product NO_3^{2-} in the second reaction is not well understood, but it is presumed that at first NO_2 is produced, which preferentially hydrates. Thus, the overall reaction is represented by

$$e_h + NO_3^- \longrightarrow H_2O\ (NO_2)_{aq} + 2OH^-.$$

The reaction of e_h with H_2S belongs to the class

$$e_h + HX \longrightarrow H + X^-;$$

or $\longrightarrow$ dissociation products.

HX^- may lose H or dissociate with probability depending on bond strength and structure. The rate of reaction with H_2S, $1.4 \times 10^{10}\ M^{-1}s^{-1}$, is diffusion-controlled, and the products are $H + HS^-$ or $H_2 + S^-$, in a $65:35$ ratio.

Halide ions have lower orbitals filled in a rare gas configuration. Their reaction rates with e_h are expected to be small, which is verified experimentally. I, I_3^-, and ClO^-, however, react with e_h at near diffusion-controlled rates.

Formally, the immediate product of an electron transfer reaction may be envisaged as an electron adduct. Rarely, however, is the immediate product unreactive (Hart and Anbar, 1970, Table V.1). Exceptions exist in the case of Cu^{2+}, MnO_4^-, and others.

Cations in general should be reactive toward e_h, but considerations of redox potential and free energy change are important. Thus, the alkali metal cations, having higher redox potentials, are unreactive toward e_h. Another example is

$$e_h + (M^{2+})_{aq} \longrightarrow M_{aq}^+.$$

With M as Be, Mg, Ca, Sr, or Ba, the reaction has a positive free energy change due to the overcompensating effect of ion hydration energy relative to electron affinity. Therefore, these ions do not react with e_h.

Stein (1971) has proposed a generalized Hammett equation correlating the rate of reaction of e_h with the dissociation constant of a Brønsted acid. However, there are exceptions, including $(H_3O^+)_{aq}$. Thomas et al. (1964) note that the rate of the reaction $e_h + M^{3+}$ decreases as the redox potential $E^0(M^{3+}/M^{2+})$ becomes more negative. Against the redox potential, Hart and Anbar (1970) plotted the rates of reaction of e_h with some lanthalide ions after correcting for diffusion according to Noyes (1961), namely,

$$k_{act} = (k_{obs}^{-1} - k_{diff}^{-1})^{-1}, \tag{6.18}$$

where k_{obs} and k_{diff} are observed and diffusion-controlled rates, respectively. A good exponential fit is obtained, which breaks down when the rate is low. Several hundred reactions of inorganic compounds and ions with e_h are known. Only a few representative ones have been discussed here. A larger selection will be found in Hart and Anbar (1970), Anbar et al. (1973), Ross (1975), and the CRC Handbook of Radiation Chemistry (1991, Sect. VI.B.3).

6.6.1c Reactions with Organic Compounds

The reactivity of organic compounds toward e_h varies systematically (Hart and Anbar, 1970), the variation often attributable to electron density of a specific functional group. Other generalizations are the following:

1. Although electron transfer reactions, kinetically e_h behaves as a classical nucleophilic reagent.
2. The immediate product of the electron transfer reaction with an organic compound is unstable.
3. Organic reactions are exothermic. In ~1 ps, radiationless transitions occur, giving the primary product and evolving energy.
4. Most bonds in organic molecules, except C—H and C—C, may be hydrogenated or cleaved upon reduction by e_{am} and by e_h.

Generally, organic compounds containing only H, C, O, and N and having no π orbitals have either negative or a small positive electron affinity. The free energy change in a reaction with e_h therefore has to be sufficiently negative so that a reasonably good rate of reaction ($\geq 10^7$ M^{-1}s^{-1}) will result. In this sense, saturated hydrocarbons, aliphatic alcohols, ethers, and amines are considered unreactive toward e_h (Hart and Anbar, 1970).

Matheson (1975) and Hart and Anbar (1970) note the following characteristics of the reaction of e_h with unsaturated organic compounds:

1. The rate of e_h reaction with ethylene is low, ~10^6 M^{-1}s^{-1}. An electron-donating group adjacent to NH$_2$ or OH makes the rate very low. Similarly, an electron-accepting group enhances the rate as in the case of pyrrole, vinyl alcohol, or ethylene derivatives, where some reactions proceed at diffusion-controlled rates.
2. The reactivity of a carbonyl compound R$_1$COR$_2$ is correlated with its σ^* value (Taft, 1965; Hart et al., 1967). Since σ^* values may be assumed to be additive, Hart et al. plotted the logarithm of the rate constant with e_h against ($\sigma_1^* + \sigma_2^*$). The result was a fairly good straight line, with a slope of -0.74. The active center of reaction is known to be the carbonyl oxygen.
3. The reactivity of oximes is less than that of carbonyls, attributed to the inductive effect of OH on N. Nevertheless, they conform to the plot of Hart et al. (1967).
4. Amides and esters give a positive slope of about 1.2 on Hart et al.'s (1967) plot, due to the reduction of the double bond character of C=O by the mesomeric effect of NH$_2$ and OH, which reduces the reactivity toward e_h.

The sulfur compounds RSH and disulfides are highly reactive, but RS$^-$ and RSR are not. Nitro compounds usually react at diffusion-controlled rates. Aromatic compounds also fit into the Hammett (1940) equation when log k is plotted against free-energy change due to polar effects $\log(k/k_0) = \sigma\rho$. Anbar and Hart (1964) studied the reactions of monosubstituted benzenes with the σ values of Van Bekkum *et al.* (1959) and obtained $\rho = 4.8$. Bromobenzene and iodobenzenes give higher reactivities than obtained from the preceding equation due to the reaction of e_h with both the halogen and the ring.

The reactivity of haloaliphatic compounds decreases in ascending order. This seems to be related to polarizability. RX^- ion has been postulated as an intermediate, but it has not been experimentally observed. Again, the reactivity is increased by adjacent electron-accepting groups.

6.6.2   REACTIONS OF OTHER SOLVATED ELECTRONS

The radiation-induced e_{am} reacts with NH_2 with a fairly high rate (see Table 6.6), which limits its lifetime to ~1 μs under typical linear accelerator doses. NH_2 radical is absent in metal-ammonia solutions, where e_{am} is highly unreactive (Matheson, 1975). The low reactivity of e_{am} has been explained by Schindewolf (1968) in terms of the transition state theory as due to a large negative activation volume arising from the cavity of e_{am}. However, Farhataziz and Cordier (1976) found a number of reactions of e_{am} with inorganic ions at 23°, the rates of which are high. In units of 10^{11} $M^{-1}s^{-1}$, these rates are given parenthetically as follows: Tl^+ (12), Cu^{2+} (8.3), Co^{3+} (7.7), Co^{2+} (5.7), Pb^{2+} (5.6), Fe^{3+} (4.9), Cd^{2+} (4.6), Hg^{2+} (3.2), Ce^{3+} (3.1), Hg^+ (2.3), Ce^{4+} (1.5), Ni^{2+} (1.1), Cr^{3+} (0.58), Zn^{2+} (0.13), and so on. Of these, only the reaction with Tl^+ is considered diffusion-controlled based on an estimate of the Debye-Smoluchowski rate taking the cations to be the same size as NH_4^+. After crudely estimating the standard free energy of the electron transfer reaction (ETR) and computing k_{act} by Eq. (6.18), log k_{act} was plotted versus ΔF_r^0. On the basis that the highest rates saturate with the largest values of $-\Delta F_r^0$, the authors favor the quantum-mechanical theory of Ulstrup and Jortner (1975) over the semiclassical theory of Marcus (1964, 1965a, b).

There is apparently no analog of the reaction $2e_h \rightarrow H_2$ in liquid ammonia, where e_{am} is very stable. The loss of paramagnetism in concentrated solutions has been interpreted to be either by formation of $(e_{am})_2$ or by association with metal cation; in neither case is the spectral shift drastic. For Na in ethylenediamine (EDA), Dye *et al.* (1972) measured the rate of $2e_s \rightarrow (e_s)_2$ as 1.7×10^9 $M^{-1}s^{-1}$, which is comparable to that of the corresponding reaction in water, 6×10^9 $M^{-1}s^{-1}$, although the products are different. A few rate constants have been measured in cesium-EDA systems, but it is not clear whether the electron or an associated form of the electron and the cation is the reactant.

The solvated electron is reactive in alcohols, both with solutes and solvents (Watson and Roy, 1972). With methanol, ethanol, and 1- and 2-propanols, somewhat different rates of e-solvent reactions have been measured by Freeman (1970) and by Baxendale and Wardman (1971). However, the (pseudo-first-order) rates are generally $\sim 10^5$ s^{-1}. Although the activation energy of these reactions are not definitely established, the slowness of the reactions is probably not due to the activation energy.

Baxendale and Wardman (1973) note that the reaction of e_s with neutrals, such as acetone and CCl_4, in n-propanol is diffusion-controlled over the entire liquid phase. The values calculated from the Stokes-Einstein relation, $k = 8\pi RT/3\eta$, where η is the viscosity, agree well with measurement. Similarly, Fowles (1971) finds that the reaction of e_s with acid in alcohols is diffusion-controlled, given adequately by the Debye equation, which is *not* true in water. The activation energy of this reaction should be equal to that of the equivalent conductivity of $e_s + ROH_2^+$, which agrees well with the observation of Fowles (1971).

Of the ethers, rate constants for e_s reactions are available for tetrahydrofuran (THF). Since the neutralization reaction, THF$^+$ + e_s, is very fast, only fast reactions with specific rates $\sim 10^{11}$–10^{12} M^{-1}s^{-1} can be studied (see Matheson, 1975, Table XXXII). Bockrath and Dorfman (1973) compared the observed rate of the reaction e_s + Na$^+$ in THF, 8×10^{11} M^{-1}s^{-1}, with that calculated from the Debye equation, $<3 \times 10^{11}$ M^{-1}s^{-1}. Although the reaction radius is not well known, the authors note on a spectroscopic basis that Na$^+$ and e_s are strongly coupled in THF. Thus, the reaction of a solute with (Na$^+$, e_s) in THF is much slower, sometimes by an order of magnitude, than the corresponding reaction with e_s only. Reaction with pyrene is an example.

6.6.3 SOME THEORETICAL FEATURES OF REACTIONS OF SOLVATED ELECTRONS

We should remember (1) that the activation energy of e_h reactions is nearly constant at 3.5 ± 0.5 Kcal/mole, although the rate of reaction varies by more than ten orders of magnitude; and (2) that all e_h reactions are exothermic. To some extent, other solvated electron reactions behave similarly. The theory of solvated electron reaction usually follows that of ETR in solution with some modifications. We will first describe these theories briefly. This will be followed by a critique by Hart and Anbar (1970), who favor a tunneling mechanism. Here we are only concerned with k_{act}, the effect of diffusion having been eliminated by applying Eq. (6.18). Second, we only consider simple ETRs where no bonds are created or destroyed. However, the comparison of theory and experiment in this respect is appropriate, as one usually measures the rate of disappearance of e_s rather than the rate of formation of a product.

In the (semi-)classical models of ETR (Marcus; the Russian school), redox orbitals of reactants overlap at a close separation, followed by swift electron transfer. The activated complex, considered in equilibrium with the reactants, consists of these overlapping orbitals. In the tunneling model, the electron penetrates the barrier at some separation and makes a transition from the solvated state to negative ion state of the acceptor without the necessity of a transition state. These two models are not necessarily exclusive (Marcus, 1964).

Figure 6.6 illustrates the potential energy diagram of the Marcus theory of ETR. The abscissa indicates a line drawn through the equilibrium positions of reactants (A) and products (A'), with configurational coordinates orthogonal to it. In the absence of reactions, the configurational coordinate fluctuates on one curve only going through S (actually a hypersurface). The redox orbital is split at each point of S by electronic interaction, and it is this interaction that induces chemical reaction. The state at S is the activated complex. If the probability of reaction per passage through S is small, then the reactants mainly stay on R, only occasionally jumping to P; such reactions are called nonadiabatic. If the same probability of reaction is ~1, then the system "almost always" adheres to the lower curves and thereby "slithers" from the reactants to products. Such reactions are called adiabatic, and their probability is given by the "first passage" through S. The configurational coordinate is a general concept, meaning that a suitable change in that coordinate will let the reactants come closer.

In addition to the usual assumptions of ETR theories, Marcus (1965a, b) notes three features specific to e_s reactions:

1. The wavefunction of e_s is sensitive to orientational fluctuations of solvent molecules.
2. The solvated electron disappears on reaction, thus reducing the number of reactants.

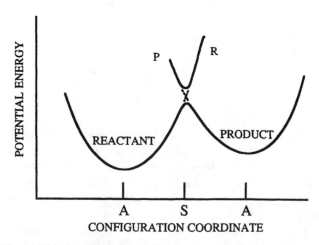

FIGURE 6.6   Potential energy diagram for the theory of electron transfer reactions. The activated complex is at S. For reasonably fast reactions, the reactant adheres to the lower curve and slithers into the product curve through the activated complex—that is, an adiabatic electron transfer occurs.

3. Unlike most ETRs, many reactions of e_s are apparently diffusion-controlled. (In light of later experimental findings, this point is debatable.)

Marcus uses the Born-Oppenheimer approximation to separate electronic and nuclear motions, the only exception being at S in the case of nonadiabatic reactions. Classical equilibrium statistical mechanics is used to calculate the probability of arriving at the activated complex; only vibrational quantum effects are treated approximately. The result is

$$k_{act} = Z\kappa\rho \exp\left(\frac{\Delta F^*}{k_B T}\right),$$
(6.19)

where Z is a collision frequency, usually $\sim 10^{11}$ $M^{-1}s^{-1}$, given by $(8\pi k_B T/m^*)^{1/2}R^2$ where m^* is the reduced mass for the reaction coordinate and R is the most probable separation of the reactants in the activated complex. For adiabatic reactions, the reaction probability per passage $\kappa \approx 1$; for nonadiabatic reactions, $\kappa < 1$. The quantity $\rho \approx 1$ is the ratio of rms fluctuation of the reactants' separation in the activated complex to the same quantity on an orthogonal surface. The most important quantity in the rate equation, (6.19), is the free energy of activation ΔF^*. A rather similar expression for ΔF^* is obtained as for ordinary ETR, namely,

$$\Delta F^* = \frac{1}{2}(w + w' + \Delta F^{0\prime}) + \frac{\lambda}{4}\left[1 + \left(\frac{\Delta F^{0\prime}+ w' - w}{\lambda}\right)^2\right],$$

where w is the work done to bring the reactants at the reaction distance R is the work done to separate the products from there, λ is the reorganization energy to conform the coordination of the electron acceptor in a manner suitable for electron transfer, and ΔF^0 is the standard free energy of ETR in the *prevailing medium*—that is, when the reactants and products are at infinity. The standard free energy of reaction at the reaction distance R in the medium is given by $\Delta F_R^{0\prime} = \Delta F^0 + w' - w$, while the free energy of activation may be expressed as

$$\Delta F^* = w + \frac{(\Delta F_R^{0\prime} + \lambda)^2}{4\lambda}.$$

Since e_s is a strong reducing agent, one may assume that $\Delta F^0 \ll 0$ and that ΔF^* is small. Thus, Marcus (1965a) obtains (see Hart and Anbar, 1970)

$$\Delta F^* = w + \frac{\lambda}{4}\left(1 + \frac{\Delta F^0 - w}{\lambda}\right)^2,$$

where ΔF^0 is the *standard* free energy of reaction in the prevailing medium corrected for the free energy loss of translation when e_s is lost in ETR; w' is equated to zero, consistent with simple ETR. In comparing theory with experiment, one must correct for diffusion as in Eq. (6.18). The aim of the Marcus theory is to correlate the e_s reaction rate with (1) the standard reduction potentials of reactants, (2) physical data of e_s, such as the absorption spectrum, and (3) rates of ordinary ETR of the reactant. The theory can explain the rate dependence on electron affinity. The slowness of many reactions of e_h, still having small activation energy, can be attributed to the reorganization energy. Quantitatively, there are several shortcomings (Hart and Anbar, 1970), including the systematic effect of electrophilic aromatic substitution or nucleophilic aliphatic substitution.

The Russian school of ETR (Levich, 1966; Dogonadze, 1971; Vorotyntsev *et al.*, 1970) treats the medium polarization by a second-quantized Hamiltonian, written as

$$ H = \frac{1}{2} \sum_k \hbar \omega_k \left(q_k^2 - \frac{\partial^2}{\partial^2 q_k} \right), $$

where q is a set of normal coordinates representing dipolar orientation, k denotes the wave index appropriate to the mode of oscillation, and $\omega/2\pi$ is the frequency of oscillation, typically $\approx 10^{12 \pm 1}$ s^{-1} in water. In the gas phase, the electron transfer reaction $A^{n+} + B^{m+} \longrightarrow A^{(n+1)+} + B^{(m-1)+}$ requires near resonance of the electronic energy levels of A^{n+} and $B^{(m-1)+}$ due to the Franck-Condon principle. In the liquid phase, some mismatch can be adjusted by the medium rearrangement energy. Thus, polarization fluctuations are extremely important, as they can bring about an otherwise restrictive reaction.

Changes in the degrees of freedom in a reaction can be classified in two ways: (1) classical over the barrier for frequencies ω such that $\hbar \omega < k_B T$; and (2) quantum mechanical through the barrier for $\hbar \omega > k_B T$. In ETR, only the electron may move by (1); all the rest move by (2). Thus, the activated complex is generated by thermal fluctuations of all subsystems (solvent plus reactants) for which $\hbar \omega < k_B T$. Within the activated complex, the electron may penetrate the barrier with a transmission coefficient determined entirely by the overlap of the wavefunctions of the quantum subsystems, while the activation energy is determined entirely by the motion in the classical subsystem.

Levich (1966) gives the specific rate of ETR in absolute units as

$$ k_{act} = \int W(R) \exp\left(-\frac{U}{k_B T} \right) dV $$

where $W(R)$ is the probability per unit time of electron transfer at separation R between the reactants, $U(R)$ is the interaction energy between the reactants at separation R, and the integration is taken over all available space. In a dilute solution, the electrostatic part of U is given in the Debye theory by $(mn/\varepsilon R) \exp(-R/L_D)$ where ε is the dielectric constant, L_D is the Debye length, and the electron is transferred from A^{n+} to B^{m+}. The most important factor in the calculation is W, which is obtained quantum mechanically. The result is given in the high- and low-temperature limits according to $k_B T \gg \hbar \omega_0$ or $k_B T \ll \hbar \omega_0$, where ω_0 is the long-wavelength limit of the frequency of optical phonons. In both cases, the Arrhenius form for W is obtained, but with different activation energies.

In the high-temperature limit, the activation energy for both adiabatic and nonadiabatic cases is given by $(\Delta E + E_s)^2/4E_s$, where ΔE is the difference in energies of the reactants and products at their respective equilibrium configurations and E_s is the medium rearrangement energy of the reaction (comparable to λ of Marcus's theory). The preexponential terms contain, among other factors, the product of exchange integrals, having somewhat different forms for adiabatic and nonadiabatic transfers. Schmidt (1973), using nonequilibrium statistical mechanics, has shown that both forms for the *rate* expression arise from a single matrix element of the Coulomb interaction that induces the ETR. In the low-temperature limit, the activation energy is simply ΔE, and Levich (1966) finds that this limit is appropriate when bonds are broken. Schmidt (1973) further finds that only for inner-sphere ETR may the rate always appear in the Arrhenius form.

Hart and Anbar (1970) state the conditions favorable for e_h reactions as (1) a low positive or negative redox potential, (2) sufficiently negative free energy change on reaction, (3) high electron affinity of the reactant, and (4) a vacant low-lying electron orbital. These authors favor reaction by electron tunneling over transition state theories. Their arguments are the following.

1. Some reactions of aromatic compounds in water and methanol and of certain other compounds in water and ice have comparable specific rates. Due to considerable differences in solvation energies of both reactants and products, a transition state would require different barriers against reaction.

2. The activation energy of most of the e_h reactions, 3.5±0.5 Kcal/mole, is much less than the hydration energy of the electron, ~40 Kcal/mole. There are other barriers against reaction, such as repulsion by electrons in molecules. This can only be an accident in the classical mechanism, but not in electron tunneling theory as long as the reaction is exothermic overall.

3. In electron tunneling, the barrier height is determined mainly by the free energy change, whereas the distance of transfer is determined by, among other things, that of closest approach to the reactant. There is evidence of significant slowing down of the reaction when an acceptor atom is surrounded by ligands. However, Marcus (1965a,b) points out the error of identifying the electron tunneling probability with Eyring's transmission coefficient. The electron does tunnel a barrier, but *not* the barrier of Figure 6.6. The simple barrier penetration probability (p) calculation is nevertheless useful in estimating the energy gap between the adiabatic (solid) curves in Figure 6.6.

REFERENCES

Adams, G. E., Baxendale, J. H., and Boag, J. W (1964), Proc. Roy. Soc. (London) A277, 549.

Allen, A. O. (1948), J. Phys. Colloid Chem. 52, 479.

Allen, A. O. (1961), The Radiation Chemistry of Water and Aqueous Solutions, Van Nostrand, Princeton, N.J.

Anbar, M. (1965), Adv. in Chem. Ser. 50, 55.

Anbar, M., and Hart, E. J. (1964), J. Am. Chem. Soc. 86, 5633.

Anbar, M., and Meyerstein, D. (1966), Trans. Faraday Soc. 62, 2121.

Anbar, M., and Neta, P. (1965), Int. J. Appl. Radial. Isotopes 16, 227.

Anbar, M., Gloria, H. R., Reinish, R. F., and St. John, G. A. (1969), NASA Project NA2-5119.

Anbar, M., Bambenek, M., and Ross, A. B. (1973), Selected Specific Rates of Reaction of Transients from Water in Aqueous Solution. 1. Hydrated Electron. NSRDS-NBS 43, National Bureau of Standards, Washington, D.C.

Arnikar, H. J., Damle, P. S., Chaure, B. D., and Rao, B. S. M. (1970), Nature 228, 357.

Asmus, K.-D., and Fendler, J. H. (1969), J. Phys. Chem. 73, 1583.

Avery, E. C., Remko, J. R., and Smaller, B. (1968), J. Chem. Phys. 49, 951.

Bales, B. L., and Kevan, L. (1974), J. Chem. Phys. 60, 710.

Barr, N. F., and Allen, A. O. (1959), J. Phys. Chem. 63, 928.

Basco, N., Kenney, G. A., and Walker, D. C. (1969), Chem. Commun. 917.

Baxendale, J. H., and Hughes, G. (1958), Z. Phys. Chem. (Frankfurt) 14, 306.

Baxendale, J. H., and Rasburn, E. J. (1974), J. Chem. Soc. Faraday 1 70, 705.

Baxendale, J. H., and Wardman, P. (1971), J. Chem. Soc. Chem. Commun. 429.

Baxendale, J. H., and Wardman, P. (1973), J. Chem. Soc. Faraday 1 69, 584.

Baxendale, J. H., Bell, C., and Wardman, P. (1971), Chem. Phys. Lett. 12, 347.

Baxendale, J. H., Bell, C., and Wardman, P. (1973), J. Chem. Soc. Faraday 1 69, 776.

Belloni, J., Cordier, P., and Delavie, J. (1974), Chem. Phys. Lett. 27, 241.

Bennett, J. E., Mile, B , and Thomas, A (1964), Nature 201, 919.

Bennett, J. E., Mile, B., and Thomas, A. (1967), J. Chem. Soc. A 1393.

Boag, J. W., and Hart, E. J. (1963), Nature 197, 45.

Bockrath, B., and Dorfman, L. M. (1973), J. Phys. Chem. 77, 1002.

Bolton, G. L., and Freeman, G. (1976), J. Am. Chem. Soc. 98, 6825.

Bonin, M. A., Lin, J., Tsuji, K., and Williams, F. (1968), Adv. Chem. Ser. 82, 269.

Bronskill, M. J., Wolff, R. K., and Hunt, J. W. (1970), J. Chem. Phys. 53, 4201.

Buckingham, A. D. (1957), Disc. Faraday Soc. 24, 151.

Burton, M. (1947), J. Phys. Colloid Chem. 51, 611.

Burton, M. (1950), Ann. Rev. Phys. Chem. 1, 113.

Busi, F., and Ward, M. D. (1973), Int. J. Rad. Phys. Chem. 5, 521.

Cady, H. P. (1897), J. Phys. Chem. 1, 707.

Catterall, R. (1976), in Electron–Solvent and Anion–Solvent Interactions (Kevan, L., and Webster, B., eds.), p. 45, Elseiver, Amsterdam.

Collinson, E., Dainton, F. S., Smith, D. R., and Tazuke', S. (April 1962), Proc. Chem. Soc. 140.

Copeland, D. A., Kestner, N. R., and Jortner, J. (1970), J. Chem. Phys. 53, 1189.

CRC Handbook of Radiation Chemistry (1991) (Tabata, Y., Ito, Y., and Tagawa, S., eds.), CRC Press, Boca Raton, Fla.

Czapski, G., and Schwarz, H. A. (1962), J. Phys. Chem. 66, 471.

Dainton, F. S., and Watt, W. S. (1963), Proc. Roy. Soc. A275, 447.

Dainton, F. S., Skwarski, T., Smithies, D., and Wezranowski, E. (1964), Trans. Faraday Soc. 60, 1068.

Davydov, A. S. (1948), Zh. Eksp. Thoer. Fiz. 18, 913.

Debierne, A. (1914), Ann. Phys. (Paris) 2, 97.

Delahay, P. (1976), in *Electron–Solvent and Anion–Solvent* Interactions (*Kevan, L., and Webster, B., eds.*), *p. 115, Elsevier, Amsterdam.*

Delahay, P., and Srinivasan, V. S. (1966), J. Phys. Chem. *70*, 420.

Dewald, R. R., and Tsina, R. V. (1967), Chem. Commun. 13, 647.

Dewald, R. R., Dye, J. L., Eigen, M., and Demaeyer, L. (1963), J. Chem. Phys. *39*, 2388.

Diegen, M. F. (1954), Zh. Eksp. Theor. Fiz. 26, 300.

Dixon, R. S., and Lopata, V. J. (1975), J. Chem. Phys. *62*, 4573.

Dogonadze, R. R. (1971), Ber. Bunsenges. Phys. Chem. *75*, 628.

Dogonadze, R. R., Kuznetsov, A. M., and Levich, V. G. (1969),

Dokl. Akad. Nauk SSSR *188*, 383.

Dorfman, L. M. (1965), Adv. Chem. Series *50*, 36.

Draganic', I., and Draganic', Z. D. (1971), *The Radiation Chemistry of Water*, ch. 5, Academic, New York.

Draganic', I. G., Nenadovic', M. T., and Draganic', Z. D. (1969), J. Phys. Chem. *73*, 2564.

Dye, J. L., Debaeker, M. G., Eyre, J. A., and Dorfman, L. M. (1972), J. Phys. Chem. *76*, 839.

Eckstrom, A. (1970), Radiat. Res. Rev. 2, 381.

Eckstrom, A., Suneram, R., and Willard, J. E. (1970), J. Phys. Chem. *74*, 1888.

Eisele, I., and Kevan, L. (1970), J. Chem. Phys. *53*, 1867.

Eisele, I., and Kevan, L. (1971), J. Chem. Phys. *55*, 5407.

Elliot, A. J., McCracken, D. R., Buxton, G. V., and Wood, N. D. (1990), J. Chem. Soc. Faraday Trans. 86, 1539.

Farhataziz (1976), private communication.

Farhataziz and Cordier, P. (1976), J. Phys. Chem. *80*, 2635.

Farhataziz, Knight, R. J., Milnex, D. J., Mozumder, A., and Povinelli, R. J., eds. (1966), *Proc. 5th Informal Conf. on Radiation Chemistry of Water* (CO0-38-519), Table IV, p. 27, Radiation Laboratory, University of Notre Dame, Ind.

Farhataziz, Perkey, L. M., and Hentz, R. R. (1974a), J. Chem. Phys. *60*, 717.

Farhataziz, Perkey, L. M., and Hentz, R. R. (1974b), J. Chem. Phys. *60*, 4383.

Feng, D.-F., Fueki, K., and Kevan, L. (1973), J. Chem. Phys. *58*, 3281.

Feng, D.-F., Kevan, L., and Yoshida, H. (1974), J. Chem. Phys. *61*, 4440.

Fielden, E. M., and Hart, E. J. (1967), Radiat. Res. *32*, 564.

Fielden, E. M., and Hart, E. J. (1968), Trans. Faraday Soc. *64*, 3158.

Fowles, P. (1971), J. Chem. Soc. Faraday Trans. *67*, 428.

Franck, J., and Scheibe, G. (1928), Z. Phys. Chem. *A139*, 22.

Freed, S., and Sugarman, N. (1943), J. Chem. Phys. *11*, 354.

Freeman, G. R. (1970), in *Actions Chimiques et Biologiques des Radiations*, v. 14 (Haissinsky, M., ed.), p. 73, Masson, Paris.

Fueki, K. (1969), J. Chem. Phys. *50*, 5381.

Fueki, K., Feng, D. F., and Kevan, L. (1970), J. Phys. Chem. *74*, 1976.

Fueki, R., Feng, D. F., Kevan, L., and Christofferson, R. (1971), J. Phys. Chem. *75*, 2297.

Fueki, R., Feng, D. F., and Kevan, L. (1973), J. Am. Chem. Soc. *95*, 1398.

Funabashi, K. (1974), Advan. Radiat. Chem. 4, 103.

Gaathon, A., and Jortner, J. (1973), in *Electrons in Fluids* (Jortner, J., and Kestner, N. R., eds.), p. 429, Springer, Heidelberg.

Gallivan, J. B., and Hamill, W. H. (1966), J. Chem. Phys. *44*, 2378.

Ganguly, A. K., and Magee, J. L. (1956), J. Chem. Phys. *25*, 129.

Gordon, S., and Hart, E. J. (1964), J. Am. Chem. Soc. *86*, 5343.

Gottschall, W. C., and Hart, E. J. (1967), J. Phys. Chem. *71*, 2102.

Grossweiner, L. I. (1968), in *Energetics and Mechanisms in Radiation Biology* (Phillips, G. O., ed.), p. 303. Academic Press, London.

Grossweiner, L. I., Zwicker, E. F., and Swenson, G. W. (1963), Science *141*, 1180.

Gutman, F., and Lyons, L. (1967), *Organic Semiconductors*, Wiley, New York.

Haissinsky, M. (1967), in *Actions Chimiques et Biologiques des Radiations*, v. 11 (Haissinsky, M., ed.), p. 131, Masson, Paris.

Hamill, W. H. (1968), in *Radical Ions* (Kaiser, E. T., and Kevan, L., eds.), p. 321, Interscience, New York.

Hammett, L. P. (1940), *Physical Organic Chemistry*, p. 186, McGraw, New York.

Hart, E. J., ed. (1965), *Solvated Electron*, Adv. Chem. Series, v. 50, ACS, Washington, D.C.

Hart, E. J. (1966), in *Actions Chimiques et Biologiques des Radiations*, v. 10 (Haissinsky, M., ed.), p. 1, Masson, Paris.

Hart, E. J., and Anbar, M. (1970), *The Hydrated Electron*, Wiley Interscience, New York.

Hart, E. J., and Boag, J. W. (1962), J. Am. Chem. Soc. *84*, 4090.

Hart, E. J., and Platzman, R. L. (1961), in *Mechanisms in Radiobiology*, v. 1 (Errera, M., and Forssberg, A., eds.), p. 184 et seq., Academic, New York.

Hart, E. J., Gordon, S., and Fielden, E. M. (1966), J. Phys. Chem. *70*, 150.

Hart, E. J., Fielden, E. M., and Anbar, M. (1967), J. Phys. Chem. *71*, 3993.

Hase, H., Warashina, T., Noda, M., Namiki, A., and Higashimura, T. (1972a), J. Chem. Phys. *57*, 1039.

Hase, H., Higashimura, T., and Ogasawara, M. (1972b), Chem. Phys. Lett. *16*, 214.

Hayon, E., and Weiss, J. (1958), *Proc. Second Inter. Conf. Peaceful Uses of Atomic Energy*, v. 29, p. 80, IAEA, Geneva.

Hentz, R. R., and Kenney-Wallace, G. A. (1972), J. Phys. Chem. *76*, 2231.

Hentz, R. R., and Kenney-Wallace, G. A. (1974), J. Phys. Chem. *78*, 514.

Hentz, R. R., and Knight, R. J. (1970), J. Chem. Phys. *52*, 2456.

Hentz, R. R., Farhataziz, Milner, D. J., and Burton, M. (1967a), J. Chem. Phys. *46*, 2995.

Hentz. R. R., Farhataziz, and Milner, D. J. (1967b), J. Chem. Phys. *47*, 4865.

Hentz, R. R., Farhataziz, and Milner, D. J. (1967c), J. Chem. Phys. *47*, 5381.

Hentz, R. R., Farhataziz, and Hansen, E. M. (1972), J. Chem. Phys. *57*, 2959.

Higashimura, T., Noda, M., Warashina, T., and Yoshida, H. (1970), J. Chem. Phys. *53*, 1152.

Hills, G. J., and Kinnibrugh, D. R. (1966), J. Electrochem. Soc. *113*, 1111.

Hochanadel, C. J., and Ghormley, J. A. (1962), Radiat. Res. *16*, 653.

Howat, G., and Webster, B. (1972), J. Phys. Chem. *76*, 3714.

Hughes, G., and Roach, R. J. (1965), Chem. Commun. 13, *600*.

Hunt, J. W., Wolff, R. D., Bronskill, M. J., Jonah, C. D., Hart, E. J., and Matheson, M. S. (1973), J. Phys. Chem. *77*, 425.

Iguchi, K. (1968), J. Chem. Phys. *48*, 1735.

Ishimaru, S., Yamabe, T., Fukui, K., and Kato, H. (1973), J. Phys. Chem. *77*, 1450.

Jha, K. N., Bolton, G. L., and Freeman, G. R. (1972), J. Phys. Chem. *76*, 3876.

Jonah, C. D. (1974), J. Phys. Chem. *78*, 2103.

Jonah, C. D., Hart, E. J., and Matheson, M. S. (1973), J. Phys. Chem. *77*, 1838.

Jonah, C. D., Matheson, M. S., Miller, J. R., and Hart, E. J. (1976), J. Phys. Chem. *80*, 1267.

Jortner, J. (1959), J. Chem. Phys. *30*, 839.

Jortner, J. (1962), Mol. Phys. *5*, 257.

Jortner, J. (1964), Radiation Res. Suppl. *4*, 24.

Jortner, J. (1970), in *Actions Chimiques et Biologiques des Radiations*, v. 14 (Haissinsky, M., ed.), p. 7, Masson, Paris.

Jortner, J., and Noyes, R. M. (1966), J. Phys. Chem. *70*, 770.

Jortner, J., and Rabani, J. (1962), J. Phys. Chem. *66*, 2081.

Jortner, J., and Stein, G. (1955), Nature *175*, 893.

Jortner, J., Ottolenghi, M., and Stein, G. (1962), J. Phys. Chem. *66*, 2029.

Jortner, J., Ottolenqhi, M., and Stein, G. (1963), J. Phys. Chem. *67*, 1271.

Jortner, J., Ottolenghi, M., and Stein, G. (1964), J. Phys. Chem. 68, 247.

Jou, F. Y., and Dorfman, L. M. (1973), J. Chem. Phys. 58, 4715.

Keene, J. P. (1960), British Empire Cancer Campaign 38(2), 498.

Keene, J. P. (1964), Radiation Res. 22, 1.

Kestner, N. R. (1976), in Electron–Solvent and Anion–Solvent Interactions (Kevan, L., and Webster, B., eds.), p. 1, Elsevier, Amsterdam.

Kestner, N. R., and Jortner, J. (1973), J. Phys. Chem. 77, 1040.

Kestner, N. R., and Logan, J. (1972), Metal-Ammonia Solutions, Colloque Weyl III, Kibbutz Hanita, Israel.

Kestner, N. R., Jortner, J., and Gaathon, A. (1973), Chem. Phys. Lett. 19, 328.

Kevan, L. (1968), in Radiation Chemistry of Aqueous Systems (Stein, G., ed.), p. 21, Weizmann Science, Jersualem.

Kevan, L. (1972), J. Chem. Phys. 56, 838.

Kevan, L. (1974), Advan. Radiat. Chem. 4, 181.

Kraus, C. A. (1908), J. Am. Chem. Soc. 30, 1323.

Kuppermann, A. (1961), in Actions Chimiques et Biologiques des Radiations, v. 5 (Haissinsky, M., ed.), p. 85, Masson, Paris.

Kupperman, A. (1967), in Radiation Research (Silini, G., ed.), p. 212, North-Holland, Amsterdam.

Kupperman, A. (1974), in Physical Mechanisms in Radiation Biology (Cooper, R. D., and Wood, R. W., eds.), p. 155. Technical Information Center, USERDA, Springfield, Virginia.

Kuppermann, A., and Belford, G. G. (1962), J. Chem. Phys. 36, 1427.

Laitinen, H. A., and Nyman, C. J. (1948), J. Am. Chem. Soc. 70, 3002.

Lam, K. Y., and Hunt, J. W. (1974), J. Phys. Chem. 78, 2414.

Land, R. H. and O'Reilly, D.E. (1967), J. Chem. Phys. 46, 4496.

Landau, L. (1933), Phys. Z. fur Sowjetunion 3, 664.

Lefort, M., and Tarrago, X. (1959), J. Phys. Chem. 63, 833.

Levich, V. G. (1966), Advan. Electrochem. Engg. 4, 249.

Magee, J. L. (1953), in Basic Mechanisms in Radiobiology, p. 51, USNRC publication no. 305 Washington, D.C.

Magee, J. L. (1964), Radiation Res. Suppl. 4, 20.

Magee, J. L. (1975), private communication.

Magee, J. L., and Mozumder, A. (1973), in Advances in Radiation Research, v. 1 (Duplan, J. F., and Chapiro, A., eds.), p. 15, Gordon and Breach, New York.

Mahlman, H. A., and Sworski, T. J. (1967), in The Chemistry of Ionization and Excitation (Johnson, G. R. A., and Scholes, G., eds.), p. 259, Taylor and Francis, London.

Marcus, R. A. (1964), Ann. Rev. Phys. Chem. 15, 155.

Marcus, R. A. (1965a), Adv. Chem. Ser. 50, 138.

Marcus, R. A. (1965b), J. Chem. Phys. 43, 3477.

Matheson, M. S. (1975), in Physical Chemistry, v. 7 (Eyring, H., Henderson, D., and Jost, W., eds.), p. 533, Academic, New York.

Matheson, M. S., and Rabani, J. (1965), J. Phys. Chem. 69, 1324.

Michael, B. D., Hart, E. J., and Schmidt, K. H. (1971), J. Phys. Chem. 75, 2798.

Miller, J. R. (1972), J. Chem. Phys. 56, 5173.

Minday, R. M., Schmidt, L. D., and Davis, H. T. (1971), J. Chem. Phys. 54, 3112.

Mittal, J. P. (1971), Nature 230, 160.

Moissan, H. (1889), C. R. Acad. Sci. (Paris) 128, 26.

Mozumder, A., and Magee, J. L. (1975), in Physical Chemistry, an Advanced Treatise, v. 7 (Eyring H., Henderson, D., and Jost, W., eds.), p. 742, Academic Press, New York also see, Mozumder, A. (1977), Khimiya Visokikh Energii. 11, 182.

Mozumder, A., and Tachiya, M. (1975), J. Chem. Phys. 62, 979.

Naleway, C. A., and Schwartz, M. E. (1972), J. Phys. Chem. 76, 3905.

Narayana, P. A., and Kevan, L. (1976), J. Chem. Phys. 65, 3379.

Natori, M. (1968), J. Phys. Soc. Japan 24, 913.

Natori, M. (1969), J. Phys. Soc. Japan 27, 1309.

Natori, M., and Watanabe, T. (l966), J. Phys. Soc. Japan 21, 1573.

Newton, M. D. (1973), J. Chem. Phys. 58, 5833.

Newton, M. D. (1975), J. Phys. Chem. 79, 2795.

Nielsen, S. O., Pagsberg, P., Hart, E. J., Christensen, H., and Nilsson, G. (1969), J. Phys. Chem. 73, 3171.

Nielsen, S. O., Michael, B. D., and Hart, E. J. (l976), J. Phys. Chem. 80, 2482.

Noyes, R. M. (1961), in *Progress in Reactions Kinetics* (Porter, G., ed.), v. 1, p. 137, Pergamon, Oxford.

Ottolenghi, M., Rabani, J., and Stein, G. (1967), Israel J. Chem. 5, 309.

Platzman, R. L. (1953), in *Basic Mechanisms in Radiobiology* (Magee, J. L., Kamen, M. D., and Platzman, R. L., eds.), NAS-NRC publication no. 305, p. 22, National Research Council, Washington, D.C.

Platzman, R. L. (1967), in *Radiation Research* (Silini, G., ed.), p. 20, North-Holland, Amsterdam.

Platzman, R. L., and Franck, J. (1954), Z. Phys. 138, 411.

Rabani, J., Mulac, W. A., and Matheson, M. (1965), J. Phys. Chem. 69, 53.

Raff, L., and Pohl, H. A. (1965), Adv. Chem. Ser. 50, 173.

Renaudiere, J. Fradin de la, and Belloni, J. (1973), Int. J. Rad. Phys. and Chem. 5, 31.

Rentzepis, P. M., Jones, R. P., and Jortner, J. (1973), J. Chem. Phys. 59, 766.

Richards, J. T., and Thomas, J. K. (1970), J. Chem. Phys. 53, 218.

Risse, O. (1929), Strahlentherapie 34, 578.

Ross, A. B. (1975), *Selected Specific Rates of Reactions of Transients from Water in Aqueous Solution. Hydrated Electron, Supplemental data*, NSRDS-NBS 43, Supplement, National Bureau of Standards, Washington, D.C.

Rzad, S. J., and Schuler, R. H. (1973), J. Phys. Chem 77, 1926.

Samuel, A. H., and Magee, J. L. (1953), J. Chem. Phys. 21, 1080.

Sauer, M. C., Jr., Arai, S., and Dorfman, L. M. (1964), J. Chem. Phys. 42, 708.

Sawai, T., and Hamill, W. H. (1969), J. Phys. Chem. 73, 2750.

Schindewolf, U. (1968), Angew. Chem. (Inter. Ed.) 7, 190.

Schindewolf, U., Kohrmann, H., and Lang, G. (1969), Angew. Chem. International Ed., 8, 512.

Schmidt, K. H. and Anbar, S. (1969), J. Phys. Chem. 73, 2846.

Schmidt, K. H., and Buck, W. L. (1966), Science 151, 70.

Schmidt, K. H., and Hart, E. J. (1968) in *Radiation Chemistry I* (E. J. Hart and R. F. Gould eds.) Advances in Chemistry Series 81, p. 267, Amer. Chem. Soc., Washington, D.C.

Schmidt, P. P. (1973), J. Phys. Chem. 77, 488, and references therein.

Schmidt, W. F., and Bakale, G. (1972), Chem. Phys. Lett. 17, 617.

Schwarz, H. A. (1969), J. Phys. Chem. 73, 1928.

Scott, N. D., Walker, J. F., and Hansley, V. L. (1936), J. Am. Chem. Soc. 58, 2442.

Seddon, W. A., Fletcher, J. W., Sopchyshyn, F. C, and Jevcak, J. (1974), Can. J. Chem., 52, 3269.

Sokolov, U., and Stein, G. (1966), J. Chem. Phys. 44, 2189.

Spinks, J. W. T., and Woods, R. J. (1976), *An Introduction to Radiation Chemistry*, p. 260, Wiley, New York.

Springett, B. E., Jortner, J., and Cohen, M. H. (1968), J. Chem. Phys. 48, 2720.

Stein, G. (1952a), Disc. Faraday Soc. 12, 227.

Stein, G. (1952b), Disc. Faraday Soc. 12, 289.

Stein, G. (1969), in *Actions Chimiques et Biologiques des Radiations*, v. 13, (Haissinsky, M., ed.), p. 119, Masson, Paris.

Stein, G. (1971), Israel J. Chem. 9, 413.

Stein, G., and Treinin, A. (1959), Trans. Faraday Soc. 55, 1087, 1091.

Swallow, A. J. (1965), in *Pulse Radiolysis* (Ebert, M., Keene, J. P., and Swallow, A. J., eds.), p. 54, Academic, London.

Tachiya, M. (1972), J. Chem. Phys. *56*, 6269.

Tachiya, M. (1974), J. Chem. Phys. *60*, 2275.

Tachiya, M., and Mozumder, A. (1974a), J. Chem. Phys, *60*, 3037.

Tachiya, M., and Mozumder, A. (1974b), J. Chem. Phys, *61*, 3890.

Taft, R. W. (1956), in *Steric Effects in Organic Chamistry* (Newman, M. S., ed.), p. 556, Wiley, New York.

Taub, I. A., and Gillis, H. A. (1969), J. Am. Chem. Soc. *91*, 6507.

Taub, I. A., Sauer, M. C., Jr., and Dorfman, L. M. (1963), Disc. Faraday Soc. *36*, 206.

Taub, I. A., Hurter, D. A., Sauer, M. C., Jr., and Dorfman, L. M. (1964), J. Chem. Phys. *41*, 979.

Thomas, J. K., Gordon, S., and Hart, E. J. (1964), J. Phys. Chem. *68*, 1524.

Ulstrup, J., and Jortner, J. (1975), J. Chem. Phys. *63*, 4358.

Van Bekkum, H., Verkade, P. E., and Wepater, B. M. (1959), Rec. Trav. Chim. *78*, 815.

Vorotyntsev, M. A., Dogonadze, R. R., and Kuznetsov, A. M. (1970), Dokl. Akad. Nauk SSSR *195*, 1135.

Walker, D. C. (1966), Can. J. Chem. *44*, 2226.

Walker, D. C. (1967), Can. J. Chem. *45*, 807.

Walker, D. C. (1968), Adv. Chem. Ser. *81*, 49.

Wallace, S. C., and Walker, D. C. (1972), J. Phys. Chem. *76*, 3780.

Ward, B. (1968), Adv. Chem. Series *81*, 601.

Watson, C., Jr., and Roy, S. (1972), *Selected Specific Rates of Reactions of the Solvated Electrons in Alcohols*, NSRDS-NBS 42, U.S Government Printing Office, Washington, D.C.

Webster, B. C., and Howat, G. (1972), Radiat. Res. Revs. *4*, 259.

Weiss, J. (1944), Nature *153*, 748.

Weiss, J. (1953), Ann. Revs. Phys. Chem. *4*, 143.

Weiss, J. (1960), Nature *186*, 751.

Weissmann, M., and Cohan, N. V. (1973), J. Chem, Phys. *59*, 1385.

Westermark, T, and Grapengiesser, B. (l960), Nature *188*, 325.

Weyl, W. (1864), Ann. Phys. *121*, 601.

Willard, J. E. (1968), in *Fundamental Processes in Radiation Chemistry* (Ausloos, P., ed.), p. 599, Interscience, New York.

Wolff, R. K., Broskill, M. J., Aldrich, J. E., and Hunt, J. W. (1973), J. Phys. Chem. *77*, 1350.

Yoshida, H., Feng, D. F., and Kevan, L. (1973), J. Chem. Phys. *58*, 3411.

CHAPTER 7

Spur Theory of Radiation Chemical Yields: Diffusion and Stochastic Models

7.1 Early Studies
7.2 Multiradical Diffusion Models
7.3 Stochastic Kinetics
7.4 Comparison with Experiment: The Case of Water Radiolysis
7.5 Kinetics of Electron-Ion Recombination in Irradiated Dielectric Liquids

7.1 EARLY STUDIES

The spur diffusion model arose from the necessity of explaining the variation of molecular and radical yields in water radiolysis with particle LET and scavenger concentration (Samuel and Magee, 1953; Ganguly and Magee, 1956; Schwarz, 1955). There is no *a priori* reason for the diffusion model to be a valid theory. Its success is to be judged by the ability to encompass a great variety of experimental results using relatively few adjustable parameters (Schwarz, 1969). For low-LET radiations, the reactive intermediates are thought to be formed in *spurs,* which are small localized regions (on the order of a few nanometers in extent) of energy deposition by the primary particle and by secondary electrons. In the high-LET case, the spurs coalesce to form cylindrical tracks (Samuel and Magee, 1953). This basic consideration is modified somewhat by details of energy deposition pattern (see Chapter 3).

The diffusion model assumes that (1) Fick's law is valid as modified by reactions; (2) reactant concentrations are interpreted as probability densities; (3) specific rates correspond to otherwise homogeneous reactions—Monchick *et al.* (1957) show that this implies neglect of interparticle correlation, which is not always justified; (4) the rate of irradiation is low; and (5) reactions proceed under quasi-thermal conditions. The model implicitly ignores reactions in the prediffusion regime. Subject to these assumptions, the space-time evolution of the concentration of the ith species is given by

$$\frac{\partial c_i}{\partial t} = D_i \nabla^2 c_i - c_i \sum_j k_{ij} c_j + \sum_{j,k \neq i} k_{jk} c_j c_k - c_i \sum_s k_{si} c_s; \; i, j = 1, 2, ..., n. \quad (7.1)$$

Here D_i is the diffusion coefficient of the ith species, which reacts with the jth species at a specific rate k_{ij}, and s denotes a homogeneously distributed solute called a scavenger. The first term on the right-hand side of (7.1) represents pure diffusion; the second and fourth terms give disappearance of the ith species via reaction with jth species ($j = i$ is allowed) and scavengers, respectively; and the third term denotes the regeneration of the ith species from other species by reaction.

In earlier studies, the system of equations (7.1) was solved by the *method of prescribed diffusion*, meaning that the spatial part of c_j was written *ad hoc* as a gaussian function for a freely diffusing nonreactive particle. The time-dependent normalizing factor, denoting the species survival probability over all space, was obtained by inserting the gaussian functions in (7.1) and solving the resultant system of first-order nonlinear simultaneous equations.

Later, Kuppermann and Belford (1962a, b) initiated computer-based numerical solution of (7.1), giving the space–time variation of the species concentrations; from these, the survival probability at a given time may be obtained by numerical integration over space. Since then, this method has been vigorously followed by others. John (1952) has discussed the convergence requirement for the discretized form of (7.1), which must be used in computers; this turns out to be $\Delta\tau/(\Delta\rho)^2 \leq 1/2$, where ρ and τ are respectively the normalized forms of r and t. Often, $\Delta\tau/(\Delta\rho)^2 = 1/6$ is used to ensure better convergence. Of course, any procedure requires a reaction scheme, values of diffusion and rate coefficients, and a statement about initial number of species and their distribution in space (*vide infra*).

7.1.1   THE SAMUEL-MAGEE MODEL

Samuel and Magee (1953) employed a 1-radical model to find the relative forward yield in water radiolysis as a function of radiation quality. In such models, no distinction is made between reactive radicals or molecular products. The products of radiolysis are called *forward* (F) to denote *observable* molecular yield or *radical* (R), denoting yield of scavenger reaction at small concentration. The aim of the theory is to calculate the relative forward yield $G(F)/[G(F) + G(R)]$, where the G values refer to the respective yields for 100 eV energy absorbed in water. The model further assumes that (1) the initial spur volume is proportional to its radical content, a feature that has been maintained in most subsequent models; and (2) the ionized electron is recaptured, giving H and OH as the primary reactive species. Thus, the reaction scheme is

$$H_2O \rightarrow \frac{1}{2} H_2 + \frac{1}{2} H_2O_2 , \tag{F}$$

$$H_2O \rightarrow H + OH. \tag{R}$$

The second assumption has been effectively invalidated by the discovery of the hydrated electron. However, the effects of LET and solute concentration on molecular yields indicate that *some* kind of radical diffusion model is indeed required. Kuppermann (1967) and Schwarz (1969) have demonstrated that the hydrated electron can be included in such a model. Schwarz (1964) remarked that Magee's estimate of the distance traveled by the electron at thermalization (on the order of a few nanometers) was correct, but his conjecture about its fate was wrong. On the other hand, Platzman was correct about its fate—namely, solvation—but wrong about the distance traveled (tens of nanometers).

In the spirit of prescribed diffusion, Samuel and Magee write the *normalized* distribution of radicals at time t as

$$\rho(r) = \left(\frac{\beta^3}{\pi^{3/2}} \right) \exp(-\beta^2 r^2). \tag{7.2}$$

Here $\beta^2 = 1/Lv(t + \tau)$, L is the mean free path of radicals at thermal velocity v, and the initial spur radius r_0 and the fictitious time τ are related by $r_0^2 = Lv\tau$. On random scattering, the probability per unit time of any two radicals colliding in volume dv will be $\sigma v/dv$, where σ is the collision cross section. The probability of finding these radicals in dv at the same time t is $N(N - 1)\rho^2 \, dv^2$, giving the rate of reaction in that volume as $\sigma v N(n - 1)\rho^2 \, dv$. Thus,

$$-\frac{dN}{dt} = \sigma v N(N - 1) \int \rho^2 \, dv = \frac{\sigma v N(N - 1)}{(8\pi^3)^{1/2}[Lv(t + \tau)]^{3/2}}. \tag{7.3}$$

The final result is obtained by using Eq. (7.2) and the definition of β. Integrating (7.3) with the initial condition $N = N_0$ at $t = 0$ gives the surviving number of radicals at infinite time as

$$\frac{N_\infty}{N_0} = [N_0 - (N_0 - 1) \exp(-x N_0^{-1/3})]^{-1}, \tag{7.4}$$

where $x = \sigma/(2\pi^3)^{1/2} La$ and $r_0 = a N_0^{1/3}$, in conformity with the initial spur volume being proportional to N_0. Equation (7.4) shows that for a *spherical spur*, a certain fraction of radicals remain extant at infinite time. This can be picked up by a scavenger at a very small concentration, and it is interpreted as the radical fraction. The fraction giving *observable* molecular yield is given by

$$F(N_0) = \left(1 - \frac{N_\infty}{N_0}\right)\left(\frac{N_0}{2} - 1\right)(N_0 - 1), \qquad (7.5)$$

where allowance has been made for re-formation of water through the factor $(N_0/2 - 1)/(N_0 - 1)$. This factor is obviously correct for $N_0 = 2$ and ∞, and it is assumed to be valid for intermediate values of N_0. To compare with experiments for the forward or radical fraction using Eqs. (7.4) and (7.5) entails an averaging over the distribution of N_0, the initial number of radicals in the spur. Samuel and Magee use relative ionization frequency observed in a cloud chamber for this purpose, which is open to criticism. In any case, they use the normalized distribution

$$f(N_0) = 0.65 \exp\left(-\frac{N_0}{4}\right). \qquad (7.6)$$

It is unfortunate that in many later applications of the model the spur size distribution was ignored and an average spur size was used, inconsistent with the statistics of energy loss events.

Samuel and Magee take $L = 1$ Å, $\sigma = 4\pi A^2$, $a = 5$ A and evaluate x. With this value of x, G_R and G_F are calculated to be proportional respectively to $\Sigma N_0 f(N_0)(N_\infty/N_0)$ and $(1/2)\Sigma N_0 f(N_0)F(N_0)$. From (7.5) and (7.6), they then get $G_F/(G_F + G_R) = 0.231$, which compares favorably with experimental values under ^{60}Co-γ radiolysis; these are 0.21 (Argonne) and 0.25 (Oak Ridge). The fact that the correct value of x corresponds to reasonable values of L, σ, and a speaks strongly in favor of the model.

In summary, the Samuel-Magee model of low-LET tracks consists of isolated spherical spurs distributed exponentially in energy. No distinction is made between primary and secondary tracks; inherent slowing down of the particle is also ignored.

The Samuel-Magee model can be extended to α-particle tracks, considered as cylindrical columns formed by excessive spur overlap due to high LET. To a good approximation, the length l of the cylinder remains constant while its radius grows by diffusion. In this geometry, the normalized radical distribution is given by

$$\rho(r) = \frac{\beta^3}{\pi l} \exp(-\beta^2 r^2). \qquad (7.7)$$

Since N, the number of radicals in the column, is much greater than 1, one obtains

$$-\frac{dN}{dt} = \sigma v \int N^2 \rho^2(r)\, dv = \frac{\sigma v N^2}{2\pi L l v(t + \tau)}, \qquad (7.8)$$

which on integration gives, with $N = N_0$ at $t = 0$,

$$\frac{N_0}{N} = 1 + \frac{\sigma N_0}{2\pi L l} \ln\left(\frac{t + \tau}{\tau}\right). \tag{7.9}$$

Equation (7.9) implies that nothing survives radical-radical recombination for an isolated cylindrical track—that is, at zero dose. At a finite dose rate, tracks will overlap at a large, but finite time and, for even a very small solute concentration, a scavenger reaction may precede intertrack overlap. Thus, the relevant question is the determination of the maximum time to be used in Eq. (7.9) consistent with a given dose rate.

Samuel and Magee note that $(t + \tau)/\tau$ represents the ratio of track volumes per unit length at times t and 0. Considering generation, diffusion, and reaction, Magee (1951) derived the a relation between the dose rate I (reps/s) and x_m, the maximum volume ratio-$x_m(x_m - 1) = 2.5 \times 10^{-13} DN_0 /Ilv_0^2$, where D (cm²s⁻¹) is the radical diffusion coefficient and v_0 (cm³/cm) is the initial track volume per unit length. Taking $N_0 /i = 8.5 \times 10^7$ cm⁻¹ for a 6-MeV α-particle in water, $v_0 = \pi \times 10^{-14}$, $D = 8 \times 10^{-5}$ cm²/s for the mutual diffusion coefficient and $I = 200$, equivalent to 1.1×10^{16} eV/g.s, x_m is estimated to be approximately 10^8 and Eq. (7.9) reduces to $N_0 /N = 1 + 1.7 \ln x$. From this, it is seen that at $x = 10^4$ when only 0.01% of track life has elapsed, 97% of all recombinations have taken place.

Thus, although there is no unequivocal value of $G_F /(G_F + G_R)$ for cylindrical tracks, Samuel and Magee argue that the track life may be divide into two stages. The first stage is relatively very short but most recombinations occur in it. Fortunately, the demarcation is not critical; $x = 10^4$ is used for it. The ratio G_F /G_R is then proportional to $(1/2)(1 - N/N_0)(N_0/N)$, the factor 1/2 accounting for the *observable* molecular yield when $N \gg 1$. Samuel and Magee then obtain $G_F /(G_F + G_R) = 0.887$, which is close to the observed value.

Samuel and Magee (1953) also applied their model to the radiolysis of tritiated water in solution. Electrons from β-decay of tritium have energy in the interval 0–18 KeV, with a peak at 5.5 KeV. Samuel and Magee consider the spur spacing on tritium tracks (~15 nm) about ten times the initial spur diameter and note that the average spur contains 5.08 radicals. Initially, the spurs expand independently in their respective spherical geometry up to such a time t_1 when the spurs have expanded to ten times their initial linear dimension. Then, they begin to overlap in an approximately cylindrical geometry, and eventually in the longer time scale, a prolate spheroidal geometry energies. Of the 380 radicals initially present in an average tritium-β track, Samuel and Magee estimate that 57 detectable molecules are formed by combination during the first stage, which is similar to situation in γ-radiolysis; 27 detectable molecules form during the later stages, leaving 183 radicals uncombined in the longest time scale. Thus, the calculated value of

$G_F/(G_F + G_R) = (57 + 27)/(57 + 27 + 183) = 0.315$ for tritium radiolysis of water, which compares favorably with Hart's (1952) experimental value of 0.30. However, in this calculation, the particle slowing down was ignored and the interspur distance was taken to what corresponds to the beginning of the track.

7.1.2   THE GANGULY-MAGEE MODEL

The Ganguly-Magee model is a 1-radical-1-scavenger model where competition between recombination and scavenging reaction is represented as follows:

$$R + R \rightarrow R_2 ,$$
$$R + S \rightarrow RS.$$

Further refinements of the model (Ganguly and Magee, 1956) include interspur overlap in the same track. The track is seen as a "string of beads." The beads are spurs, the average spacing of which corresponds to the local LET of the particle. The free path of the incident particle between spur centers is distributed exponentially with a mean value given by

$$l = \frac{\varepsilon \eta R}{E_0} \left(1 - \frac{\rho}{R} \right)^{(\eta-1)/\eta} , \tag{7.10}$$

where R is the range of the particle with starting energy E_0, ρ is the axial track length, η is the power index relating residual energy and range at ρ, and ε is the spur energy, assumed constant. The index η is evaluated from experimental data, or by extending the Bethe theory into the low-energy regime. Thusly evaluated, η generally lies between 1.4 and 1.8, typical values being 1.6 for 7.7-MeV α, 1.57 for 1-MeV protons, and 1.4 for 0.5-MeV electrons. Taking the z direction of a cylindrical coordinate system as the track axis, the radical concentration at a given point (r, z) at time t may be written as

$$c(r, z, t) = \frac{\exp(-r^2/4Dt')}{(4\pi Dt')^{3/2}} \sum_{i=1}^{m} v_i \exp\left[-\frac{(z - z_i)^2}{4Dt'} \right], \tag{7.11}$$

where $m = E_0/\varepsilon$ is the total number of spurs on the track, z_i is the location of the ith spur center having v_i radicals extant at time t, D is the radical diffusion coefficient, and $t' = t + \tau$. The fictitious time τ is related to the initial spur radius r_0 by $r_0^2 = 4D\tau$.

With the form of c given by (7.11), the diffusion model gives its space-time dependence as follows:

$$\frac{\partial c}{\partial t} = D\nabla^2 c - kc^2 - k_s c_s c. \tag{7.12}$$

Here c_s is scavenger concentration, and k and k_s are the specific rates of the recombination and scavenging reactions, respectively. Integrating (7.12) over all volume, noting that the integral of the Laplacian vanishes, and writing , the number of radicals in the track at t, one gets

$$n(t) = \sum_{i=1}^{m} v_i(t),$$

the number of radicals in the track at t, one gets

$$\frac{dn}{dt} = -k \int c^2 \, dv - \lambda n, \tag{7.13}$$

where $\lambda = k_s c_s$. Ganguly and Magee define an effective track volume at t by $V(t) = n(t)/\int c^2 \, dv$ and use prescribed diffusion (see Eq. 7.11) to show that

$$\int c^2 \, dv = (8\pi Dt')^{-3/2} \sum_{i,j}^{m} v_i v_j \exp\left(-\frac{z_{ij}^2}{8Dt'}\right), \tag{7.14}$$

where $z_{ij} = z_i z_j$. They then convert this double summation into an integral using the exponential distribution for z_i with the mean value of Eq. (7.10). They further assume that (1) all spurs contain the same number of radicals at t—that is, $v_j(t) = v$; and (2) only nearest-neighbor spurs overlap. While the second assumption seems reasonable in view of rapid decline of radical concentration with distance from spur center, the first assumption is hard to justify, except in an undefined average sense, due to random variation of interspur separation. In any case, as a result of these manipulations, Ganguly and Magee obtain, from (7.14),

$$V(t) = m(8\pi Dt')^{3/2}\left[1 + \frac{m(8\pi Dt')^{1/2}}{\eta(2 - \eta)R}\right]^{-1}; \qquad t' = t + \frac{r_0^2}{4D}. \tag{7.15}$$

Equation (7.15) shows that at small times the track develops as a collection of m isolated spheres, whereas at long times, $V(t)$ increases linearly with t, which is characteristic of cylindrical diffusion. Using the definition of $V(t)$ the solution of Eq. (7.13) may be given as

$$n(t) = \exp(-\lambda t)\left[n(0)^{-1} + k \int_0^t V(t')^{-1} \exp(-\lambda t') \, dt'\right]^{-1}, \tag{7.16}$$

where $n(0) = mv_0$ is the initial number of radicals in the track and $V(t)$ is given by Eq. (7.15).

In the infinite time limit, no radical survives. They either combine to give molecular products or undergo scavenging reactions. The probability of the latter is given by

$$S = \frac{\lambda}{n(0)} \int_0^{\infty} n(t) \, dt.$$

Ganguly and Magee take $\varepsilon = 100$ eV corresponding to 6 radicals, $D = 2 \times 10^{-5}$ cm^{2}s^{-1}, $\tau = 1.25 \times 10^{-10}$ s, $k = 10^{-11}$ cm^{3}s^{-1} and plot the surviving fraction $1 - S$ as a function of log q, where $q = \lambda\tau$, for α-particles, protons, and high-energy electrons in water. In the light of present knowledge, the spur energy content was overestimated, whereas the G values of radicals were underestimated. Nevertheless, the model has been useful in interpreting scavenging experiments in aqueous solutions semiquantitatively. Figure 7.1 reproduces some of their results. Considering k_s and τ as constants, it can be used to evaluate the reduction of molecular yield with scavenger concentration. At very small concentrations, the yield is nearly constant, being characteristic of radiation quality. This part corresponds to the Samuel-Magee calculation. The molecular yield decreases with increasing scavenger concentration, slowly at first and then more rapidly when $q \sim 1$. Also, at a given concentration, the effect of a scavenger increases with particle energy.

In the early days of water radiolysis, it was empirically established in several instances that the reduction of molecular yield by a scavenger was proportional to the cube root of its concentration (Mahlman and Sworski, 1967). Despite attempts by the Russian school to derive the so-called cube root law from the diffusion model (Byakov, 1963; Nichiporov and Byakov, 1975), more rigorous treatments failed to obtain that (Kuppermann, 1961; Mozumder, 1977). In fact, it has been shown that in the limit of small concentration, the reduction of molecular yield by a scavenger should be given by a square root law in the orthodox

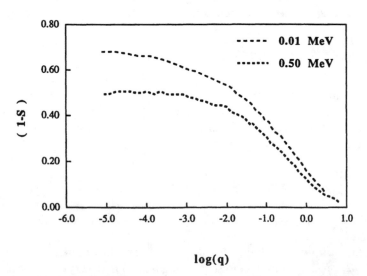

FIGURE 7.1   Plot of radical recombination probability, ordinate, $1 - S$ vs. log q for electron tracks of incident energy 0.5 and 0.01 MeV. Here S is the scavenger reaction probability and $q = k_s c_s \tau$, with k_s the rate constant of scavenging, c_s the scavenger concentration, and τ a fictitious time representing initial radical concentration. See text for explanation. From Ganguly and Magee (1956), by permission of Am. Inst. Phys.©

diffusion theory (Mozumder, 1971, 1977). Nonetheless the cube root relationship can be helpful in organizing a body of experimental results.

7.1.3 OTHER ONE-RADICAL MODELS AND LIMITATIONS

Fricke (1955), Flanders and Fricke (1958), and Schwarz (1955) have used somewhat different versions of the 1-radical model employing numerical or semianalytical techniques and arriving at similar conclusions. Schwarz's (1955) treatment gave for the first time a universal relation connecting the suppression of molecular yield and the scavenger concentration in water radiolysis. Figure 7.2 shows the Schwarz (1955) plot, which satisfactorily explains the data at low and intermediate concentrations. As pointed out by Kuppermann (1961), Schwarz does not solve Eq. (7.12) directly but calculates the relative molecular yield in presence of a scavenger by evaluating

$$(k/2)\int_0^\infty \int_0^\infty c^2 \, dv \, dt.$$

In so doing, he assumes $c(r, t) = c^0(r, t) \exp(-k_s c_s t)$ where c^0 is the prescribed diffusion solution of the problem in the absence of scavenger. This adds two radicals whenever a combination reaction occurs, but not for the scavenging reaction. Thus, the molecular yield is overcalculated both in the presence and absence of scavenger, resulting in some hopeful cancellation in the relative yield $G(R_2)/G^0(R_2)$. Figure 7.2 shows that the procedure works reasonably well at low and intermediate concentrations. At high concentration, there is probably some other complication arising from direct effects of radiation. The reason for obtaining a universal curve when the abscissa is multiplied by the ratio of scavenging rate constants lies in the fact that in the diffusion equation (see Eq. 7.12) k_s and c_s always comes as a product $\lambda = k_s c_s$. Kuppermann and Belford (1962a, b) solved the partial differential equation (7.12) numerically using the 1-radical model. From the solution $c(r, t)$, they compute the relative yields of molecular product, scavenging reaction, and radicals extant at time t respectively from the integrals

$$\frac{1}{2} k \int_0^t \int_0^\infty c^2(r, t') \, dv \, dt',$$

$$k_s c_s \int_0^t \int_0^\infty c(r, t') \, dv \, dt',$$

$$\text{and} \int_0^\infty c(r, t') \, dv$$

where $dv = 4\pi r^2 \, dr$ in spherical geometry and $dv = 2\pi r \, dr$ per unit length in cylindrical geometry. These authors demonstrate that the effects of spur size,

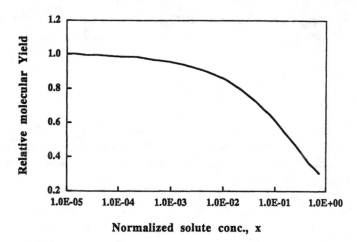

FIGURE 7.2   Relative molecular yields of H_2 and H_2O_2 in the radiolysis of aqueous solutions (Schwarz, 1955). Full curve is theoretical for solute concentrations (X axis) in different experiments multiplied by suitable normalization factors. Notice that even after normalization experimental points remain considerably higher than theoretical at high concentrations ($x > 0.1$). From Schwarz (1955), with permission of Am. Chem. Soc.©

diffusion, and rate constants are as expected. They also show that any initial distribution of radicals become gaussian by diffusion in about a 1-ns interval during which ~80% of radical reaction is over.

Sworski (1964, 1965) introduced a modified spur diffusion model for explaining the H_2 yield in water radiolysis. According to this model, H_2 is produced both by intra- and interspur reactions. The latter is diffusion-controlled but considered less important. The former is homogeneous with "excited water" as a precursor. The reduction of the H_2 yield with scavenger concentration is claimed to be consistent with experiment. No peroxide yield is calculated, and the nature of "excited water" remains elusive. Burns and Barker (1965) extended the diffusion model to organic liquids and concluded that the spur size must be bigger in these media than in water. Mozumder and Magee (1966a–c) introduced a spur size distribution based on a synthesized dipole oscillator strength distribution of the water molecule. However, as observed by Draganic' and Draganic' (1971), their diffusion calculation, within the context of a 1-radical model, was directed more toward the correctness of the physical picture of energy deposition than toward better agreement with experiment.

The main problem with 1-radical models is that they are chemically unrealistic. The distinction between oxidizing and reducing species is not recognized, and different rate and diffusion coefficients must be lumped into one.

Re-formation of water must be considered somewhat arbitrarily. However, these models have been generally useful in delineating LET and scavenger effects on the total molecular yield.

7.2   MULTIRADICAL DIFFUSION MODELS

Before the discovery of the hydrated electron in 1962, several authors used variations of the 2-radical diffusion model with H and OH as reactive entities. Burch (1959) attempted to explain the radiation-chemical yields in the ferrous- and ceric-sulfate dosimeters in which a spectrum of local energy deposition was included. However, use of a cylindrical track model for all types of radiation largely invalidates many of his conclusions. Filinovskii and Chizmadzev (1958) considered cylindrical tracks and ignored diffusion altogether. According to Kuppermann (1961), this procedure overcalculates molecular yields, but in principle it can distinguish between the Samuel-Magee and the Lea-Gray-Platzman models of radiation action.

Dyne and Kennedy (1958, 1960) applied the 2-radical diffusion model for both spherical spurs (γ-radiolysis) and cylindrical tracks (α-radiolysis). Designating H and OH by $i = 1$ and 2, respectively, they obtained a numerical solution of the two simultaneous equations (see 7.1); from this, they computed the yields of H_2, H_2O_2, re-formed H_2O, H, and OH through space-time integration in a manner similar to 1-radical computations (*vide supra*). They take $D_2 = 2 \times 10^{-5}$ cm²/s, $D_1/D_2 \sim 4$ (scaled inversely as the square root of the mass ratio), $k_{11} = 1.2 \times 10^{-11}$ cm³/s, $k_{12} = 3 \times 10^{-11}$ cm³/s, $k_{22} = 0.9 \times 10^{-11}$ cm³/s, and all scavenger specific rates as 1×10^{-11} cm³/s in accordance with Fricke (1955). Calculations were performed for $c_s = 10^{-3}$ M and visually extrapolated to 10 ns to give the "infinite-time" yields. Using $G(-H_2O) = 3.7$ somewhat arbitrarily and a spur energy 50–70 eV (which is rather implausible), they find, in agreement with Samuel and Magee, best agreement with γ-radiolysis experiment for a small-size spur, ~1 nm initial radius. Their infinite-time computed yields are $G(H_2) = 0.45$, $G(H_2O_2) = 0.7$, and $G(H_2O) = 0.9$. The observable molecular yields agree very well with experiment, but the respective combination rate constants were effectively adjusted to achieve that. Dyne and Kennedy calculated the scavenger effect in reducing molecular yields, but the result agreed with experiment no better than Schwarz's (1955) calculation. When applied to cylindrical α tracks, the best agreement was found using 20 eV per radical pair.

Kuppermann and Belford (1962b) also performed 2-radical calculations using a reaction scheme and numerical procedure similar to Dyne and Kennedy; however, their stated purpose was not the obtaining of agreement with experiment but to progressively develop the quantitative aspects of a multiradical diffusion model. On the basis of their calculation, Kuppermann and Belford tend to favor the Samuel-Magee initial distribution rather the Lea-Gray-Platzman form. The latter would give excessive suppression of molecular yields at a relatively small scavenger concentration.

7.2.1 KUPPERMANN'S MODEL

Discovery of the hydrated electron and pulse-radiolytic measurement of specific rates (giving generally different values for different reactions) necessitated consideration of multiradical diffusion models, for which the pioneering efforts were made by Kuppermann (1967) and by Schwarz (1969). In Kuppermann's model, there are seven reactive species. The four primary radicals are e_h, H, H_3O^+, and OH. Two secondary species, OH^- and H_2O_2, are products of primary reactions while these themselves undergo various secondary reactions. The seventh species, the O atom was included for material balance as suggested by Allen (1964). However, since its initial yield is taken to be only 4% of the ionization yield, its involvement is not evident in the calculation.

The reaction scheme (Kuppermann, 1967, Table 1) contains three scavenging reactions, one each for H, OH, and e_h. There specific rates are nominally taken as 6×10^9 $M^{-1}s^{-1}$ but can be altered in a given situation. There are seven primary reactions generating H_2, H_2O_2 (only by the OH + OH reaction), OH^-, and H_2O with a certain quantity of H atoms formed by the reaction $e_h + H_3O^+$; of the three secondary reactions, two are between H_2O_2 and H and e_h and the third is the neutralization reaction OH^- and H_3O^+. The rates of these reactions are obtained from pulse radiolysis while all reactions involving O atoms are assigned a specific rate 6×10^9 $M^{-1}s^{-1}$. Of the diffusion coefficients, only that of $e_h = 4.75 \times 10^{-5}$ cm^2s^{-1} is available from experiment (Schmidt and Buck, 1966). The values for others were taken as follows: (1) those for OH, OH^-, and O were equated to $D(H_2O)$, the self-diffusion coefficient of water—namely, 2×10^{-5} cm^2s^{-1}; (2) that of H_2O_2 was taken as $D(H_2O)/2^{1/2}$ in approximate mass consideration; (3) that of H was taken from the experimental value of $D(H_2) = 3.65 \times 10^{-5}$ cm^2s^{-1} multiplied by the ratio of sizes of H_2 and H according to the Stokes–Einstein relation; and (4) that of H_3O^+ was computed to be 1×10^{-4} cm^2s^{-1} using a proton jump mechanism.

Physical arguments regarding the formation of primary radicals led the investigators to take two initial size parameters for the gaussian distributions: one for e_h and one for the rest, here called r_2 and r_1, respectively. These values and the initial (i.e., at $\sim 10^{-11}$ s) number of radicals in the spur were adjusted to give best agreement between calculation and experiment for ^{60}Co-γ radiolysis. Thus, for an "average spur" Kuppermann obtains initially

$r_1 = 6.25$ Å,[1] $r_2 = 18.75$ Å, $N(OH) = 2.10$, $N(e_h) = N(H_3O^+) = 2.08$, $N(OH) = 0.18$, and $N(O) = 0.08$, giving a sum total of 6.52 radicals. After this, no additional adjustment was made, and the calculated variations of the yields with LET and scavenger concentration were to be viewed as predictions to be compared with experiment.

At first, Kuppermann calculates *relative* yields for isolated spurs and tracks having same the initial radii but different linear radical densities. These are then combined for a given quality of radiation upon weighing the results of spurs and tracks according to the energy deposition spectrum of Burch (1957). Finally, the relative yields are converted to absolute G values by normalization, requiring that $G(OH) = 2.22$ for γ-radiolysis. Draganic' and Draganic' (1971) point out that this value of $G(OH)$ is now considered too low. Figure 7.3 shows the variations of radical and molecular yields in water radiolysis as functions of track-averaged $\langle LET \rangle$ according to Kuppermann (1967), which agree well with the experiments of Schwarz *et al.* (1959).

Kuppermann's calculation of the reduction of molecular yield by scavenger reactions generally follows the pattern of Figure 7.2. Except at high scavenger concentration, the results for different scavenging reactions can be made to fall on a universal curve with concentration scaled by relative rate constants of scavenging. For this purpose, Kuppermann uses the absolute rate constant of the $e_h + NO_2^-$ reaction as given experimentally—that is, 4.6×10^9 M^{-1}s^{-1}.

For the reduction of H_2 yield by radical scavenging, the calculations give good agreement with experiment till ~0.3 M concentration, beyond which theory gives a decrease faster than that observed experimentally. Kuppermann offers two explanations. The first relates to the variation of the initial radical density with spur size, which cannot be verified independently. The other refers to an alternative source of H_2 without a free radical precursor. Kuppermann suggests that the latter could be a first-order process, also quenched by radical scavengers though less efficiently. It should also be pointed out that at high scavenger concentration there is always the possibility of direct effects that cannot be accounted for in the diffusion model.

Another finding of the Kuppermann model is the increase of R_2 yield with the scavenging of the *other* kind of radical R'. The explanation is that the scavenger interferes with the $R + R'$ reaction. Thus, some of the R radicals that would have otherwise re-formed water are now available to form R_2. Draganic' and Draganic' (1969) reported an increase in the yield of H_2O_2 with the increase of e_h scavenger concentration. At low and intermediate concentrations, this increase follows approximately Kuppermann's calculation. At high concentration, the calculated increase is much greater than observed experimentally.

---

[1]Kuppermann writes the initial distribution as ~$\exp(-r^2/2r_{jj}^2)$. Therefore, for the *same* distribution, his r_i value is smaller than that of the standard form by the factor $2^{1/2} = 1.414$ (see Eq. 7.2).

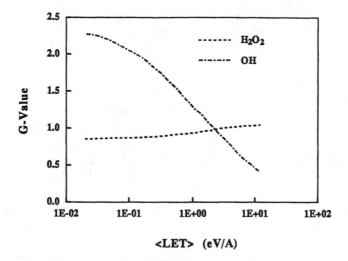

FIGURE 7.3   Variation of H_2O_2 molecular yield and OH radical yield with track-averaged $\langle LET \rangle$ according to Kupperman's (1967) calculation. Generally, the experimental values lie somewhat lower than calculated.

Kuppermann vigorously defends the diffusion model of water radiolysis. According to him, a model based on second-order homogeneous kinetics would be inconsistent with experiments because (1) a very small solute concentration would eliminate molecular yields; (2) at a given solute concentration, the dose-rate effect would be orders of magnitude greater than what is observed; and (3) the calculated curvature of the molecular yield vs. scavenger concentration curve would be much greater than observed or given by diffusion kinetics. To these, one may add that no significant LET dependence of molecular yield is expected in a homogeneous model.

7.2.2   SCHWARZ'S MODEL

Schwarz's model is a multiradical extension of the Ganguly–Magee model with some additional improvements, to be described later. Schwarz assumes that initially—that is, $\sim 10^{-11}$ s after the act of energy deposition in water—there appear five species, namely e_h, H, OH, H_3O^+, and H_2. Their initial yields, indicated by superscript zero, are related by charge conservation and material balance. Thus, there are three independent initial yields, taken to be those of e_h, H, and H_2. The initial yield of H_2 is identified with the unscavengable molecular hydrogen yield. No mechanism of its production is speculated, except that it is *not* formed by radical recombination. For the gaussian distribution of the radicals, two initial size parameters are needed: one for the hydrated electron, and one for the rest symbolized by OH. Following Samuel and Magee (1953), the initial spur-size parameter is taken to be proportional to the cube root of the number N of molecules dissociated in the spur. That is,

$$r_{OH} = r_{OH}{}^0 N^{1/3},$$

where $r_{OH}{}^0$ is the size parameter of the unit OH spur. The electron spur size is assumed to be $(r_e^{02} + r_{OH}{}^2 - r_{OH}{}^{02})^{1/2}$ where r_e^0 is the size parameter of the unit electron spur. Thus, the model has five adjustable parameters—three initial yields and two unit sizes. These are obtained by requiring agreement between calculation and experiment for some standard experiments, to be discussed later. Rate constants are taken from pulse radiolysis experiments, and diffusion coefficients are either taken from experiment or calculated. The diffusion coefficient of H, while high, is shown not to be a critical parameter.

The reaction scheme of Schwarz, with the specific rates, is shown in Table 7.1. Comparison with later compilations (Anbar *et al.*, 1973, 1975; Farhataziz and Ross, 1977) indicates that most of these rates are reasonable within the bounds of experimental error. Some of the rates are pH-dependent, and when both reactants are charged there is a pronounced ionic strength effect; these have been corrected for by Schwarz. He further notes that the second-order rates are not accurate for times less than ~1 ns if the reaction radius

TABLE 7.1   Reaction Scheme in Schwarz's Diffusion Model

Reaction[a]	Specific rate[b]
$e_h + e_h \rightarrow H_2 + 2OH^-$	0.55[c]
$e_h + H \rightarrow H_2 + OH^-$	2.5
$e_h + H_3O^+ \rightarrow H$	1.7[d]
$e_h + OH \rightarrow OH^-$	2.5
$e_h + H_2O_2 \rightarrow OH + OH^-$	1.3
$H + H \rightarrow H_2$	1.0[c]
$H + OH \rightarrow H_2O$	2.0
$H + H_2O_2 \rightarrow H_2O + OH$	0.01
$H_3O^+ \rightarrow OH + 2H_2O$	10[d]
$OH + OH \rightarrow 2H_2O_2$	0.6[c]

[a]In addition, there are scavenging reactions for the primary species as appropriate to the solution.

[b]In units of 10^{10} M^{-1}s^{-1}.

[c]Rate of actual reaction. The reactant species disappear at twice this rate.

[d]Corrected by a factor 0.75 for ionic strength effect in spur.

is on the order of a few angstroms and the diffusion coefficients are ~10^{-5} cm^{2}s^{-1}. A time-dependent specific rate, actually valid for diffusion-controlled reactions, is used:

$$k(t) = k_d\left(1 + \frac{k_d}{4\pi^{1/2}D^{3/2}t^{1/2}}\right) \qquad (t > 1 \text{ ps}),$$

Here k_d is the infinite-time specific rate, and D is the mutual diffusion coefficient of the reactants. Using the time-dependent specific rates, Schwarz reports an increase of molecular yields that is ~2% at low concentration, and ~5% at high concentration of solutes.

The diffusion coefficients for e_h, H_3O^+, OH_{aq}^-, H, OH, and H_2O_2, in units of 10^{-5} cm^{2}s^{-1}, are taken respectively as 4.5, 9.0, 5.0, 7.0, 2.8, and 2.2. Of these, the first three are for charged species taken from experiment. D_{OH} is taken the same as for self-diffusion of water. $D_{H_2O_2}$ is derived from self-diffusion of water using Stokes law to correct for the size effect. D_{OH} is obtained from the diffusion of He.

The set of equations considered by Schwarz is the same as Eq. (7.10) except that a radical is scavenged by a single solute, although a single solute may scavenge more than one species. Seven simultaneous equations are considered—five for the primary species, and two for the secondary species OH_{aq}^- and H_2O_2. In the spirit of the Ganguly-Magee model, taking a track as a collection of n spurs one obtains, by integrating Eq. (7.1),

$$n\frac{dN_i}{dt} = -\sum_j k_{ij}\int_0^\infty c_i c_j \, dv + \sum_{j,k\neq i} k_{jk}\int_0^\infty c_j c_k \, dv - nk_s c_s N_i, \qquad (7.17)$$

where $k_s \equiv k_{si}$ and N_i is the number of species i extant in the spur at time t. Taking the track as a "string of beads," the concentration of the ith species at a point is given by

$$c_i = \pi^{-3/2}b_i^{-3}\exp\left(-\frac{r^2}{b_i^2}\right)\sum_{m=1}^n N_{im}\exp\left[-\frac{(z-z_m)^2}{b_i^2}\right], \qquad (7.18)$$

where z_m is the location of the mth spur, (r, z) are the cylindrical coordinates of the point with z axis along the track, and b_i is defined as the gaussian width parameter for the ith species. From (7.18), Schwarz obtains

$$\int_0^\infty c_i c_j \, dv = nN_i N_j f_{ij}, \qquad (7.19)$$

where

$$f_{ij} = [\pi(b_i^2 + b_j^2)]^{-3/2}\{1 + [\pi(b_i^2 + b_j^2)]^{1/2}Z\} \qquad (7.20)$$

and Z is the energy-averaged spur density per unit track length, defined as

$$Z = (E_0\varepsilon)^{-1} \int_0^\infty \left(-\frac{dE}{dx}\right) dE.$$

Here E_0 is the initial particle energy, $-dE/dx$ is the stopping power, and ε is the mean energy content of the spur. From Eqs. (7.17) and (7.19), one obtains

$$\frac{dN_i}{dt} = -\sum_j k_{ij} N_i N_j f_{ij} + \sum_{j,k \neq i} k_{jk} N_j N_k f_{jk} - k_s c_s N_i. \qquad (7.21)$$

Note that f_{ij} is time-dependent through the b factors. In the absence of reactions, b^2 varies linearly with time as $b^2 = 2r_0^2 + 4Dt$. Because of bimolecular reactions, it must increase somewhat faster to account for the reduction of central density. Schwarz takes this effect into account by writing $db_i^2/dt = 4D_i + \beta_i(t)$ and devises an approximate procedure to evaluate β by separating the contributions due to different reactions—namely,

$$\beta_i = \sum_j \beta_{ij}.$$

He argues that solute reaction does not contribute to β, but creation of a species by reaction of radicals or a solute does. With this understanding and comparing with numerical calculation for the 2-radical, 1-solute case, he finally suggests

$$\beta_{ij} = \alpha b_i^2 k_{ij} N_j f_{ij} \frac{b_i^2}{b_j^2}. \qquad (7.22)$$

However, b_i/b_j was taken as 1 whenever $b_i > b_j$ in (7.22) to prevent overcorrection. Extensive comparison with Kuppermann's calculation indicated a value of $\alpha = 0.67$, for which the results of the two sets of calculations agreed within 1%. The rest of Schwarz's calculations were performed with $\alpha = 0.67$ by solving Eqs. (7.20)–(7.22) by a "march of steps" method until no further significant reaction occurred. The results so obtained were further averaged over the distribution of spurs in energy according to Mozumder and Magee (1966c). This averaging is implied in what follows.

Nitrite ion in aqueous solution depresses the yields of both H_2 and H_2O_2. The reduction of H_2 is insensitive to parameters other than $G_{H_2}^0$ and r_e^0. These are then fixed by requiring that the calculated variation of $G(H_2)$ with $[NO_2^-]$ agree closely with experiment, giving $r_e^0 = 23$ Å and $G_{H_2}^0 = 0.15$. Without the initial molecular hydrogen yield, $G(H_2)$ would be too low at high NO_2^- concentration. Next, the initial H atom yield was adjusted until good agreement was obtained between theory and experiment for the variation of $G(H)$ with $[OH^-]$, giving $G_H^0 = 0.62$. Comparison of theoretical and experimental curves for the suppression of the H_2O_2 yield by NO_2^- gave $r_{OH}^0 = 7.5$ Å. Finally, $G^0_{e_h}$ was determined to be 4.78 by requiring that the calculated total water decomposition yield $[G(e_h) + G(H) + 2G(H_2)]$ equal the experimental number 4.2 at a low solute concentration.

While fitting five adjustable parameters to four sets of experimental data may not seem surprising, the strength of the diffusion model lies in predicting a much wider body of experimental results. Of these, the most important are the variations of molecular yields with LET and solute concentration. Since these calculated variations agree quite well with experiment, no further comment is necessary except to note that calculations often require normalization, so that only relative yields can be compared with experiment. One main reason is that the absolute yields often differ from laboratory to another for the same experiment. Thus, Schwarz's theoretical predictions have *reasonable normalization constants*, which, however, are *not* considered as new parameters. In the next subsection, we will consider some experimental features that could possibly be in disagreement with the diffusion model.

7.2.2a   Reconciliation of Apparent Contradictions in the Diffusion Model for Water Radiolysis According to Schwarz

Four observation were thought to be in disagreement with the diffusion model: (1) the lack of a proportional relationship between the electron scavenging product and the decrease of H_2 yield; (2) the lack of significant acid effect on the molecular yield of H_2 ; (3) the relative independence from pH of the isotope separation factor for H_2 yield; and (4) the fact that with certain solutes the scavenging curves for H_2 are about the same for neutral and acid solutions. Schwarz's reconciliation follows.

1. Dainton and Logan (1965) found that in the radiolysis of N_2O solution containing a small amount of nitrite to scavenge the OH radicals, the N_2 yield is in excess of twice the reduction of H_2 yield. The discrepancy remains even if two molecules of N_2 are assumed to be formed by electron scavenging, instead of one by the overall reaction $e_h + N_2O \rightarrow N_2 + O^-$. On the other hand, using normalization factors 0.98 and 1.10 respectively for the N_2 and H_2 yields, Schwarz's calculation agreed rather well with both the observed yields over the entire concentration range of N_2O. The explanation lies in the fact that removal of e_h by reaction with OH has four times the reaction radius of the reaction $2e_h \rightarrow H_2$. The effect is accentuated by the narrow distribution of OH compared with that of e_h. Elimination of OH by the nitrite ion makes many more e_h available for reaction with N_2O.

2. In the presence of 10 μM peroxide, the yields of H_2, H_2O_2, and of $H + e_h$ are about the same in neutral and 0.4 M acid solutions. Since H atoms produced by the reaction of acid with hydrated electrons have different reaction rates and sequences of reaction, a much greater difference of the

yields were expected between neutral and acid solutions. Schwarz, however, points out that the so-formed H atoms have a wider distribution, inherited from that of e_h, than that of H atoms produced directly by radiolysis. When this effect was incorporated, theoretical calculations agreed much better with experiment.

3. Anbar and Meyerstein (1966) found that in solutions containing 25% D the overall H/D isotope factor for H_2 was relatively independent of pH in the range 1–10, this factor being between 2.2 and 2.3. Of the three reactions giving H_2, $e_h + e_h$ has the largest isotope factor (4.7), that of the H + e_h is the lowest (1.65), and H + H has intermediate values ranging from 2.7 and 3.4 in neutral and 0.2 M acid solutions, attributable to the two different sources of H atoms. No isotope factor for the initial H_2 yield being available, Schwarz uses the value 1.8 to get agreement with Anbar and Meyerstein's data at neutral pH. Further predictions from the diffusion model are in agreement with observation—that is, calculation shows no great dependence of the isotope separation factor on pH. With increasing acidity, the H + H reaction gradually takes over at the expense of other reactions. This, combined with the intermediate value of the isotope factor for this reaction, explains the observation.

4. Mahlman and Sworski (1967) found that curves for scavenging of H_2 by NO_3^- have nearly the same shape for neutral and 0.1 M acid solutions, over the concentration range 1 mM to 0.4 M of the solute, despite the fact that nitrate reacts much faster with e_h than with H. Schwarz points out, however, that when two solutes (H^+ and NO_3^-) compete for the same intermediate (e_h), the extrapolation to zero scavenger concentration is not a valid procedure. The calculation of the diffusion model agrees with experiment over the entire NO_3^- concentration range. Further, the model predicts a lower H_2 yield at *very* low scavenger concentrations.

7.2.2b Evolution of the Yield of the Hydrated Electron[2]

It has been suggested that a sensitive test of the diffusion model would be found in the evolution of the e_h yield (Schwarz, 1969). Early measurements by Hunt and Thomas (1967) and by Thomas and Bensasson (1967) revealed ~6% decay within the first 10 ns and ~15% decay in 50 ns. The diffusion theory of Schwarz predicts a very substantial decay (~30%) in the first nanoseconds for instantaneous energy deposition. Schwarz (1969) tried to mitigate the situation by first integrating over pulse duration (~4.2 ns) and then over the detector response time (~1.2 ns). This improved the agreement between theory and experiment somewhat, but a hypothesis of no decay *in this time scale* would also agree with experiment. Thus, it was decided that a crucial test of the diffusion theory would

---

[2]It is advisable that the rest of this section be read along with Sect. 6.2.2.

be provided by the evolution of the e_h yield in the subnanosecond region. On the other hand, the vestiges of spur reactions are not over in 10 ns, but continue till ~100 ns. In this longer time scale, Schwarz found excellent agreement between experiment and calculation.

The first subnanosecond experiments on the e_h yield were performed at Toronto (Hunt et al., 1973; Wolff et al., 1973). These were followed by the subnanosecond work of Jonah et al. (1976) and the subpicosecond works of Migus et al. (1987) and of Lu et al. (1989). Summarizing, we may note the following: (1) the initial (~100 ps) yield of the hydrated electron is 4.6±0.2, which, together with the yield of 0.8 for "dry" neutralization, gives the total ionization yield in liquid water as 5.4; (2) there is ~17% decay of the e_h yield at 3 ns, of which about half occurs at 700 ps; and (3) there is a relatively fast decay of the yield between 1 and 10 ns. Of these, items (1) and (3) are consistent with the Schwarz form of the diffusion model, but item (2) is not. In the time scale of 0.1–10 ns, the experimental yield is consistently greater than the calculated value. The subpicosecond experiments corroborated this finding and determined the evolution of the absorption spectrum of the trapped electron as well.

Following Schwarz (1969), we may draw the following conclusions regarding the diffusion model of water radiolysis:

1. The model explains satisfactorily the large part of molecular yields and their variations with solute concentration and LET, *with the probable exception of phenomena at high solute concentration.*
2. The model predicts correctly the evolution of the e_h yield beyond 10 ns.
3. The model predicts a very small initial H atom yield (0.62), smaller than the corresponding yield in the vapor phase. It may be conjectured that some of the incipient H atoms convert to e_h in the liquid phase, but the mechanism is unknown.

7.2.3   OTHER DIFFUSION MODELS

There are several other diffusion kinetic models of water radiolysis designed for specific purposes; some of these will be briefly discussed here. Trumbore et al. (1978) considered an initial radical distribution that included a dose effect of the Linac pulse. Inherent interspur overlap produced a non-gaussian distribution. The calculated results are claimed to agree with the decay of the e_h yield over the range of applicable dose. However, the model has a built-in inconsistency in the limit of zero dose. Girija and Gopinathan (1980, 1981, 1982) consider a gaussian distribution of size parameters of a spur for a given number of dissociations, the meaning of which is not clear. They claim good agreement between their calculation and the evolution of e_h yield, as well as with the yield of molecular products (Gopinathan and Girija, 1983). Yamaguchi (1987, 1989) modified Schwarz's model to include δ-ray effects through a defined restricted energy loss. The model, intended to compute the LET effect on the yield of Fe^{3+} in the Fricke dosimeter, has built-in high- and low-LET components of electron tracks. Differential G values calculated for track segments are assembled to get the integral G values, which are in agreement with available experiments.

Burns and Curtiss (1972) and Burns et al. (1984) have used the Facsimile program developed at AERE, Harwell to obtain a numerical solution of simultaneous partial differential equations of diffusion kinetics (see Eq. 7.1). In this procedure, the changes in the number of reactant species in concentric shells (spherical or cylindrical) by diffusion and reaction are calculated by a march of steps method. A very similar procedure has been adopted by Pimblott and LaVerne (1990; LaVerne and Pimblott, 1991). Later, Pimblott et al. (1996) analyzed carefully the relationship between the electron scavenging yield and the time dependence of e_h yield through the Laplace transform, an idea first suggested by Balkas et al. (1970). These authors corrected for the artifactual effects of the experiments on e_h decay and took into account the more recent data of Chernovitz and Jonah (1988). Their analysis raises the yield of e_h at 100 ps to ~4.8, in conformity with the value of Sumiyoshi et al. (1985). They also conclude that the time dependence of the e_h yield and the yield of electron scavenging conform to each other through Laplace transform, but that neither is predicted correctly by the diffusion-kinetic model of water radiolysis.

7.3 STOCHASTIC KINETICS

The procedures discussed so far take as fundamental variables the species concentration and specific rates, the latter obtained from homogeneous experiments. Such procedures are called deterministic—that is, admitting no fluctuation in the number of reactant species—as opposed to stochastic methods where statistical variation is built in.

The difference is clearly seen for a spur initially containing two dissociations of AB molecules into radicals A and B (Pimblott and Green, 1995). Considering the same reaction radii for the reactions $A + A$, $A + B$, and $B + B$ and the same initial distributions of radicals, the statistical ratio of the products should be $1 : 4 : 1$ for $A_2 : AB : B_2$, since there is one each of A–A and B–B distances but there are four A–B distances. For n dissociations in the spur, this combinatorial ratio is $n(n - 1)/2 : n^2 : n(n - 1)/2$, whereas deterministic kinetics gives this ratio always as $1 : 2 : 1$. Thus, deterministic kinetics seriously underestimates cross-recombination and overestimates molecular products, although the difference tends to diminish for bigger spurs. Since smaller spurs dominate water radiolysis (Pimblott and Mozumder, 1991), many authors stress the importance of stochastic kinetics *in principle*. Stochasticity enters in another form in radiolysis—namely in entire track simulation, when it becomes necessary to consider interspur overlap (Turner et al., 1983; Zaider and Brenner, 1984).

McQuarrie (1967) has considered, via the stochastic method, homogeneous reactions involving a few reactants. His result for second-order pairwise reactions can be applied to a 1-radical spur in a straightforward manner, giving the surviving number of particles as a sum over terms that are products of combinatorial factors and a suitable power of Ω the pair survival function (see Clifford et al., 1982a,b). It appears, however, that stochastic kinetics was first applied to radiation chemistry by Bota'r and Vido'czy (1979, 1984). They used a rudimentary random walk model on a lattice to compute the scavenging reaction probability. No kinetics was calculated, but the result indicated a scavenging reaction probability varying roughly as the cube root of scavenger concentration (see the last paragraph of Sect. 7.1.2).

The Oxford group has vigorously pursued the stochastic kinetic method for spur reactions in water radiolysis (Clifford et al., 1982a, b; 1985, 1986, 1987a,b; Green et al., 1989a,b). After some preliminary studies with random walks on a lattice, these authors proposed three procedures for calculating spur reactions using stochastic kinetic philosophy. These are (1) Monte Carlo simulation, (2) the master equation approach, and (3) the independent reaction time (IRT) model. Here, we will only give a summary of these procedures; details will be found in the original literature and in the review of Pimblott and Green (1995).

7.3.1   MONTE CARLO (MC) SIMULATION

In *Monte Carlo simulation*, the life histories of the diffusing-reacting species are generated by random flight simulation in three dimensions and the probabilities of reaction, diffusion, and separation are obtained by averaging over a large number of trials (typically $\sim 10^4$–10^5). During a time interval δt, the diffusive jump of the particle i is simulated as $\delta \vec{r}_i = (2D_i\,\delta t)^{1/2}\vec{R}(0, 1) + D_i\vec{F}_i(\vec{r}_i)\,\delta t/k_B T$, where D_i is the diffusion coefficient of the ith species having original position $\vec{r}_i$, $\vec{R}(0, 1)$ is a three-dimensional normal random vector of zero mean and unit variance, $\vec{F}_i$ is the deterministic force on that species (e.g., the coulombic force on charged species), and $k_B T$ is the Boltzmann factor at absolute temperature T.

After the jump, the particle is taken to have reacted with a given probability if its distance from another particle is within the reaction radius. For fully diffusion-controlled reactions, this probability is unity; for partially diffusion-controlled reactions, this reaction probability has to be consistent with the specific rate by a defined procedure. The probability that the particle may have reacted while executing the jump is approximated for binary encounters by a *Brownian bridge*—that is, it is assumed to be given by $\exp[-(x - a)(y - a)/D'\,\delta t]$, where a is the reaction radius, x and y are the interparticle separations before and after the jump, and D' is the mutual diffusion coefficient of the reactants. After all particle jumps have been executed in time δt, the procedure is repeated for successive intervals of time until either all relevant reactions have taken place or a sufficiently long time has elapsed to terminate the trial.

In principle, Monte Carlo simulation can generate a huge volume of output data, giving details of diffusion and reaction, some of which are not experimentally accessible. It is computationally intensive and is vulnerable to time discretization errors, calling for validation by comparison, for example, with the exact solution for a geminate pair (Green *et al.*, 1989). Therefore, its use is recommended only for spurs containing very few reactants or for validating other stochastic methods in special cases.

7.3.2   THE MASTER EQUATION (ME) APPROACH

The *master equation approach* considers the *state* of a spur at a given time to be composed of N_i particles of species i. While N_i is a random variable with given upper and lower limits, transitions between states are mediated by binary reaction rates, which may be obtained from bimolecular diffusion theory (Clifford *et al.*, 1987a,b; Green *et al.*, 1989a,b, 1991; Pimblott *et al.*, 1991). For a 1-radical spur initially with N_0 radicals, the probability P_N that it will contain N radicals at time t satisfies the master equation (Clifford *et al.*, 1982a)

$$\frac{dP_N}{dt} = -\frac{1}{2}\mu(t)N(N-1)P_N(t) + \frac{1}{2}\mu(t)(N+1)(N+2)P_{N+2}, \qquad (7.23)$$

where $\mu(t)$ is the (time-dependent) probability of reaction per unit time and the right-hand side of Eq. (7.23) considers the decay of P_N by the transition $N \rightarrow N - 2$ and its regeneration by the transition $N + 2 \rightarrow N$ with respective numbers of pairs involved. From (7.23), the mean number of radicals

$$\langle N \rangle = \sum_{N=2}^{N_0} NP_N$$

satisfies

$$\frac{d\langle N \rangle}{dt} = -\frac{1}{2}\mu(t)\langle N(N-1) \rangle. \qquad (7.24)$$

Equation (7.24) can be compared with the corresponding prescribed-diffusion equation, namely $d\langle N \rangle/dt = -(1/2)\mu(t)\langle N \rangle(\langle N \rangle - 1)$ (Clifford *et al.*, 1982a). These two equations would be equivalent if $\langle N^2 \rangle = \langle N \rangle^2$—that is, the variance of N would be zero. This implies that all spurs would have exactly the same number of radicals at a given time. Since stochasticity denies this, a considerable difference is expected between the results of these two methods; however, this difference tends to decrease with the spur size N_0 (Clifford *et al.*, 1982a; Pimblott and Green, 1995).

The master equation methodology can be readily generalized to multiradical spurs, but it is not easy to include the reactions of reactive products (Green *et al.*, 1989; Pimblott and Green, 1995). This approach is therefore limited to spur reactions where the reaction scheme is relatively simple.

7.3.3   THE INDEPENDENT REACTION TIME MODEL

The *independent reaction time (IRT) model* was introduced as a shortcut Monte Carlo simulation of pairwise reaction times without explicit reference to diffusive trajectories (Clifford *et al.*, 1982b). At first, the initial positions of the reactive species (any number and kind) are simulated by convolving from a given (usually gaussian) distribution using random numbers. These are examined for immediate reaction—that is, whether any interparticle separation is within the respective reaction radius. If so, such particles are removed and the reactions are recorded as static reactions.

For the surviving particles, the interparticle separations are used to generate random reaction times by inverting a *first passage time* problem. That is, we seek to answer this question: given a reaction probability W chosen by a uniformly distributed random number between 0 and 1, at what time was the reactive configuration first reached? The reaction probability $W = W(r, a; t)$ depends on the initial separation r, reaction radius a, and time t, and it may involve other factors, like the probability of reaction at an encounter for partially diffusion-controlled reactions and an interparticle field such as the coulombic field for ionic reactions. In any case, W is a monotonically increasing function of t. Therefore, given its form and the value of W, it is relatively straightforward to invert it numerically to obtain the reaction time t (Clifford *et al.*, 1985; Green *et al.*, 1987).

Great simplification is achieved by introducing the hypothesis of *independent reaction times (IRT)*: that the pairwise reaction times evolve independently of any other reactions. While the fundamental justification of IRT may not be immediately obvious, one notices its similarity with the molecular pair model of homogeneous diffusion-mediated reactions (Noyes, 1961; Green, 1984). The usefulness of the IRT model depends on the availability of a suitable reaction probability function $W(r, a; t)$. For a pair of neutral particles undergoing fully diffusion-controlled reactions, W is given by $(a/r)\,\mathrm{erfc}[(r - a)/2(D't)^{1/2}]$ where D' is the mutual diffusion coefficient and erfc is the complement of the error function.

Green and Pimblott (1989) have extended the IRT model to partially diffusion-controlled reactions between neutrals. They derive an analytical expression that involves an additional parameter, namely the reaction velocity at encounter. For reactions between charged species, W generally cannot be given analytically but must be obtained numerically. Furthermore, numerical inversion to get t then becomes computationally expensive. However, for spur reactions in water radiolysis, most reactions are either between neutrals or between ions in a high-permittivity medium, for which a good approximation to W can be obtained by comparing with W for neutrals and using a scaled distance (see Pimblott and Green, 1995). With a suitable form of W, a table is drawn for the first independent reaction times for all possible pairs of radicals. Of these, the smallest is chosen to have occurred, and all pairs involving any of the reactants are removed (see the next subsection). IRT simulation is now continued with the remaining radicals, and the procedure repeated until all species have reacted or a defined terminal time has been reached to collect the unreacted radicals as escaped.

7.3.4 REACTIVE PRODUCTS

We have already commented that the master equation method is not suitable, at present, to handle reactive products because, *inter alia*, the dimensionality of the problem increases with reactions of products. There is no difficulty, in principle, to including reactive products in Monte Carlo simulation, since the time of reaction and the positions of the products can be recorded. In practice, however, this requires a greatly expanded computational effort, which is discouraging.

In the IRT model, reactions of products can be incorporated indirectly and approximately by one of the following procedures (Green *et al.*, 1987): (1) the diffusion approach, (2) the time approach, or (3) the position approach. The *diffusion approach* is conceptually the simplest. In it, the fundamental entity is the interparticle distance, which evolves by diffusion independently of other such distances along with IRT. Thus, if the interparticle distance was $\bar{r}$ at $t = 0$, that at time t is simulated as $\bar{r}' = \bar{r} + R_3$, where R_3 is a three-dimensional normally distributed random number of zero mean and variance $2D't$. When reaction occurs at t, the product inherits the position of one of the parents taken at random. The procedure is then repeated with new interparticle distances so obtained.

The *time approach* relies entirely on independent diffusion-reaction time without reference to distances. The reaction product inherits the time sequence of one of the parents chosen at random; however, its *residual time* to react with another species is scaled inversely relative to *its mutual diffusion coefficient*. A heuristic correction is also made for the change of reaction radius (Clifford *et al.*, 1986).

The *position approach* strives to get the positions of the reactive particles explicitly at the reaction time t obtained in the IRT model. While the nonreactive particles are allowed to diffuse freely, the diffusion of the reactive particles is *conditioned* on having a distance between them equal to the reaction radius at the reaction time. Thus, following a fairly complex procedure, the position of the reactive product can be simulated, and its distance from other radicals or products evaluated, to generate a new sequence of independent reaction times (Clifford *et al.*, 1986).

It is clear that the different procedures of handling reactive products are based on different approximations; therefore, somewhat different results are expected. On the whole, since the IRT methodology is based on the conceived independence of pairwise bimolecular reactions, it needs validation by comparison with well-known examples for which Monte Carlo results are available. Such validations have in fact been made (see Figures 4 and 7 in Clifford et al., 1986, and Figures 12, 27, and 28 in Pimblott and Green, 1995).

7.3.5   RESULTS

For a pair of particles, neutral or ionic, very good agreement has been found between random flight Monte Carlo (MC) simulation and exact solution with regard to the time-dependent survival probability; the agreement with the result of IRT simulation is also quite good (Pimblott and Green, 1995). Similarly good agreement has been found between random flight Monte Carlo and IRT methods for the time-dependent survival and reaction probabilities for a 2-pair ionic spur in a medium of high permittivity (Green et al., 1989a,b).

Figure 7.4 compares the results of MC and IRT simulations for a spur initially containing two pairs of neutral radicals (A, B). The initial distributions of both radicals are gaussian with standard deviation 1 nm, all reaction radii are assumed to be 1 nm, and both diffusion constants are taken as 5×10^{-5} cm^{2}s^{-1}. From the decay of A, growth of AA, and regeneration of AB, it is apparent that the agreement is very good, the difference being barely perceptible at the longest times. Figure 7.5 shows the same situation, except that reaction of products is included with the same reaction radius. Comparing the growth of AAB obtained by MC and IRT, including different approximations for the reactions of products, it is seen that the time approach comes closest to MC simulation. The diffusion and position approaches consistently overestimate product reactions, although not excessively.

Clifford et al. (1987a,b) considered acid spurs (primary radicals H and OH) and computed the evolution of radical and molecular products by the master equation (ME) and IRT methods. *Reasonable* values were assumed for initial yields, diffusion constants, and rate constants, and a distribution of spur size was included. To be consistent with experimental yields at 100 ns, however, they found it necessary that the spur radius be small—for example, the radius of H distribution (standard deviation in a gaussian distribution) for a spur of one dissociation was only in the 0.4–0.75 nm range. Since in acid spurs H atoms inherit the distribution of e_h, this is considered too low. This preliminary finding has later been revised in favor of spurs of much greater radius.

7.3.6   OTHER STOCHASTIC MODELS

Zaider and Brenner (1984) have developed computer code for fast chemical reactions on electron tracks; Zaider et al. (1983) have performed MC simulation of

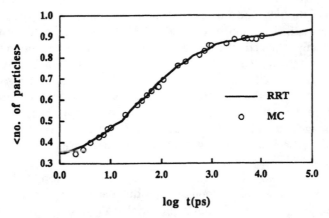

FIGURE 7.4   Comparison of Monte Carlo (MC) and independent reaction time (RRT) simulations with respect to the product *AB* for a spur of two neutral radical-pairs. See text for explanation. From Clifford *et al* (1986), by permission of The Royal Society of Chemistry©

proton tracks. In these calculations, entire tracks are simulated with given or assumed elastic and inelastic cross sections of incident charged particles. Thus, both spur structure and proximity functions of radicals in the spur are evaluated. Zaider and Brenner find that the initial radical distribution in the spur is not well represented by a gaussian. In any case, they perform MC simulations in

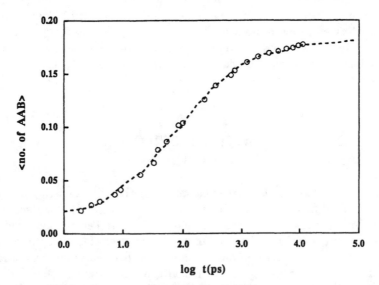

FIGURE 7.5   Same as in Figure 7.4, where reaction of products is included with equal reaction radius. MC and RRT simulations (time approach) for the product *AAB*. From Clifford *et al.* (1986), by permission of The Royal Society of Chemistry©.

the spirit of the IRT model using the diffusion constants, rate coefficients, and initial spur radii as given by Short *et al.* (1981). In this sense, the results are *not* parametrically adjusted to agree with experiments. However, taking all radical reactions as diffusion-controlled, which they do, is questionable. Also, it is not clear how they treat the reactive products. For the time dependence of e_h decay, their calculated G values are consistently and significantly lower than experiment, which they attribute to the use of transport code for the gas phase. One is therefore urged to compare the shapes of the decay curves, not the absolute values. On the other hand, their calculation agrees fairly well with experiment for the decay of OH.

Turner *et al.* (1983, 1988) have simulated entire electron tracks and calculated the yields of radical and molecular products as functions of primary energy. The methodology here is random flight simulation starting with a discussion of events from energy deposition ($\sim 10^{-15}$ s) to spur formation ($\sim 10^{-11}$ s), and continuing to the end of spur reactions ($\sim 10^{-6}$ s). The elastic scattering cross section is taken from the gas phase as in most similar work, simply because the liquid phase elastic cross section is unavailable. Reaction radii are obtained from measured reaction rates using the Debye–Smoluchowski equation for the fully diffusion-controlled rate. For partially diffusion-controlled reactions, such as the OH + OH and e_h + e_h reactions, a probability of reaction at encounter is evaluated by comparison with measured rate. This is *not* a correct procedure. In any case, these authors obtain reaction probabilities of 1/3 and 1/6, respectively, for the two reactions and use these values in their calculations. Transient G values of radicals are claimed to be consistent with experiment. With these G values, they compute $G(Fe^{3+})$ in the aerated Fricke dosimeter to be 12.9 and 12.1 per 100 eV for 5- and 1-KeV electrons, respectively.

A great deal of work has been done by Hummel and co-workers on electron-ion recombination in hydrocarbon liquids using random flight MC simulation. This will be discussed in Sect. 7.5.

7.4   COMPARISON WITH EXPERIMENT: THE CASE OF WATER RADIOLYSIS

Several authors have made restricted comparisons between experiment and calculations of diffusion theory. Thus, Turner *et al.* (1983, 1988) considered $G(Fe^{3+})$ in the Fricke dosimeter as a function of electron energy, and Zaider and Brenner (1984) dealt with the *shape* of the decay curve of e_h (*vide supra*). These comparisons are not very rigorous, since many other determining experiments were left out. Subsequently, more critical examinations have been made by LaVerne and Pimblott (1991), Pimblott and Green (1995), Pimblott *et al.* (1996), and Pimblott and LaVerne (1997). These authors have compared their calculations with a *wide* spectrum of experimental results. Some of their findings are briefly reported in what follows; the interested reader is referred to the original references for details.

LaVerne and Pimblott (1991) basically use the reaction scheme of Schwarz (1969) with somewhat different rate constants for some of the reactions in view of later measurements(see Table 7.1). Thus, the rates of reaction of e_h with H^+, OH, and H_2O_2 are taken respectively as 2.3, 3.0, and 1.1×10^{10} $M^{-1}s^{-1}$; those of $H_3O^+ + OH^- \rightarrow 2H_2O$, $H + H_2O_2 \rightarrow H_2O + OH$, and $OH + OH \rightarrow H_2O_2$ as 1.4×10^{11}, 9.0×10^7, and 5.5×10^9 $M^{-1}s^{-1}$, respectively. Rates of other reactions remain the same, whereas the three reactions of O_2 with e_h, H, and H_3O^+ are incorporated with specific rates 1.9, 2.1, and 3.8×10^{10} $M^{-1}s^{-1}$, respectively. In this work, deterministic diffusion kinetic methodology is used in the form of Facsimile code (Burns et al., 1984). Initial yields of e_h (4.78) and OH (5.50) are taken from shortest-time measurement, and the radii of their initial distributions (cf. standard deviation of a gaussian distribution) are taken respectively as 2.23 and 0.85 nm. The diffusion constants of the species are either taken from experiment for the charged species or are rationalized. LaVerne and Pimblott (1991) found good agreement between their calculation and experiment for the scavenging of e_h, OH, H_2, and H_2O_2 by various scavengers using a spur energy of 62.5 eV, which gave the best agreement. The experimental results on e_h scavenging may be expressed empirically, by comparison with electron-ion recombination in liquid hydrocarbons, as

$$G(s) = G_{esc} + (G_0 - G_{esc})F(s), \qquad (7.25)$$

where G_0 is the initial yield (4.80), G_{esc} is the long-time or escape yield (2.56), and $F(s)$ is an analytical function involving the scavenger concentration c_s.

Early work of Balkas et al. (1970) suggested $F(s) = (\alpha c_s)^{1/2}/[1 + (\alpha c_s)^{1/2}]$, where α is a fitted constant. LaVerne and Pimblott find better agreement with $F(s) = [(\alpha c_s)^{1/2} + \alpha c_s/2]/[1 + (\alpha c_s)^{1/2} + \alpha c_s/2]$ where, with the known rate constant of scavenging, the value of α implied a decay time of e_h in pure water of 2.77 ns. However, the inverse Laplace transform (ILT) of these forms of $F(s)$, which can be found analytically, did not agree with the real-time decay of e_h in pure water (Jonah et al., 1976; Chernovitz and Jonah, 1988), as it was expected to do. The discrepancy was traced by Pimblott et al. (1996) as originating from an incorrect value of the reaction rate of e_h with methyl chloride (1.1×10^9 $M^{-1}s^{-1}$), against which all other scavenging rates were standardized. When this rate was changed to its correct value of 4.7×10^8 $M^{-1}s^{-1}$ (Schmidt et al., 1995), the problem was resolved and the scavenging yield and the time dependence of e_h yield in the pure solvent could be related by the Laplace transform method. Also within the context of diffusion kinetics, the calculated stationary yields of radical and molecular products agreed well with experiment. The time-dependent growth of H_2 and H_2O_2 were evaluated, but there were no experiments for comparison.

Figure 7.6 shows the effect of a hydrated electron scavenger on the scavenging yield. The abscissa is the product of scavenger concentration and the specific rate of scavenging. The solid line is a calculated result according to LaVerne and Pimblott (1991) and the dashed line represents experimental results compiled by LaVerne and Pimblott (1991) and by Pimblott and LaVerne (1994). The agreement is quite satisfactory. Figure 7.7, from Pimblott and Green (1995), compares calculated results of LaVerne and Pimblott (1991) with experiments for the suppression of the yield of molecular hydrogen by scavenging of e_h. Again, the agreement is reasonable.

Recently, Pimblott and LaVerne (1997) have used stochastic kinetics with the IRT methodology to calculate the time dependence of the e_h yield and the scavenging capacity dependence of the yields of e_h, OH, H_2, and H_2O_2. Both sets of results agreed quite well with experiments. A distribution of spur size was included. However, for best agreement with experiment, the gaussian size parameter (standard deviation) for the initial e_h distribution turns out to be 4.0 nm, which is considerably larger than what was found by earlier deterministic method (Schwarz, 1969). The size for the initial distribution of other radicals was found to be the same as in deterministic studies—namely, 0.75 nm. Comparing stochastic and deterministic kinetics with experimental results, the following conclusions can be drawn:

1. Deterministic methods are easier to visualize and simpler to implement. However, these represent poor physical and statistical pictures. The spur energy (62.5 eV) needed for best agreement between deterministic calculation and experiment is *not* supported by the mean energy loss of high-energy electrons in water-producing spurs, which is somewhat less than 40 eV (Pimblott *et al.*, 1990).

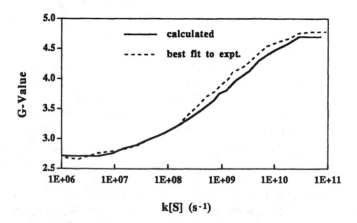

FIGURE 7.6   Effect of hydrated electron scavenger on the scavenging yield. See text for explanation. Reproduced from Pimblott and Green,1995, with permission of Elsevier©.

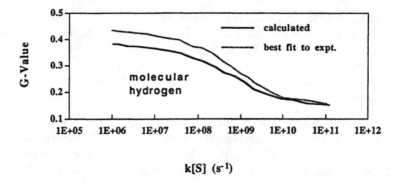

FIGURE 7.7 Suppression of molecular hydrogen yield by scavenging of the hydrated electron. See text for explanation. Reproduced from Pimblott and Green,1995, with permission of Elsevier©.

2. Stochastic kinetics requires details of individual particle reactions. It is computer-intensive and produces a huge volume of output. In this sense, it is overparameterized. However, stochastic kinetics can be made consistent with the statistics of energy deposition and reaction.

3. Using deterministic kinetics, one can force-fit the time evolution of one species—for example, e_h; but then those of other yields (e.g., OH) will be inconsistent. Stochastic kinetics can predict the evolutions of radicals correctly and relate these to scavenging yields via Laplace transforms.

7.5 KINETICS OF ELECTRON-ION RECOMBINATION IN IRRADIATED DIELECTRIC LIQUIDS[3]

7.5.1 GENERAL CONSIDERATIONS: THE GEMINATE PAIR

In the kinetics of electron–ion recombination in liquid hydrocarbons and in the associated problem of free-ion yield in media of relatively low electron mobility ($\mu_e \leq 10$ cm^{2}v^{-1}s^{-1}), the *geminate ion pair* has a dominant role. Originally, the term geminate referred to electron–ion or radical pairs formed out of the same molecule; later use has generalized it to any pairwise reaction in isolation. Although the physics of energy deposition in hydrocarbons indicates that a substantial fraction of energy (ca. 25%) is found in nongeminate situations

[3]The subject matter of this section is closely related to that of Sects. 9.2 and 9.3. It would be profitable to read these sections together.

(Mozumder and Magee, 1967), the success of the geminate pair model in kinetics (Warman *et al.*, 1969; Rzad *et al.*, 1970) and in free-ion yield determination (Hummel and Allen, 1966; Freeman, 1963a, b) has been rationalized on the basis that in multi-ion pair spurs all but the final *e*-ion pair would quickly neutralize due to intense internal coulombic interaction. Therefore, emphasis is laid in this section on the geminate pair with two caveats:

1. We are dealing here with low-LET irradiation. There is definite evidence of multiple ion-pair involvement at higher LETs (Mozumder and Magee, 1967; Holroyd and Sham, 1985; Bartczak and Hummel, 1997).
2. We are *not* dealing here with media of high electron mobility (≥ 100 $cm^2v^{-1}s^{-1}$). In such cases, the ionizations cannot be considered to be isolated even at the minimum LET (see Sect. 9.6).

7.5.2   LAPLACE TRANSFORMATION AND SCAVENGING

Recombination of the electron-ion geminate pair can be intercepted by reaction of either species with a homogeneously distributed solute, called the scavenger. Most experiments use electron scavenging, the probability of which increases with the solute concentration and, of course, with the specific rate of scavenging. Monchick (1956) first drew attention to a *Laplace transform* relationship between the probability of scavenging and the evolution of the recombination process in the pure liquid—that is, in the absence of scavenger (*vide infra*). Its corollary, emphasized by experimentalists, is equally valid. That is, given the scavenging probability as a function of scavenger concentration, one can get the *relative* kinetics of the *e*-ion recombination process in the neat liquid by *inverse Laplace transform (ILT)*. Absolute kinetics can then be evaluated if the specific scavenging rate is known. However, it should be stressed that the scavenging probability must be available over the entire range of scavenger concentration, zero to infinity, to obtain the ILT. Therefore, analytical extension of the scavenging function is needed in the experimentally inaccessible region.

Warman *et al.* (1969) used alkyl halides (CH_3Cl, CH_3Br, C_2H_5Br) in the concentration range 10^{-4} to 0.5 M to scavenge electrons produced by γ-radiolysis in cyclohexane solvent. The products were identified as results of dissociative electron capture–$e + RX \rightarrow R\bullet + X^-$, where $R\bullet$ is an alkyl radical and X is a halogen atom. The scavenging yield (the G value for 100-eV energy deposition in the solution) was shown to follow quantitatively the so-called WAS equation, namely

$$G(s) = G_{fi} + \frac{G_{gi}(\alpha c_s)^{1/2}}{1 + (\alpha c_s)^{1/2}},$$ (7.26)

where α is an adjustable constant specific to the scavenger, c_s is the scavenger concentration, G_{fi} is the extrapolated yield at zero scavenger concentration identified as the free-ion yield (see Chapter 9), and $G_{gi} + G_{fi}$ is the total ionization yield. The last is an extrapolated value at $c_s \to \infty$, and, by inference, G_{gi} is the yield destined for geminate recombination.

The WAS form of scavenging equation has found wide application not only with various combinations of hydrocarbon liquids and scavengers, but also in aqueous radiation chemistry (Balkas *et al.*, 1970). In the latter case, LaVerne and Pimblott (1991) improved the functional form somewhat without entailing any additional adjustable parameters [see the discussion of Eq. (7.25)]. For liquid n-hexane, cyclohexane, and isooctane, van den Ende *et al.* (1984) claim a better agreement with experiment by changing the power of c_s from 0.5 to ~0.6, but the difference is not great except at very low concentrations of scavenger (see also Bartczak and Hummel, 1997). In any case, the *square root dependence* has a sound theoretical basis in diffusion theory (Mozumder, 1971).

Rzad *et al.* (1970) obtained the relative lifetime distribution of electron-ion recombination in cyclohexane by ILI of Eq. (7.26). Denoting the probability that the lifetime would be between t and $t + dt$ as $f(t)\,dt$, the thusly defined scavenging function at scavenger concentration c_s is given by

$$F(c_s) = \int_0^\infty [1 - \exp(-k_s c_s t)]\, f(t)\, dt = 1 - \int_0^\infty \exp(-k_s c_s t)\, f(t)\, dt,$$

where k_s is the specific rate of scavenging. Identifying $F(c_s)$ with $[G(s) - G_{fi}]/G_{gi} = (\alpha c_s)^{1/2}/[1 + (\alpha c_s)^{1/2}]$ (see Eq. 7.26), one then gets

$$\int_0^\infty \exp(-k_s c_s t)\, f(t)\, dt = 1 - F(c_s) = [1 + (\alpha c_s)^{1/2}]^{-1}.$$

On inverse Laplace transformation, $f(t)$ is now given by

$$f(t) = \frac{k_s}{\alpha}\left[\left(\frac{\alpha}{\pi k_s t}\right)^{-1/2} - \exp\left(\frac{k_s t}{\alpha}\right)\operatorname{erfc}\left(\frac{k_s t}{\alpha}\right)\right], \tag{7.27}$$

where erfc is the complement of the error function. It is clear that k_s must be proportional to α, since $f(t)$ refers to the pure solvent. Equation (7.27) therefore is sometimes written in relative form, $f(t) = \lambda[(\pi\lambda t)^{-1/2} - \exp(\lambda t)\operatorname{erfc}(\lambda t)^{1/2}]$, where $\lambda = k_s/\alpha$. The probability $F(t)$ that the e–ion pair will remain extant at time t, over and above the escape probability, is then given by $F(t) = \exp(\lambda t)\operatorname{erfc}(\lambda t)^{1/2}$.

To evaluate the absolute kinetics of recombination from Eq. (7.27), it is necessary to obtain a reliable value of k_s in addition to α measured from scavenging experiments. For electron scavenging in cyclohexane by biphenyl, Rzad *et al.* (1970) used $k_s = 3.0 \times 10^{11}$ $M^{-1}s^{-1}$ on the basis of the α value of scavenging

(15 M^{-1}) and the ratios of mobilities of electrons and ions. Theoretical analysis by Mozumder (1971) produced a higher value, 1.0×10^{12} M^{-1}s^{-1}. Later experiments of Beck and Thomas (1972) gave $k_s = (2.2–3.0) \times 10^{12}$ M^{-1}s^{-1}, which is consistent with a recent mobility model (Mozumder, 1995; see Sect. 10.3.3).

Rzad et al.(1970) compared the consequences of the lifetime distribution obtained by ILT method (Eq. 7.27) with the experiment of Thomas et al. (1968) for the decay of biphenylide ion (10–800 ns) after a 10-ns pulse-irradiation of 0.1 M biphenyl solution of cyclohexane. It was necessary to correct for the finite pulse width; also, a factor r was introduced to account for the increase of lifetime on converting the electron to a negative ion. Taking $r = 17$ and $G_{fi} = 0.12$ in consistence with free-ion yield measurement, they obtained rather good agreement between calculated and experimental results. The agreement actually depends on λ /r, rather than separately on λ or r.

The discussion so far has been empirical in the sense that Laplace transform method has been utilized in conjunction with an experimentally determined scavenging function without a theoretical model for the recombination kinetics. A theoretical model will be attempted in the following subsections.

7.5.3   EVALUATION OF DIFFUSION MODELS

Williams (1964) derived the relation $\tau = \varepsilon k_B Tr_0^3/3De^2$, where τ is the recombination time for a geminate e-ion pair at an initial separation of r_0, ε is the dielectric constant of the medium, and the other symbols have their usual meanings. This *r-cubed rule* is based on the use of the Nernst-Einstein relation in a coulombic field with the assumption of instantaneous limiting velocity. Mozumder (1968) criticized the rule, as it connects initial distance and recombination time uniquely without allowance for diffusional broadening and without allowing for an escape probability. Nevertheless, the r-cubed rule was used extensively in earlier studies of geminate ion recombination kinetics.

7.5.3a   Prescribed Diffusion Treatment

In an early attempt, Mozumder (1968) used a prescribed diffusion approach to obtain the e-ion geminate recombination kinetics in the pure solvent. At any time t, the electron distribution function was assumed to be a gaussian corresponding to free diffusion, weighted by another function of t only. The latter function was found by substituting the entire distribution function in the Smoluchowski equation, for which an analytical solution was possible. The result may be expressed by

$$W(r_0, t) = \exp\left\{-a\, \mathrm{erfc}\left[\frac{r_0}{(4Dt)^{1/2}}\right]\right\}, \tag{7.28}$$

where $W(r_0, t)$ is the probability that the pair will remain unrecombined at t starting from an initial separation r_0, $a = r_c/r_0$, and $r_c = e^2/\varepsilon k_B T$ is the distance, called the Onsager distance, at which the coulombic interaction of the pair numerically equals the thermal energy $k_B T$. In the infinite time limit, Eq. (7.28) gives the escape probability as $\exp(-r_c/r_0)$, which agrees *exactly* with the Onsager (1938) formula. However, more accurate numerical analysis of Abell *et al.* (1972) has shown that the prescribed diffusion approximation of Eq. 7.28 only gives a temporal broadening corresponding to free diffusion, thereby underestimating the recombination rate in the important time scale, typically by a factor of 2 or so. The approximation improves greatly in the long-time limit, where it has been usefully employed (Mozumder, 1971).

In the general case, whether the e-ion pair is isolated or not, the probability density $P(\bar{r}, t)$ that an electron will remain extant at time t is given by the Smoluchowski equation

$$\frac{\partial P}{\partial t} = D[\nabla^2 P - (k_B T)^{-1}\nabla \cdot (PF)] - k_s c_s P,\qquad (7.29)$$

where F is the electrostatic field on the electron and c_s is the concentration of the scavenger that reacts with the electron with a specific rate k_s. Writing $P_0 = P \exp(k_s c_s t)$, Eq. (7.29) may be transformed to the diffusion equation in the absence of scavengers—that is,

$$\frac{\partial P_0}{\partial t} = D[\nabla^2 P_0 - (k_B T)^{-1}\nabla \cdot (P_0 F)].\qquad (7.30)$$

Equation (7.30) shows that the fundamental information on recombination kinetics is contained in the solution of the scavenger-free case, from which the recombination kinetics with a scavenger may be obtained via an exponential transformation. The scavenger reaction probability is now given by

$$N(c_s) = k_s c_s \int_0^\infty dt \int_0^\infty P\, dv = k_s c_s \int_0^\infty \exp(-k_s c_s t) F(t)\, dt,\qquad (7.31)$$

where

$$F(t) = \int_0^\infty P_0(\bar{r}, t)\, dv$$

is the survival probability at time t in the *absence* of scavenger. Equation (7.31) establishes the Laplace transform relationship in scavenging.

Mozumder (1971) calculated $F(t)$ by the prescribed diffusion method. For the isolated ion-pair case, the solution appears in (7.28); for the multiple ion-pair case, further approximation was introduced in the nature of mean force acting on an electron, by which the problem was reduced to that of a collection of isolated pairs as the recombination progressed. In all cases, the integrated survival probability in the long-time limit is given by $G_{fi} + At^{-1/2}$, where G_{fi} is the free-ion yield and A is a factor specific to the initial number and geometry of the ion-pairs.

After obtaining $F(t)$, the scavenging probability was calculated from Eq. (7.31) by numerical integration, except at the longest time scale, for which analytical integration is facilitated by the approximation just mentioned. The calculated result for an isolated ion-pair was compared with the experiments of Rzad et al. (1970), which showed good agreement up to ~10 mM concentration of scavenger; beyond this, the prescribed diffusion method overestimated the scavenging probability by ~26%. The agreement improved somewhat when multiple ion-pairs in blobs and short tracks were included to represent γ-radiolysis, but a considerable difference (~19%) still remained. Mozumder (1971) also computed, by the prescribed method, the time dependence of the biphenylide ion yield in pulse-irradiated solution of biphenyl in cyclohexane and compared it with the experiment of Thomas et al. (1968). In this computation, a self-consistent correction, due to homogeneous recombination for the estimated dose, was made and the result agreed very well with the experiment.

A special situation arises in the limit of small scavenger concentration. Mozumder (1971) collected evidence from diverse experiments, ranging from thermal to photochemical to radiation-chemical, to show that in all these cases the scavenging probability varied as $c_s^{1/2}$ in the limit of small scavenger concentration. Thus, importantly, *the square root law* has nothing to do with the specificity of the reaction, but is a general property of diffusion-dominated reaction. For the case of an isolated e-ion pair, comparing the $t \rightarrow \infty$ limit of Eq. (7.28) followed by Laplace transformation with the $c_s \rightarrow 0$ limit of the WAS Eq. (7.26), Mozumder derived

$$\alpha = r_c^2 \left(\frac{G_{fi}}{G_{gi}} \right)^2 \frac{k_s}{D},$$

giving a theoretical significance to the WAS parameter α [note that $G_{fi} \propto \exp(-a)$]. This equation can also be derived without using prescribed diffusion.

7.5.3b   The Eigenvalue Method

The limitation of the prescribed diffusion approach was removed, for an isolated ion-pair, by Abell et al. (1972). They noted the equivalence of the Laplace transform of the diffusion equation in *the absence of scavenger* (Eq. 7.30) and the *steady-state equation in the presence of a scavenger* with the initial e–ion distribution appearing as the source term (Eq. 7.29 with $\partial P/\partial t = 0$). Here, the Laplace transform of a function $f(t)$ is defined by

$$\tilde{f}(p) = \int_0^\infty f(t) \exp(-pt)\, dt.$$

Denoting the Laplace transforms of $n(r, t)$ and $I(r, t)$ respectively by $\bar{n}$ and $\bar{I}$, where n is the electron density and I is the outward electron current as seen from the positive ion, the authors derive the following equations:

$$p\bar{n} = n_0(r) - (4\pi r^2)^{-1} \frac{d\bar{I}}{dr}, \qquad (7.32a)$$

$$\bar{I} = -4\pi D \left(r^2 \frac{d\bar{n}}{dr} + r_c \bar{n} \right). \qquad (7.32b)$$

In Eqs. (7.32a, b), n_0 is the initial electron density function and $p = k_s c_s$. The boundary conditions are $\bar{n}(R, p) = 0$, where R is the reaction radius, and

$$\lim_{r \to \infty} \bar{I}(r, p) = 0.$$

The second boundary condition assures total finite existence probability at any time; the first boundary condition implies that the recombination is fully diffusion-controlled, which has been found to be true in various liquid hydrocarbons (Allen and Holroyd, 1974). [The inner boundary condition can be suitably modified for partially diffusion-controlled reactions, which, however, does not seem to have been done.]

Given n_0 and $\bar{I}(R, p)$, Eqs. (7.32a, b) can be integrated successively from $r = R$ to a large value of r. By definition $\bar{I}(R, p) = -\gamma$, where γ is the recombination probability in presence of scavenger. Only for the correct value of γ do the solutions of (7.32a, b) smoothly vanish asymptotically as $r \to \infty$; otherwise, they diverge. Thus, the mathematics is reduced to a numerical *eigenvalue problem* of finding the correct value of $\bar{I}(R, p)$.

Abell et al. (1972) have shown how to calculate γ very accurately by setting numerical upper and lower bounds within close tolerance. Thus, scavenging curves $(1 - \gamma)$ were generated for different values of p and for various initial δ-function distributions, thereby verifying the square root law at small concentration. Comparing with the experiments of Rzad et al. (1970) and taking $k_s/D = 1.2 \times 10^{14}$ liters/mole$\propto$cm^2, these authors found best agreement for an initial separation ~90 Å (the effect of the initial distribution is discussed presently). From the scavenging curve, the recombination kinetics in the pure liquid was obtained by ILT. For this, the scavenging probability $1 - \gamma$ was expanded in powers of $z = \beta p^{1/2}/(1 + \beta p^{1/2})$, where β is a suitable scaling parameter, and the result was inverted term by term. Abell et al.'s procedure is free from arbitrary assumptions, is quite accurate, and only involves moderate computer time.

The eigenvalue method was extended by Abell and Funabashi (1973) to investigate the effect of the initial distribution. This only required an integration over that distribution. However, the authors also used the effect of an external field on the free-ion yield as a further probe of the initial distribution. The latter was calculated using Onsager's (1938) equation (see Chapter 9). After experimenting with different initial density distributions, the authors concluded that in all liquid hydrocarbons the exponential distribution of the form $\sim r^{-2}$ $\exp(-r/b)$ describes both effects best, although in many cases the gaussian distribution was also acceptable. The value of b ranged from 50 to 60 Å for hexane and cyclohexane to about 250 Å for neopentane, generally increasing with the electron mobility in the liquid.

Figure 7.8 shows the comparison of calculated scavenging yields for initial gaussian and exponential e-ion distributions with an "observed" scavenging function. Here "observed" means Eq. (7.26) with the value of α that best describes electron scavenging in cyclohexane (Rzad et al., 1970). Although the exponential distribution describes the scavenging and free-ion yield experiments very well, Abell and Funabashi's theoretical demonstration for its validity is less convincing. Essentially, they invoke the phenomenon of the spreading of a wavefunction in the presence of absorption or traps, much like the propagation of light from a point source in an absorbing medium. Thus, they mix classical and quantum concepts.

7.5.3c   The Hong-Noolandi Treatment

Hong and Noolandi (1978a) first gave an analytical solution for the diffusion equation of an e-ion pair in the absence of an external field—that is, of Eq. (7.30) with $F = -e^2/\varepsilon r^2$, where ε is the dielectric constant of the medium. They then extended their solution in the presence of an external field of arbitrary strength (Hong and Noolandi, 1978b). Since the method involves fairly complex mathematical manipulations, we will only present its outline and some important conclusions.

Hong and Noolandi first transform the time-dependent density function to an auxiliary function h by writing

$$n(r,t) = \frac{1}{4\pi\sqrt{rr_0}} \exp(r^{-1} - r_0^{-1}) h(r,t),$$

where r_0 is the initial e–ion separation. Substitution into the Smoluchowski equation with the coulombic field (Eq. 7.30) then gives the partial differential equation for $h(r, t)$. Upon Laplace transformation, an ordinary differential equation is obtained for $\bar{h}$. This equation is solved in terms of two subsidiary functions, y_1 and y_2, which are themselves expanded in series of modified Bessel function products I and K. Finally, if desired, an inverse Laplace transformation may be taken to obtain the time dependence. Thus, although the method has been called exact, it nevertheless entails considerable numerical work. On the other hand, many limiting cases—for example, $t \to \infty$ or R, reaction radius, $\to 0$–can be obtained exactly and analytically, offering simplification.

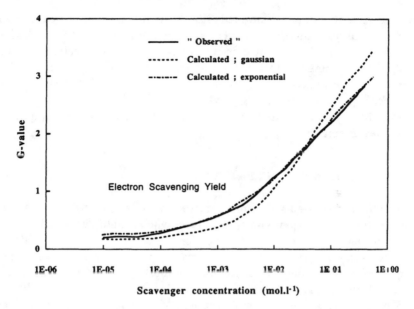

FIGURE 7.8   Comparison of initial gaussian and exponential distributions for the scavenging yield. Apparently, the exponential distribution agrees better with experiments; see text for details. Reproduced from Abell and Funabashı (1973), with permission of Am. Inst. Phys.©

Another virtue of the procedure is that it can explicitly take into account a partially diffusion-controlled recombination reaction in the form of Collins-Kimball radiation boundary condition—namely, $j(R, t) = -\kappa n(R, t)$ where $j(R, t)$ is the current density at the reaction radius and κ is the reaction velocity; $\kappa \rightarrow \infty$ implies a fully diffusion-controlled reaction. Thus, the time dependence of e-ion recombination in high-mobility liquids can also be calculated by the Hong-Noolandi treatment.

Standard limiting results have been reproduced by the Hong-Noolandi treatment—for example, the square root law of scavenging in the small concentration limit and the Onsager probability of escaping recombination for $\kappa \rightarrow \infty$ and $R \rightarrow 0$. In the long-time limit, the survival probability given by the prescribed diffusion method (Eq. 7.28) is validated. However, to achieve this, we must have $t \gg r_c^2/D$. For shorter times, the prescribed diffusion approximation was found to overestimate the survival probability in the same manner as determined by the eigenvalue method of Abell et al. (1972). The long-time survival probability given by Hong and Noolandi can be expressed as $\Omega(t) = [U(r_0)/U(\infty)]\{1 + r_c/[U(\infty)(\pi Dt)^{1/2}]\}$ where r_0 is the initial separation and $U(r) = \exp(-r_c/r) + (Dr_c/\kappa R^2 - 1) \exp(-r_c/R)$.

A surprising prediction of the Hong-Noolandi (1978b) theory is the existence of a critical field F_c above which the long-time decay becomes purely exponential—that is, rate-controlled rather than diffusion-controlled. For ordinary hydrocarbon liquids at room temperature, this field is ~20 KV/cm, a fairly modest value. For $F < F_c$, the long-time density distribution behaves as $\exp[-(F/2)^2 t]/t^{1+v}$ where v is in general a function of F, only approaching 1/2 for $F = 0$; here, t is expressed in unit of $r_c^2/4D$. For $F > F_c$, however, this long-time density changes to $\exp(-|s_0|t)$ where $|s_0| < (F/2)^2$. As yet, there does not seem to be any experimental evidence in radiation chemistry to support this theoretical prediction.

7.5.4   STOCHASTIC TREATMENTS

The methods discussed so far are essentially limited to isolated ion-pairs or, in the admittedly crude approximation, to cases when a multiple ion-pair spur can be considered to be a collection of single ion-pairs. Additionally, it is difficult to include an external field, as that will destroy the spherical symmetry of the problem. Stochastic treatments can incorporate both multiple ion-pairs and the effects of an external field.

The methodology of stochastic treatment of e–ion recombination kinetics is basically the same as for neutrals, except that the appropriate electrostatic field term must be included (see Sect. 7.3.1). This means the coulombic field in the dielectric for an isolated pair and, in the multiple ion-pair case, the field due to all unrecombined charges on each electron and ion. All the three methods of stochastic analysis—random flight Monte Carlo (MC), independent reaction time (IRT), and the master equation (ME)—have been used (Pimblott and Green, 1995).

MC simulation for multiple ion-pair case is straightforward in principle. A recombination, if necessary with a given probability, is assumed to have taken place when an e-ion pair is within the reaction radius. Simulation is continued until either only one pair is left or the uncombined pairs are so far apart from each other that they may be considered as isolated. At that point, isolated pair equations are used to give the ultimate kinetics and free-ion yield.

To implement the IRT or ME method, it is necessary to know, usually numerically, the recombination kinetics of an isolated ion-pair (vide infra). The general conclusion is the same as for neutral radicals—that is, the IRT method comes close to MC, requiring much less computer effort. It is surprisingly accurate under a variety of conditions and in the presence of a combination of an external field and that due to extant charges (Green et al., 1989b; Pimblott, 1993). In many cases of practical interest, ME provides acceptable result; although more approximate, it is computationally the least expensive (Green and Pimblott, 1990; Pimblott, 1993; Pimblott and Green, 1995).

For an isolated ion-pair with $r_c = 29$ nm, r_0 (initial separation) = 6.0 nm, $D = 2.5 \times 10^{-5}$ cm^{2}s^{-1}, and R (reaction radius) = 1.0 nm, all appropriate to n-hexane, random flight MC simulation reproduces accurately the kinetics of recombination as given by the numerical solution of the Debye-Smoluchowski equation. The IRT simulation also gives a very good approximation, especially at short and long time scales; at intermediate times, it tends to give slightly slower kinetics. Again using MC simulation, the variation of the escape yield with the external field is predicted to be essentially the same as given by the Onsager theory at low fields. In this calculation, the same data were used for n-hexane while the initial separation was sampled from a gaussian distribution of standard deviation 6.5 nm (Pimblott and Green, 1995, Figures 12 and 13).

The IRT method has been validated for the recombination kinetics of one and two pairs of ions by comparison with MC simulation (Green et al., 1989b). Using data suitable for low-dielectric-constant media such as n-hexane, good agreement is found between IRT and MC simulated results. The method has been extended to high-dielectric-constant media such as water by a distance scaling, $r \rightarrow r_{eff} = r_c /[1 - \exp(-r_c /r)]$ and thereafter treating the problem as for neutral radicals (Clifford et al., 1987b). Good agreement has been claimed for the kinetics of an ion-pair.

Figure 7.9 shows the recombination kinetics of two cation-anion pairs over the time scale $1-10^5$ ps, taking the initial distribution from identically distributed gaussians of standard deviation 8.0 nm, $D = 1.0 \times 10^{-4}$ cm^{2}s^{-1}, and other data as given before. The IRT simulation gave slightly slower kinetics than MC, but it was a very good approximation. The MC simulation for normally distributed random flights agreed quite well with the uniformly distributed random flight model of Bartczak and Hummel (1986).

Green and Pimblott (1991) criticize the truncated distributions of Mozumder (1971) and of Dodelet and Freeman (1975) used to calculate the free-ion yield in a multiple ion-pair case. In place of the truncated distribution used by the earlier authors, Green and Pimblott introduce the *marginal distribution for all ordered pairs*, which is statistically the correct one (see Sect. 9.3 for a description of this distribution).

In the ME model of recombination kinetics in a multiple ion-pair spur, the probability P_N that N ion pairs will remain extant at time t is given by (Green and Pimblott, 1990)

$$\frac{dP_N}{dt} = (N + 1)^2 \lambda(t)P_{N+1} - N^2\lambda(t)P_N,$$

since the rate of reaction in the state N is proportional to the total number of reactive distances, N^2 (see Eq. 7.23); here, $\lambda(t)$ is the time-dependent rate coefficient for an isolated pair. From the solution of this equation, the average number of ion pairs extant at time t is given quasi-analytically by

$$\langle N \rangle = \sum_{n=1}^{N_0} \frac{2n[\Gamma(N_0 + 1)]^2 \, \Pi(t)^{n^2}}{\Gamma(N_0 + n + 1) \, \Gamma(N_0 - n + 1)},$$

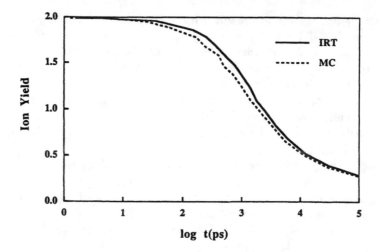

FIGURE 7.9   Recombination kinetics of two cation–anion pairs using Monte Carlo (MC) and independent reaction time (IRT) simulations. Initial distributions are identical gaussians of standard deviation 8.0 nm. Agreement between the simulations is good with IRT, giving slightly slower kinetics. Reproduced from Green *et al.* (1989a), with permission of Elsevier©.

where N_0 is the initial number of ion pairs, Γ is the gamma function, and $\Pi(t)$ is the survival function for an isolated ion-pair. Since $\Pi(t)$ is not available analytically, either a numerical solution given by Hummel and Infelta (1974) is used or the Hong-Noolandi (1978a, b) procedure could be followed. In either case, a convolution over the initial distribution is necessary. The free-ion yield is particularly simple to compute in this model, since the Onsager formula can be used for the infinite-time surviving function. The ME results of Green and Pimblott for the recombination in a multiple ion-pair spur show somewhat faster kinetics than MC or IRT simulations; it is nevertheless a useful approximation. The free-ion yield calculated by this method has been compared with the prescribed diffusion method of Mozumder (1971), giving varying degrees of agreement.

Pimblott (1993) has used MC and ME methods for the external field (E) dependence of the escape probability (P_{esc}) for multiple ion-pair spurs. At low fields, P_{esc} increases linearly with E with a slope-to-intercept ratio (S/I) very similar to the isolated ion-pair case as given by Onsager (1938). Therefore, from the agreement of the experimental S/I with the Onsager value, one cannot conclude that only isolated ion-pairs are involved. However, the near equality of S/I is contingent on small P_{esc}, which is not expected at high fields.

Bartczak *et al.* (1991; Bartczak and Hummel, 1986, 1987, 1993, 1997) have used random flight MC simulation of ion recombination kinetics for an isolated pair, groups of ion-pairs, and entire electron tracks. The methodology is similar to MC simulation described earlier, but the calculations are geared to comparison with experiments with electron tracks. Each step of the simulation involves a time differential ~10^{-5} to 5×10^{-5} in units of r_c^2/D, where $D = D_+ + D_-$, the terms on the right-hand side being the diffusion coefficients of the positive and negative ions, respectively. The motion of the charges is the result of diffusion and drift in the combined field due to all other charges and an external field, if present. The drift of the ith ion along any coordinate direction x is given by $\Delta x_i = \mu_i E_{xi} \Delta t$, where Δt is the time step, E_{xi} is the total electric field in the x direction on the ion, and μ_i is its mobility. The diffusive jump length in any direction is taken to be uniformly distributed between zero and a maximum value l_{max} consistent with the ion's diffusion coefficient: $\langle l^2 \rangle = 6D \Delta t$. Relying on the central limit theorem for a large number of simulations, this procedure has been found to be approximately equivalent to taking a gaussian distribution of jump length (vide supra).

In their early work, only a δ-function distribution was used by Bartczak et al. for the initial inter-ion separations. Later, exponential and gaussian distributions were used for the initial e-ion separation (r_-) of the same original pair. These authors first place the positive ions and then distribute the complementary electrons according to the chosen distribution. The initial inter-positive-ion separation (r_{++}) is taken to be fixed, 3.0 to 5.0 nm, for small values of N, the initial number of ion-pairs in the spur or track. For larger values of N (entire track), the local inter-positive-ion separation is taken from the 50% transmission range of Paretzke (1988). The reaction radius is taken to be ~0.5 to 1.5 nm, and reactions are checked after each step of simulation. The procedure is repeated until either one pair is left or the remaining pairs are so far apart that they can be considered to be isolated. After a large number of realizations, averages are computed for the surviving fraction at a given time and for the defined escape probability.

With $r_c = 28.45$ nm, $r_{++} = 3.0$ nm, and $r_- = 8.39$ nm, Bartczak and Hummel (1986) compute the escape probability $P_{esc} = 0.0336, 0.0261$, and 0.0230 respectively for $N = 1, 2$, and 3. While the first is comparable to the Onsager value, the latter are new results. The kinetics of recombination for the isolated pair, found by Bartczak and Hummel (1987) using MC, is very similar to that obtained by Abell et al. (1972). For $N > 1$, these authors found the recombination kinetics to be faster than that for the isolated pair. For two pairs, the calculated escape probability increased with the external field, but not as strongly as for $N = 1$.

Bartczak et al.'s entire track calculation is contingent on the specific detail of track structure. They argue that branch tracks of energy ≤50 or 100 KeV should be treated as a single entity whereas those of higher energy could be broken into their constituent spurs and tracks. In this manner, Bartczak and Hummel (1993) found that P_{esc} as a function of electron energy shows a minimum at about 2 KeV. Such a minimum should be expected since P_{esc} decreases with N for small N, but then at high energy P_{esc} would be dominated by that for the isolated ion-pair case. Bartczak et al.

(1991) also investigated the effect of the relative diffusion coefficients of the positive and negative ions, D_+/D_-, on P_{esc} for small values of N. While the escape probability decreased with N, it increased with the D_+/D_- ratio for a given N.

Bartczak and Hummel (1997) have reviewed their MC simulated results with regard to the effects of initial distribution, electron energy, long-time recombination kinetics, and so forth. On the basis of their calculations, these authors did not find convincing evidence to distinguish between exponential and gaussian initial distribution of e-ion pair. As before and as in every other simulation of the kind, the escape probability was found to decrease with N for small values of the initial number of ion-pairs. However, the authors computed P_{esc} over a wide span of electron energy, showing a clear minimum at ~3 KeV. Good agreement has been claimed with the available experiments, but this must be viewed with caution since for experiments with relatively low electron energy (~20 KeV or less), it is rather difficult to establish the zero-field free-ion yield. With regard to the long-time recombination kinetics, the authors find that the range of validity of the $t^{-1/2}$ law of decay is rather limited. Instead, they find better agreement with the equation $P(t)/P_{esc} = 1 + 0.6\tau^{-0.6}$, where $\tau = Dt/r_c^2$ and $P(t)$ is the probability of remaining uncombined at time t. This form was first suggested in the experiments of van den Ende et al. (1984). Figure 7.10 compares the asymptotic decay for the $\tau^{-0.5}$ and for $\tau^{-0.6}$ forms. It should be remembered, though, that the $t^{-1/2}$ law of decay represents pure diffusion in the absence of any field. It is entirely possible that the coulombic field will alter it somewhat.

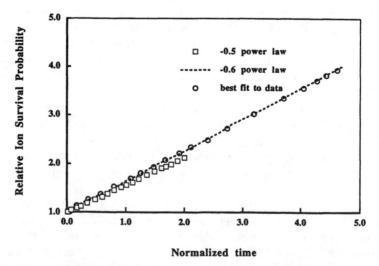

FIGURE 7.10   Comparison of long-time decay kinetics for ion-pair recombination. The authors find $\tau^{-0.6}$ decay describes the kinetics over a wider range of time than $\tau^{-0.5}$. Here τ is normalized time. See text for details. Reproduced from Bratczak et al. (1997), with permission of Elsevier©.

REFERENCES

Abell, G. C., and Funabashi, K. (1973), J. Chem. Phys. *58*, 1079.

Abell, G. C. , Mozumder, A., and Magee, J. L. (1972), J. Chem. Phys. *56*, 5422.

Allen, A. O. (1964), Radiat. Res. Suppl. *4*, 54.

Allen, A. O., and Holroyd, R. A. (1974), J. Phys. Chem. *78*, 796.

Anbar, M., and Meyerstein, D. (1966), Trans. Faraday Soc. *62*, 2121.

Anbar, M., Bambaneck ,M., and Ross, A. B. (1973), *Selected Specific Rates of Reactions of Transients from Water in Aqueous Solution. I. Hydrated Electron*, NSRDS-NBS *43*, National Bureau of Standards, Washington, D. C.

Anbar, M., Farhataziz, and Ross, A. B. (1975), *Selected Specific Rates of Reactions of Transients from Water in Aqueous Solution. II. Hydrogen Atom*, NSRDS-NBS *51*, National Bureau of Standards, Washington, D.C.

Balkas ,T. T., Fendler, J. H., and Schuler, R. H. (1970), J. Phys. Chem. *74*, 4497.

Bartczak, W. M., and Hummel, A.(1986), Radiat. Phys. Chem. *27*, 71.

Bartczak, W. M., and Hummel, A.(1987), J. Chem. Phys. *87*, 5222.

Bartczak, W. M., and Hummel, A. (1993), J. Phys. Chem. *97*, 1253.

Bartczak, W. M., and Hummel, A. (1997), Radiat. Phys. Chem. *49*, 675.

Bartczak, W. M., de Haas, M. P., and Hummel, A. (1991), Radiat. Phys. Chem. *37*, 401.

Beek, G. and Thomas, J. K. (1972), J. Chem. Phys. *57*, 3649

Bota'r, L., and Vido'czy, T. (1979), React. Kinet. Catal. Lett. *12*, 485.

Bota'r, L., and Vido'czy, T. (1984), Chem. Phys. Lett. *104*, 16.

Burch, P. R. J. (1957), Radiat. Res. *6*, 289.

Burch, P. R. J. (1959), Radiat. Res. *11*, 489.

Burns, W. G., and Barker, R. (1965), in *Progress in Reaction Kinetics* (Porter, G., ed.), p. 305, Pergamon, London.

Burns, W. G., and Curtiss, A. R. (1972), J. Phys. Chem. *76*, 3008.

Burns, W. G., Sims, H. E., and Goodall, J. A. B. (1984), Radiat. Phys. Chem. *23*, 143.

Byakov, V. M. (1963), Doklady Akad. Nauk (Fys. Khimii), USSR *153*, 1356.

Chernovitz, A. C., and Jonah, C. D. (1988), J. Phys. Chem. *92*, 5946.

Clifford, P., Green, N. J. B., and Pilling, M. J. (1982a), J. Phys. Chem. *86*, 1318.

Clifford, P., Green, N. J. B., and Pilling, M. J. (1982b), J. Phys. Chem. *86*, 1322.

Clifford, P., Green, N. J. B., and Pilling, M. J. (1985), J. Phys. Chem. *89*, 925.

Clifford, P., Green, N. J. B., Oldfield, M. J., Pilling, M. J., and Pimblott, S. M. (1986), J. Chem. Soc. Faraday Trans 1 *82*, 2673.

Clifford, P., Green, N. J. B., Pilling, M. J., Pimblott, S. M., and Burns, W. G. (1987a), Radiat. Phys. Chem. *30*, 125.

Clifford, P., Green, N. J. B., Pilling, M. J., and Pimblott, S. M. (1987b), J. Phys. Chem. *91*, 4417.

Dainton, F. S., and Logan, S. R. (1965), Trans. Faraday Soc. *61*, 715.

Dodelet, J.-P., and Freeman, G. R. (1975), Radiat. Phys. Chem. *7*, 183.

Draganic', I. G., and Draganic', Z. D. (1971), *The Radiation Chemistry of Water*, ch. 6, Academic Press, New York.

Draganic', Z. D., and Draganic', I. G. (1969), J. Phys. Chem. *73*, 2571.

Dyne, P. J., and Kennedy, J. M. (1958), Can. J. Chem. *36*, 1518.

Dyne, P. J., and Kennedy, J. M. (1960), Can. J. Chem. *38*, 61.

Ershler, B. V., and Byalov, V. M. (1962), Zh. F. Kh. *36*, 913.

Farhataziz and Ross, A. B. (1977), *Selected Specific Rates of Reactions of Transients from Water in Aqueous Solution. III. Hydroxyl Radical and Perhydroxyl Radical and Their Radical Ions*, NSRDS-NBS *59*, National Bureau of Standards, Washington, D.C.

Filinovskii, V. Y., and Chizmadzev, Y. A. (1958), Proc. 1st. All-Union Conf. Rad. Chem., Akad. Nauk S.S.S.R., Moscow, p. 19.

Flanders, D. A., and Fricke, H. (1958), J. Chem. Phys. 28, 1126.

Freeman, G. R. (1963a), J. Chem. Phys. 38, 1022.

Freeman, G. R. (1963b), J. Chem. Phys. 39, 988.

Fricke, H. (1955), Ann. N.Y. Acad. Sci. 59, 567.

Ganguly, A. K., and Magee, J. L. (1956), J. Chem. Phys. 25, 129.

Girija, G., and Gopinathan, C. (1980), Radiat. Phys. Chem. 16, 245.

Girija, G., and Gopinathan, C. (1981), Radiat. Phys. Chem. 17, 41.

Girija, G., and Gopinathan, C. (1982), Radiat. Phys. Chem. 19, 107.

Gopinathan, C., and Girija, G. (1983), Radiat. Phys. Chem. 21, 209.

Green, N. J. B. (1984), Chem. Phys. Lett. 107, 485.

Green, N. J. B., Clifford, P., Pilling, M. J., and Pimblott, S. M.(1987), Faraday Diss. No. 83, Royal Society of Chemistry, Cambridge, England.

Green, N. J. B., and Pimblott, S. M. (1989), J. Phys. Chem. 93, 5462.

Green, N. J. B., and Pimblott, S. M. (1990), J. Phys. Chem. 94, 2922.

Green, N. J. B., and Pimblott, S. M. (1991), Radiat. Phys. Chem. 37, 161.

Green, N. J. B., Pilling, M. J., and Pimblott, S. M.(1989a), Radiat. Phys. Chem. 34, 105.

Green, N. J. B., Pilling, M. J., Pimblott, S. M., and Clifford, P. (1989b), J. Phys. Chem. 93, 8025.

Green, N. J. B., Pimblott, S. M., and Brocklehurst, B. (1991), J. Chem. Soc. Faraday Trans. 87, 2427.

Hart, E. J. (1952), J. Phys. Chem. 56, 594.

Holroyd, R. A., and Sham, T. K. (1985), J. Phys. Chem. 89, 2909.

Hong, K. M., and Noolandi, J. (1978a), J. Chem. Phys. 68, 5163.

Hong, K. M., and Noolandi, J. (1978b), J. Chem. Phys. 69, 5026.

Hummel, A., and Allen, A. O. (1963), J. Chem. Phys. 44, 3426.

Hummel, A., and Allen, A. O. (1966), J. Chem. Phys. 44, 3426.

Hummel, A., and Infelta, P. P. (1974), Chem. Phys. Lett. 24, 559.

Hunt, J. W., and Thomas, J. K. (1967), Radiat. Res. 32, 149.

Hunt, J. W., Wolff, R. K., Bronskıll, M. J., Jonah, C. D., Hart, E. J., and Matheson, M. S. (1973), J. Phys. Chem. 77, 425.

John, F. (1952), Communs. Pure and Applied Math. 5, 155.

Jonah, C. D., Matheson, M. S., Miller, J. R., and Hart, E. J. (1976), J. Phys. Chem. 80, 1267.

Kuppermann, A. (1961), in Actions Chimiques et Biologiques des Radiations,.v. 5 (Haissinsky, M., ed.), p. 85, Masson, Paris.

Kuppermann, A. (1967), in Radiation Research (Silini, G., ed.), p. 212, North-Holland, Amsterdam.

Kuppermann, A., and Belford, G. G. (1962a), J. Chem. Phys. 36, 1412.

Kuppermann, A., and Belford, G. G. (1962b), J. Chem. Phys. 36, 1427.

LaVerne, J. A., and Pimblott, S. M. (1991), J. Phys. Chem. 95, 3196.

Lu, H., Long, F. H., Bowman, R. M., and Eisenthal, K. B. (1989), J. Phys. Chem. 93, 27.

Magee, J. L. (1951), J. Am. Chem. Soc. 73, 3270.

Mahlman, H. A., and Sworski, T. J. (1967), in The Chemistry of Ionization and Excitation (Johnson, G. R. A., and Scoles, G., eds.), p. 259, Taylor and Francis, London.

McQuarrie, D. A. (1967), J. Appl. Prob. 4, 143.

Migus, A. ,Gauduel, Y., Martin, J. L., and Antonetti, A. (1987), Phys. Rev. Lett. 58, 1559.

Monchick, L. (1956), J. Chem. Phys. 24, 381.

Monchick, L., Magee, J. L., and Samuel, A. H. (1957), J. Chem. Phys. 26, 935.

Mozumder, A. (1968), J. Chem. Phys. 48, 1659.

Mozumder, A. (1971), J. Chem. Phys. 55, 3026.

Mozumder, A. (1977), Khimiya Visokikh Energii 11, 182.

Mozumder, A. (1995), J.Phys. Chem. 99, 6557.

Mozumder, A., and Magee, J. L. (1966a), Radiat. Res. 28, 203.

Mozumder, A., and Magee, J. L. (1966b), Radiat. Res. 28, 215.

Mozumder, A., and Magee, J. L. (1966c), J. Chem. Phys. 45, 3332.

Mozumder, A., and Magee, J. L. (1967), J. Chem. Phys. 47, 439.

Nichiporou, F. G. and Byakoo, V. M. (1975), Khimiya Visokikh Energü 1, 199

Noyes, R. M. (1961), Prog. React. Kinet. 1, 129.

Onsager, L. (1938), Phys. Rev. 54, 554.

Paretzke, H. G. (1988), Report Gesellschaft fur Strahlen und Umweltforschung mbH, Munchen, GSF-Bericht 24/88.

Pimblott, S. M. (1993), J. Chem. Soc. Faraday 89, 3533.

Pimblott, S. M., and Green, N. J. B. (1995), in Research in Chemical Kinetics, v. 3 (Compton, R. G., and Hancock,G., eds.), p. 117, Elsevier, Amsterdam.

Pimblott, S. M., and LaVerne, J. A. (1990), Radiat. Res. 122, 12.

Pimblott, S. M., and LaVerne, J. A. (1994), Radiat. Prot. Dosim. 52, 503.

Pimblott, S. M., and Mozumder, A. (1991), J. Phys. Chem. 95, 7291.

Pimblott, S. M., LaVerne, J. A., Mozumder, A., and Green, N. J. B. (1990), J. Phys. Chem. 94, 488.

Pimblott, S. M., Green, N. J. B., and Brocklehurst, B. (1991), J. Chem. Soc. Faraday Trans. 87, 3601.

Pimblott, S. M., LaVerne, J. A., Bartels, D. M., and Jonah, C. D. (1996), J. Phys. Chem. 100, 9412.

Pimblott, S. M., and LaVerne J. A., (1997), J. Phys. Chem. 101, 5828.

Rzad, S. J., Infelta, P. P., Warman, J. M., and Schuler, R. H. (1970), J. Chem. Phys. 52, 3971.

Samuel, A. H., and Magee, J. L. (1953), J. Chem. Phys. 21, 1080.

Schmidt, K. H., and Buck, W. I. (1966), Science 151, 70

Schmidt, K H., Han, P., and Bartels, D. (1995), J. Phys. Chem. 99, 10530.

Schmidt, W. F., and Allen, A. O. (1970), J. Chem. Phys. 52, 4788.

Schwarz, H. A. (1955), J. Am. Chem. Soc. 77, 4960.

Schwarz, H. A. (1964), in Advances in Radiation Biology, v. 1 (Augenstein, L. G., Mason, R., and Quastler, H., eds.), p. 1, Academic, New York.

Schwarz, H. A. (1969), J. Phys. Chem. 73, 1928.

Schwarz, H. A., Caffery, J. M., and Scholes, G. (1959), J. Am. Chem. Soc. 81, 1801.

Short, D. R., Trumbore, C. N., and Olson, J. H. (1981), J. Phys. Chem.85, 2328.

Sumiyoshi, T., Tsugaru, K., Yamada, T., and Katayama, M., Bull. Chem. Soc. Jpn. 58, 3073.

Sworski, T. J. (1964), J. Am. Chem. Soc. 86, 5034.

Sworski, T. J. (1965), Adv. Chem. Ser. 50, 263.

Thomas, J. K., and Bensasson, R. V. (1967), J. Chem. Phys. 46, 4147.

Thomas, J K., Johnson, K., Klippert, T. and Lowers, R.J. (1968), J. Chem. Phys. 48, 1608.

Trumbore, C. N., Short, D. R., Fanning, J. E., and Olsen, J. H. (1978), J. Phys. Chem. 82, 2762.

Turner, J. E., Magee, J. L., Wright, H. A., Chatterjee, A., Hamm, R. N., and Ritchie, R. H. (1983), Radiat. Res. 96, 437.

Turner, J. E., Hamm, R. N., Wright, H. A., Ritchie, R. H., Magee, J. L., Chatterjee, A., and Bloch, W. E. (1988),Radiat. Phys. Chem. 32, 503.

van den Ende, C. A. M., Warman, J. M., and Hummel, A. (1984), Radiat. Phys. Chem. 23, 55.

Warman, J. M., Asmus, K.-D., and Schuler, R. H. (1969), J. Phys. Chem. 73, 931.

Williams, Ff. (1964), J. Am. Chem. Soc., 86, 3954.

Wolff, R. K., Bronskill, M. J., Aldrich, J. E., and Hunt, J. W., (1973), J. Phys. Chem. 77, 1350.

Yamaguchi, H. (1987), Radiat. Phys. Chem. 30, 279.

Yamaguchi, H. (1989), Radiat. Phys. Chem. 34, 801.

Zaider, M., and Brenner, D. J. (1984), Radiat. Res. 100, 245.

Zaider, M., Brenner, D J., and Wilson, W. E. (1983), Radiat. Res. 95, 231.

CHAPTER 8

Electron Thermalization and Related Phenomena

Most radiation-chemical reactions are thermal in nature; those considered in the diffusion-kinetic scheme are essentially thermal reactions (see Chapter 7). In polar media, electron thermalization is presumed to occur before solvation (Mozumder, 1988). However, ionization processes usually involve transfer of energy in excess of the ionization potential (see Chapter 4). Therefore, mechanisms of thermalization are important for radiation-chemical effects.

Platzman (1967) estimated that in the radiolysis of water the positive ion is left, on average, with an excitation energy of ~8 eV; this estimate was later lowered to ~4 eV by Pimblott and Mozumder (1991). In any case, the chemical consequences of such excess energy of the positive ion is unknown, and it will be assumed that, at least in the condensed phase, the positive ion is thermalized locally.

On the other hand, electron thermalization, although fast on the scale of thermal reactions, can still be discerned experimentally. In the gas phase, it exhibits itself through the evolution of electron energy via time-dependent reaction rates. In the liquid phase, the thermalization distance in the field of the positive ion is the all-important quantity that determines the probability of free-ion generation (see Chapter 9). In this chapter, we will deal exclusively with electron thermalization.

8.1   DEGRADATION MECHANISMS
OF SUBEXCITATION AND
SUBVIBRATIONAL ELECTRONS

As long as the energy of an electron remains above the electronic excitation potential of the medium, it loses energy very quickly. Its energy rapidly falls below the excitation potential—that is, the electron becomes subexcitational. According to Platzman (1955), *subexcitation electrons* are produced in two ways: directly in an ionization event, or by the gradual energy loss of a higher-energy electron.

Subexcitation electrons lose energy relatively slowly, the dominant mode of energy loss being the excitation of vibrations (Chen and Magee, 1962; Chen, 1964; Herzenberg and Mandl, 1962). Such vibrations can be excited directly, or via the formation a temporary negative ion. The cross section has a broad resonance character due to (1) various degrees of freedom of vibration in a polyatomic molecule and (2) excitation of multiple quanta of vibration within the same degree of freedom. Typical cross sections are ~1 $Å^2$. Thus, Mozumder and Magee (1967) estimate that in *n*-hexane, with 54 degrees of freedom, the total cross section is ~40 $Å^2$, which with a density of 0.667 g/cm^2 gives a mean free path of 5 Å between vibrational encounters. This kind of excitation process goes on efficiently till the electron energy approaches the quantum of CH vibration— that is, ~0.4 eV (*vide infra*). Assuming a subexcitation electron energy of 6.0 eV, a typical vibrational quantum ~0.2 eV, and a mean vibrationally excited state between the second and the third, the required number of free paths to degrade the energy to 0.4 eV is $(6.0 - 0.4)/(0.2 \times 2.5)$ or only 11. In an isotropic scattering model, the rms distance traveled is ~$11^{1/2} \times 5$ or about 16–17 Å.

Electrons of still lower energy have been called *subvibrational* (Mozumder and Magee, 1967). These electrons are hot (epithermal) and must still lose energy to become thermal with energy $(3/2)k_BT = 0.0375$ eV at $T = 300$ K. Subvibrational electrons are characterized not by forbiddenness of *intra*molecular vibrational excitation, but by their low cross section. Three avenues of energy loss of subvibrational electrons have been considered: (1) elastic collision, (2) excitation of rotation (free or hindered), and (3) excitation of *intermolecular* vibration (including, in crystals, lattice vibrations).

Elastic collision, determined by mass ratio, is a very inefficient process. By default, it is the only available mechanism in rare gases. Rotations are not easily excited in nonpolar molecules, especially in the condensed phase. They can be a contributing factor in molecular gases (*vide infra*). In polar media, rotations are an important degradation mechanism (Frohlich and Platzman, 1953).

Intermolecular vibration involving nearest-neighbor molecules are easily excited in liquid hydrocarbons, since the de Broglie wavelength of the electron at a few tenths of electron-volt energy is comparable to the intermolecular separation. The quantum for this vibration lies in the far IR and can be observed indirectly by Raman spectra. Raman shifts in many hydrocarbon liquids have been observed in the 30–80 cm^{-1} range. Similar values of intermolecular vibrational quanta have also been inferred from thermal diffusion data. The basic quantum used by Mozumder and Magee (1967) for liquid n-hexane is 0.01 eV (80 cm^{-1}), but its exact value is not important. In their theory, it always appears through its product with the probability of excitation per scattering mean free path (see Sect. 8.3.2).

Frohlich and Platzman (1953) developed a detailed electromagnetic theory for the *rate* of energy loss of a subexcitation electron in a polar medium due to dielectric loss. Their final result may be expressed as

$$-\frac{dE}{dt} \approx \frac{\pi e^2}{4d} \frac{\varepsilon_s - \varepsilon_{ir}}{\tau n^4},$$

(8.1)

where d is a length of the order of intermolecular separation, ε_s and ε_{ir} are respectively the static and infrared dielectric constants of the medium, τ is the dielectric relaxation time, and n is a suitably averaged optical refractive index. In this approximation, the denominator (n^4) arises mainly from the dispersion due to ionic oscillations and the numerator from the dielectric absorption at the same frequencies.

Note that Eq. (8.1) is remarkably independent of electron velocity. The stopping power $-dE/dx = v^{-1}(-dE/dt)$ does depend on velocity, but Frohlich and Platzman preferred not to use the stopping power due to a lack of knowledge of the actual tortuous path. Taking $\varepsilon_s = 80$, $\varepsilon_{ir} = 5$, $d = 3.3\text{Å}$, and $\tau = 10^{-11}$ s for water at 20°, they computed $-dE/dt \approx 10^{13}$ s^{-1}, which is about three orders of magnitude less than that for excitation and ionization at higher energies.

Platzman and Frohlich's estimate is a lower limit to the rate of energy loss, as it does not include energy loss to molecular vibrations, which is approximately of the same order of magnitude (emission of a quantum of 0.1 eV every 10^{-14} s). Taking the initial energy as a few eV, the estimated order of magnitude of slowing time is again ~10^{-13} s. Going to ice (–0.1°C), only the relaxation time is greatly modified (0.2 μs), which means that $-dE/dt$ is reduced to only 10^7 eV·s^{-1}.

Magee and Helman (1977) have extended the treatment of Frohlich and Platzman, considering path deviations of the electron on collisions and infrared-active vibrations. They used the numerical technique of fast Fourier transform to construct the frequency spectrum and adopted a Monte Carlo procedure for ensemble averaging. The frequency spectrum was nearly the same as that of Frohlich and Platzman except at very low or high frequencies, where the contribution to energy loss is minimal. Thus, they obtained about the same rate of energy loss as computed by Frohlich and Platzman. Only when they included "hesitation" in their random walk did they obtain significantly higher rate of energy loss. This hesitation may be interpreted in terms of temporary negative ion formation. The authors also considered nonpolar media such as benzene and polyethylene. The rate of energy loss of slow electrons in these media due to excitation of infrared vibrations is estimated to be approximately 3×10^{12} eV·s^{-1} or less.

8.2   ELECTRON THERMALIZATION
IN THE GAS PHASE

Shizgal *et al.* (1989) have listed a large number of processes that require an understanding of electron thermalization in the gas phase. These range from radiation physics and chemistry to radiation biology, and connect such diverse fields as electron transport, laser systems, nuclear fusion, and plasma chemistry. Certainly, this list is not exhaustive.

8.2.1   REVIEW OF RELEVANT EXPERIMENTS

In a nonattaching gas electron, thermalization occurs via vibrational, rotational, and elastic collisions. In attaching media, competitive scavenging occurs, sometimes accompanied by attachment-detachment equilibrium. In the gas phase, *thermalization time* is more significant than *thermalization distance;* because of relatively large travel distances, thermalized electrons can be assumed to be homogeneously distributed. The experiments we review can be classified into four categories: (1) microwave methods, (2) use of probes, (3) transient conductivity, and (4) recombination luminescence. Further microwave methods can be subdivided into four types: (1) cross modulation, (2) resonance frequency shift, (3) absorption, and (4) cavity technique for collision frequency.

In the cross modulation experiments (Mentzoni and Row, 1963; Mentzoni and Rao, 1965), an electron plasma is briefly heated by a microwave pulse while a weak microwave signal probes the mean electron energy. Assuming no electron loss and insignificant ambient gas heating, these authors derived the following equation for the relaxation of electron Maxwellian temperature T_e toward the ambient temperature T:

$$-\frac{dT_e}{dt} = \left(\frac{16}{3}\right)\left(\frac{2}{\pi}\right)^{1/2} qBNm^{1/2}\left(k_BT_e\right)^{-1/2}\left(T_e-T\right). \tag{8.1}$$

Here m is electron mass, N is the number density of gas molecules, B is the rotational constant, and $q = (8/15)\pi a_0{}^2Q^2$, a_0 and Q being respectively the Bohr radius and the quadrupole moment of the molecule. The experimental energy loss rate for nitrogen agreed well with Eq. (8.1) over the ambient temperature range 300–735 K. Typical values are ~0.5 µs at 300 K and 6 torr, and ~1 µs at 735 K and 4 torr. The variation of relaxation time with gas temperature and pressure are also well predicted. For oxygen, Mentzoni and Rao (1965) measure relaxation times ~160–350 ns for T = 300–900 K and at 3 torr.

In a microwave cavity containing an ionized gas, the resonant frequency shifts in proportion to the electron density n (Slater, 1946). This effect has been used by Warman and Sauer (1970, 1975) to measure n as a function of time following a pulse of ionizing radiation. It is known from separate experiments (Blaunstein and Christophorou, 1968; Christophorou et al., 1971) that, while thermal electron attachment of CCl_4 has a high rate (3.5×10^{-3} cm^3 s^{-1}), the rate constant k falls rapidly with energy; $k(e + CCl_4)$ as a function of electron energy can be obtained from these experiments. This fact has been exploited by Warman and Sauer to measure time-dependent k from the complex decay of the electron density n. Thus, the decay of the (mean) electron energy as a function of time has been indirectly obtained in Ar (which itself has the longest thermalization time) in the presence of CCl_4 and a moderating gas. Hence, thermalization times, defined operationally when mean electron energy $\langle E \rangle$ is ~10% above thermal, have been obtained. Warman and Sauer (1975) show that $\langle E \rangle$ decays exponentially in the long-time limit with a rate that varies widely with the moderating gas. The measured values for Ar and acetone are respectively 1.3×10^{-13} and 1.5×10^{-8} cm^{3}s^{-1}. Table 8.1 summarizes the measurements of Warman and Sauer together with a few other values.

The determination of electron concentration by the frequency shift method is limited to time resolution greater than a few hundred nanoseconds and is therefore not applicable to liquids. The microwave absorption method can be used virtually down to the pulse width resolution. Under conditions of low dose and no electron loss, and assuming Maxwellian distribution at all times, Warman and deHaas (1975) show that the fractional power loss is related to the mean electron energy $\langle E \rangle$ by

$$\frac{\Delta P\left(\langle E \rangle\right)}{\Delta P_{\text{th}}} = C\left(\frac{k_B T}{\langle E \rangle}\right)^n ,$$

Here the left-hand side is the ratio of power loss at time t, when the mean electron energy is $\langle E \rangle$, to that at thermalization, and C and n are determinable constants. This idealized equation is not expected to be valid in presence of the Ramsauer effect, but Warman and deHaas apply it anyway to N_2, Ar, and He at atmospheric pressure. The method relates the gradual decrease of collision frequency to an increase in conductivity, which finally rides to a plateau interpreted to be the thermal conductivity. The time needed to reach 90% of the thermal conductivity is called the thermalization time (see Table 8.1).

The foregoing equation can be used to give the evolution of $\langle E \rangle$. For N_2, exponential decay of $\langle E \rangle$ was seen, indicating efficient thermalization; but this was not observed for He. For highly efficient moderators such as ethane, the absorption signal essentially follows the pulse shape.

The method is also applicable to liquids and solids (Sowada and Warman, 1982; Sowada et al., 1982); for the condensed rare gases, a correction is needed for recombination. Sowada et al. (1982) obtained the following values of thermalization times, within 20% accuracy, given here in parenthesis as (phase, temperature

TABLE 8.1   Experimental Electron Thermalization Times in Various Gases at ~300K

Gas	τ ($\mu s \cdot torr$)[a]	Reference
He	26, 30	Warman and Sauer (1975), Warman and deHass (1975)
Ne	670	Warman and Sauer (1975)
Ne[b]	350	Dean et al. (1974)
Ar	1300, 228	Warman and Sauer (1975), Warman and deHass (1975)
Ar[b]	135	Dean et al. (1974)
H_2	1.5	Warman and Sauer (1975)
N_2	7.6, 15.2	Warman and Sauer (1975), Warman and deHass (1975)
O_2	1.7	Warman and Sauer (1975)
CH_4	0.2	Warman and Sauer (1975)
C_2H_6	0.15	Warman and Sauer (1975)
n-Hexane	0.08	Warman and Sauer (1975)
Neopentane	0.08	Warman and Sauer (1975)
C_2H_4	0.065	Warman and Sauer (1975)
C_2H_6	0.038	Warman and Sauer (1975)
N_2O	0.037	Warman and Sauer (1975)
CO_2	0.029	Warman and Sauer (1975)
NH_3	0.028	Warman and Sauer (1975)
Kr[b]	100	Dean et al. (1974)
C_2H_5OH	0.023	Warman and Sauer (1975)
$(C_2H_5)_2O$	0.022	Warman and Sauer (1975)
$(CH_3)_2CO$	0.01	Warman and Sauer (1975)

[a]Pressure normalized values.

[b]Relaxation time for electron cooling in the final stage.

[K], t_{th} [ns]): Ar (s, 82, 0.5), Ar (l, 85, 0.9), Kr (s, 113, 2.2), Kr (l, 117, 4.4), Xe (s, 157, 4.4), and Xe (l, 163, 6.5). In LAr, LKr, and Lxe, t_{th} shows a maximum ~7 ns as a function of liquid density, which is similar to the effect of density on drift mobility.

Shimamori and Hatano (1976) describe a Febetron-injected microwave cavity apparatus for measuring electron concentration following pulse irradiation. Its application to thermalization in Ar and CH_4 is similar to the method of Warman and Sauer (1975). In a related experiment, Hatano et al. (private communication) measure the electron collision frequency directly.

Dean et al. (1974) use a Langmuir probe technique in a rare gas repetitive afterglow plasma. The electron temperature is extracted from the semi-log plot of probe current versus retarding voltage. Provided that the radio-frequency discharge is weak, the electron-electron collision frequency is high enough to ensure a Maxwellian distribution. The probe measurements can give a time-resolved electron energy, which is interpreted mainly in terms of electron-neutral collision. After ensuring that there is no significant heating of the ambient gas nor any substantial spatial gradient of electron temperature, Dean et al. observed a fast initial relaxation of electron energy followed by a constant, slower relaxation. Some of their results are shown in Table 8.1.

Sowada and Warman (1982) have described a dc conductivity method for Ar gas at 295 K and 45 atm. Following a 20-ns pulse of irradiation, the conductivity rises to a peak at ~50 ns, due to the Ramsauer effect, before settling to a plateau, which is ascribed to thermal conductivity since the collecting field is very low. Since there is little electron loss, the conductivity profile is proportional to the mobility profile; this in turn can be considered a kind of image of collision frequency as a function of electron energy. The time to reach the conductivity plateau, ~150 ns, is the measure of thermalization time in the present case. At a density of ~9 × 10^{21} cm^{-3}, the conductivity maximum vanishes, indicating the disappearance of the Ramsauer minimum according to Sowada and Warman.

Takasaki et al. (1982a, b) use the delayed luminiscence in the rare gases and their mixtures to probe electron thermalization in these systems. The emissions are from the rare gas molecular excited states($^1\Sigma_u^+$ and $^3\Sigma_u^+$) to the dissociative ground state. There are well-characterized prompt emissions due to direct excitation, and delayed emission due to electron-ion recombination. Since it is known that the recombination rate of electrons with dimerized positive ions varies roughly as (electron energy)$^{-0.7}$, it may be assumed that most excited states are formed upon thermal electron recombination. Therefore, the observed delay of luminescence is roughly equated to the thermalization time. Takasaki et al. find that on admixture with N_2 the time delay is significantly reduced, indicating efficient moderation by the molecule. It is noted that ~1–2% of N_2 is sufficient to reduce the thermalization time in the rare gases by about a factor of 2, whereas the effect on the thermal electron mobility is not nearly as great.

8.2.2   THEORETICAL METHODS

Early theoretical models were based on fractional energy loss $2m/M$ per elastic collision (for details, see LaVerne and Mozumder, 1984, Sect. 3, and references therein). Thus, frequently, the energy loss rate was written as $-d\langle E\rangle/dt = (2m/M)(\langle E\rangle - 3k_B T/2)v_c$, where v_c is the collision frequency and $\langle E\rangle$ is the mean electron energy over an unspecified distribution. The heuristic inclusion of the term $3k_B T/2$ allowed the mean energy to attain the asymptotic thermal value, although the form was not derived. Since inelastic collisions were not explicitly included, the energy relaxation times computed from such equations tended to be overestimates.

Two other attempts, without the use of a distribution function, are worth mentioning, as these are operationally related to experiments and serve to give a rough estimate of the thermalization time. Christophorou et al. (1975) note that in the presence of a relatively weak external field E, the rate of energy input to an electron by that field is $\omega = eEv_d$, where v_d is the drift velocity in the stationary state. Under equilibrium, it must be equal to the difference between the energy loss and gain rates by an electron's interaction with the medium. The mean electron energy is now approximated as $\langle E \rangle = (3eD_\perp)/(2\mu)$, where $\mu = v_d/E$ is the drift mobility and $D_\perp$ is the perpendicular diffusion coefficient (this approximation is actually valid for a Maxwellian distribution). Thus, from measurements of μ and $D_\perp$, the thermalization time is estimated to be

$$\tau_{th} = \int_{3k_BT/2}^{<E>_i} \frac{d\langle E \rangle}{\omega(\langle E \rangle)}.$$

Using this equation, Christophorou et al. found that at 298 K: (1) thermalization time in gases varies greatly from one polyatomic molecule to another, and (2) this time is insensitive to initial energy above ~0.4 eV but falls rapidly below ~0.1 eV. These are consistent with the findings of Mozumder and Magee (1967) for hydrocarbon liquids. Warman (1981) extended the idea to the liquid phase. However, in the absence of measured perpendicular diffusion coefficient's in most liquids, he writes the energy-change rate in the presence of an external field E as

$$\frac{d\langle \varepsilon \rangle}{dt} = eEv_d - \nu\left(\langle \varepsilon \rangle - \frac{3k_BT}{2}\right),$$

where ν, the so-called energy exchange frequency, is taken as a constant. From this equation, the thermalization time, defined when the mean energy is $1.1 \times 3k_BT/2$, is given by

$$\tau_{th} = \nu^{-1} \ln\left[10\left(\frac{2E_0}{3k_BT} - 1\right)\right],$$

where E_0 is the initial energy. Combining the energy equation in the stationary state, $\nu = eEv_d/(\langle E \rangle - k_BT)$, with the nth power of dependence of mobility on the electron energy (n may be positive or negative), $\mu = \mu_0(2\langle E \rangle/3k_BT)^n$, Warman derives $\nu \approx 20|n|e\mu_0E_{10}^2/3k_BT$ where $|n|$ is the magnitude of n, μ_0 is the zero-field mobility, and E_{10} is the field strength at which the mobility departs from the thermal value by 10%. Substituting into the equation for thermalization time, Warman gets

$$\tau_{th} \approx 3k_B T \frac{\ln[10(2E_0/3k_B T - 1)}{20\,|n|\,e\mu_0 E_{10}^2}.$$

The thusly-obtained thermalization time depends weakly on the initial energy, for which a value ~1 eV has been used in the irradiation case. Taking $|n| = 1$ gives $\tau_{th} = 3.0$, 1.5, and 0.5 ns respectively for LXe, LKr, and LAr and the values 10.0, 0.9, and 0.6 ps respectively for methane, neopentane, and tetramethylsilane, all liquids at their triple points. In these estimates, Schmidt's (1977) data were used for μ_0 and E_{10}. However, taking $|n| = 1$ can be very crude, as certain theories and experiments give $n = -0.5$. On the other hand, the use of 10% nonlinearity of mobility may seem arbitrary, but it has partial compensation in the definition of E_{10}.

Later theoretical approaches in the gas phase can be divided into four categories: (1) the displaced pseudo-Maxwellian (DPM) approximation (Mozumder, 1980a, b; Tembe and Mozumder, 1983a, b, 1984; Knierim et al., 1981); (2) Monte Carlo simulation (Koura, 1982, 1989); (3) the eigenvalue method of the Fokker-Planck equation (Shizgal, 1981, 1983; McMahon et al., 1986; Shizgal et al., 1989); and (4) the electron degradation spectrum based on the Spencer-Fano equation (Inokuti, 1975; Dillon et al., 1988; Douthat, 1975, 1983). The DPM method, applied by Mozumder et al. to the rare gases, diatomic molecules, and mixtures of N_2 with rare gases, is similar in principle to the procedures of Knierim et al. (1981) based on the moment method of solution of the Boltzmann equation. In it, the momentum distribution is assumed to be a displaced Maxwellian:

$$(\vec{p}, t) = \left(2\pi m k_B \langle T \rangle\right)^{-3/2} \exp\left(-\frac{\left(\vec{p} - m\langle \vec{u} \rangle\right)^2}{2mk_B T}\right),$$

Where the directed velocity $\langle \vec{u} \rangle$ and the effective temperature $\langle T \rangle$ are time-dependent, satisfying kinetic equations involving elastic and inelastic cross sections. Electron cooling gives a gradual decrease of $\langle T \rangle$, whereas $\langle \vec{u} \rangle$ is either a memory of an injected velocity or a drift velocity in the external field. In the absence of a field, $\langle u \rangle$ can be taken as zero since the velocity direction always relaxes much faster than any other physical process (Mozumder, 1980a, b). The position distribution satisfies a diffusion equation with a time-dependent diffusion coefficient for which an expression has been derived. The key quantity is the evolution of $\langle T \rangle$, which is obtained from the net rate of energy loss. In the later stages, which contribute most significantly to the thermalization time, equilibrium can only be established by equality between direct (energy loss) and inverse (energy gain) collision processes. For subvibrational electrons in a monatomic gas the net fractional energy loss at gas temperature T is given by Gilardini (1972) as

$$\frac{2m}{M}\left(1 - \langle\cos\theta\rangle\right)\left[\frac{1}{2}mv^2 - \frac{3}{2}k_BT\left(\frac{4}{3} + \frac{1}{3}\frac{d\ln\sigma_m}{d\ln v}\right)\right],$$

where $\langle\cos\theta\rangle$ is the average of the cosine of the scattering angle. The Gilardini expression is consistent with *detailed balancing*. Since the collision frequency is $Nv\sigma_{el}$ and $(1 - \langle\cos\theta\rangle) = \sigma_m/\sigma_{el}$, where σ_m and σ_{el} are respectively the momentum transfer and total elastic scattering cross sections, the net energy loss rate at any given time is given by the average of the expression $(2m/M)N\sigma_m v[\cdots]$ over the momentum distribution at that time. Here $[\cdots]$ represents the factor within square brackets in the Gilardini expression. The electron cooling rate is now given by

$$-\frac{d\left(k_B\langle T\rangle\right)}{dt} = \frac{2}{3\pi^{1/2}}\frac{m}{M}N\left(\frac{2m}{k_B\langle T\rangle}\right)^{3/2}\int_0^\infty dv\,\sigma_m v^3 \exp\left(-\frac{mv^2}{k_B\langle T\rangle}\right)[\cdots]. \quad (8.2)$$

In monatomic gases, the thermalization time can be calculated from Eq. (8.2) if the momentum transfer cross section is available as a function of velocity. One starts with an initial temperature $\langle T\rangle_0 = mu_0^2/3k_B$, where u_0 is the initial, directionally randomized electron velocity, and ends with a defined final temperature—for example, $1.1T$. Figure 8.1 shows the evolution of electron temperature in He at 290 K against density-normalized time (s∝cm^{-3}) with a starting velocity $u_0 = 4.8v_{th}$ where v_{th} is the thermal velocity at that temperature (1.148×10^7 cm∝s^{-1}). In these calculations, experimentally determined cross sections were used (Mozumder, 1980a). The diffusion coefficient (not shown in figure) remains high at short times because of greater particle velocity. In the long-time limit, it relaxes to the thermal value D_{th}. Thermal mobility, computed from D_{th}, agrees well with experimental determination extrapolated to zero electric field. Similar agreement is found in other rare gases except for Ne, for which the experimental cross sections at low energy are still unclear. The calculated density-normalized thermalization times decrease with ambient temperature in He and Ne, which is normal, but show a complex behavior in Ar, Kr, and Xe due to the Ramsauer effect (Mozumder, 1980b). In the long-time limit, Eq. (8.2) gives an exponential decay of the excess temperature, $\langle T\rangle - T$, with a relaxation time τ_n given through

$$\tau_m^{-1} = \left(\frac{2}{\pi}\right)^{1/2}\frac{2m}{3M}\frac{mN}{(k_BT)^2}\left(\frac{m}{k_BT}\right)^{3/2}\int_0^\infty dv\,\sigma_m v^5[\cdots]\exp\left(-\frac{mv^2}{k_BT}\right).$$

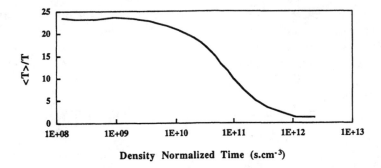

FIGURE 8.1   Evolution of effective electron temperature $\langle T \rangle$ in helium at 290 K vs. density–normalized time. Reproduced from LaVerne and Mozumder (1984), with the permission of Elsevier©.

This relaxation time decreases with T in all rare gases, particularly rapidly in He and Ne, a typical value being 10 μs∝torr in He at 300 K. Comparison with the experiments of Warman and Sauer (1975) and with those of Warman and deHaas (1975) shows excellent relative agreement for all rare gases, but the absolute values are overcalculated by a factor of 2–3. This discrepancy is not due to the distribution of initial energy, since t_{th} is insensitive to it. It is probably due jointly to the approximate nature of the theory and the uncertainty of final energy measurement. Note that there is considerable difference between t_{th} measured by different techniques. Also, when comparison is possible at an intermediate energy ~0.1 eV, the agreement is much better between experiment and calculation. In all rare gases, epithermal scavenging has been found to be ubiquitous (LaVerne and Mozumder, 1984), presumably because of long thermalization times. For example, with CCl_4 in Ar at 300 K, the scavenger concentration must be <1 ppb to avoid significant epithermal scavenging. Since such stringent purification is difficult to achieve, time-dependent reaction rates must be used to correct for these effects.

Molecular gases provide additional cooling mechanism through inelastic (vibrational and rotational) collisions. The direct and inverse collisions are related by microscopic reversibility, $p_i^2 g_i \bar{\sigma}_{if} = p_f^2 g_f \bar{\sigma}_{fi}$, where p is the electron momentum, g is molecular state degeneracy including nuclear spin, $\bar{\sigma}$ is the degeneracy-averaged cross section, and i and f refer respectively to the initial and final states. The rate of loss of mean energy due to inelastic collisions may be written as

$$-\frac{d\langle E \rangle}{dt} = \sum_{i, f > i} \varepsilon_{if} \int d\Gamma_{if} \,,$$

where ε_{if} is the energy of molecular transition and $d\Gamma_{if}$ is the difference between direct and inverse collision rates. Assuming a Maxwellian distribution, utilizing the relationship between direct and inverse cross sections as previously given, and incorporating energy conservation, one gets (see LaVerne and Mozumder, 1984)

$$-\frac{d\langle E \rangle}{dt} = \sum_{i, f > i} \varepsilon_{if} \rho_i \frac{g_f}{g_i} \left[\exp\left(-\frac{\varepsilon_{if}}{k_B \langle T \rangle} \right) - \exp\left(-\frac{\varepsilon_{if}}{k_B T} \right) \right] \xi_{fi}(\langle T \rangle),$$

where ρ_i is the molecular state density and the time-dependent collision rate coefficient ξ_{fi} is given by

$$\xi_{fi}\langle T \rangle = \int_0^\infty \sigma_{fi} v_f \left(2\pi m k_B \langle T \rangle \right)^{-3/2} \exp\left(-\frac{p_f^2}{2m k_B \langle T \rangle} \right) 4\pi \rho_f^2 \, dp_f \, .$$

At long times the excess temperature, $\langle T \rangle - T$, decays exponentially, as can be shown from the preceding equation. The relaxation rate has independent, additive contributions from momentum transfer collisions (as in the case of rare gases) and from each pair of states connected by inelastic collision. Thus the net relaxation rate is given by

$$\tau^{-1} = \tau_m^{-1} + \sum_{i, f > i} \tau_{if}^{-1}$$

where τ_m is as given before and τ_{if} is

$$\tau_{if}^{-1} = \frac{2\rho}{3Z k_B^2 T^2} \varepsilon_{if}^2 g_f \xi_{fi}(T) \exp\left(-\frac{\varepsilon_{if}}{k_B T} \right), \tag{8.3}$$

where Z is the partition function and ρ is number density of the gas. The smaller the relaxation time for a given process, the greater its importance to thermalization. The important factors are collision cross-sections, gas temperature, state energies and degeneracies. The procedure for obtaining the cross sections by combining theoretical formulas, and swarm and beam data has been outlined by LaVerne and Mozumder (1984) and by Tembe and Mozumder (1983a). Table 8.2 summarizes the results for H_2 using Eq. (8.3). The relaxation time is shown at different gas temperatures together with the principal contributor and other processes in decreasing order of importance. From this table, we see that, contrary to popular belief, elastic collisions are never unimportant. Also, consistently with the partition function, various vibrational and rotational processes assume different relative importance at a given temperature. The displaced pseudo-Maxwellian method has been applied to H_2, N_2, CO, and mixtures of N_2 with Ar and Xe (Tembe and Mozumder, 1983a, b). Figure 8.2 shows the

TABLE 8.2 Relative Importance of Various Processes to Electron Thermalization in H_2

T (K)	τ (μs)[a]	τ' (μs)[a,b]	In decreasing order of importance[c]
20	29	29	m,0,1,2
60	9.0	11.3	m,0,1,2
100	3.6	6.8	0,m,1–6
150	2.0	3.9	0,m,1–5,v,6
200	1.5	3.6	0,1,m,2–5,v,6
250	1.1	2.8	1,0,m,2–5,v,6
300	0.9	2.1	1,m,0,2–5,v,6
350	0.8	1.8	1,m,0,2–5,v,6
500	0.6	1.5	1,m,3,2,0,4,5,v,6
1000	0.3	1.2	3,m,1,2,v,5,4,0,6

[a]At density 3.3×10^{16} cm^{-3}.

[b]Most important contributor, first term in the list.

[c]Symbols: m, momentum transfer collision; v, vibrational collision $(0 \leftrightarrow 1)$; J (0 through 6) means rotational collision $(J \leftrightarrow J + 2)$. Other processes make negligible contribution in this temperature range.

evolution of the electron temperature in H_2 as a function of density-normalized time at different gas temperatures. As expected it is found that the thermalization time, controlled mainly by the relaxation time, is insensitive to the initial energy. The main contributors to thermalization time are rotational and vibrational processes at room temperatures and above, but elastic collisions make a nonnegligible contribution. H_2 is exceptional among diatomics in having a large rotational constant B. Around room temperature, $J \longleftrightarrow J + 2$ transitions contribute mostly to the thermalization time. Using a quadruple moment $Q = 0.62ea_0^2$ for H_2 and a relative rotational cross section given by the Gerjuoy-Stein (1955) formula, with absolute values adjusted with scattering experiments, the computed thermalization time 2.37 μs∝torr is ~37% larger than the experimental value of 1.5 μs∝torr (Warman and Sauer, 1975). While part of the discrepancy may be due to the uncertainty of final energy determination, a better agreement (2.05 μs∝torr) was obtained using the quadruple moment according to Engelhardt and Phelps (1963). Normal diatomics have much smaller B values than H_2 ; nevertheless, energy loss due to rotational excitation usually exceeds that due to elastic collisions. This is becuase high J values contribute significantly—for example, the maximum J value approaches 40 in N_2. In this gas, the computed t_{th} value, 11.5 μs∝torr, is bracketed by the experimental values, 7.5 μs∝torr by Warman and Sauer (1975) using CCl_4 as a probe and ~15μs∝torr by Warman and deHaas (1975) using microwave absorption. Thermalization in CO is influenced comparably by $J \longleftrightarrow J + 2$ transitions,

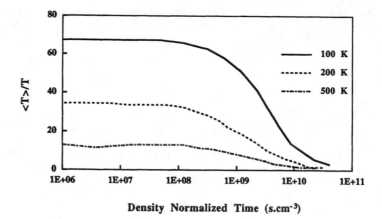

Density Normalized Time (s.cm⁻³)

FIGURE 8.2   Evolution of effective electron temperature T in H_2 as a function of density–normalized time at different gas temperatures. Reproduced from LaVerne and Mozumder (1984), with the permission of Elsevier©.

as in N_2, and by the $J \longleftrightarrow J + 1$ transitions due to the small dipole moment of the molecule. In all these gases, the thermal mobility computed from the long-time diffusion coefficient agrees excellently with experimental values.

Tembe and Mozumder (1984) applied the DPM method to calculate the time-dependent electron mobility in pulse-irradiated gaseous Ar. They used the gas kinetic formula for mobility (Huxley and Crompton, 1974),

$$\mu = -\frac{4\pi e}{3mN} \int_0^\infty \left[\frac{v^2}{\sigma_m(v)} \right] \frac{d\phi_0}{dv}\, dv$$

where ϕ_0 is the isotropic part of the electron velocity distribution function (*vide supra*). They included an energy input term to the electron population eEv_d, where v_d is the drift velocity, and showed that the DPM method should be convergent at small external fields when terms up to second order in drift velocity were retained. The calculated mobility attained a peak at ~40 ns due to the Ramsauer effect. It agreed qualitatively with the experiment of Sowada and Warman (1982), but not quantitatively. Even after convolution with the pulse width, the calculated peak appeared at ~50 ns and with a higher value at maximum compared with the experiment, in which a delayed peak is observed with a somewhat lower value. This discrepancy has been attributed to the inherent error in the assumed velocity distribution function (*vide infra*).

Shizgal *et al.* (1989) have criticized the displaced pseudo-Maxwellian (DPM) method because that approximation gives too fast an initial rise of electron temperature (velocity randomization), does not distinguish between longitudinal and transverse diffusion, and in general, the velocity distribution function does not resemble the true one sufficiently, especially at short times. Indeed, using Monte Carlo simulation, Koura (1982, 1989) finds that the Maxwellian approximation is a poor one except at very long times. In the later work Koura, (1989) considers source electrons in He with starting energy in the 100–1000 eV interval and computes the distribution function, energy, and so forth, as functions of time going through the subexcitation regime and ending with thermalization. His calculated variation of electron energy with time agrees very well with the experiment of Warman and deHaas (1975) within the entire experimental time scale.

Shizgal *et al.* start with the Boltzmann transport equation and after a number of standard approximations write it in the space-independent form as follows:

$$\frac{\partial f}{\partial t} + \frac{e\vec{E}}{m} \cdot \nabla_v f = J[f],$$

Where f is the time-dependent velocity distribution function in the presence of the external field $\vec{E}$ and J is the collision operator representing the effect of electron-moderator collisions on the distribution function. Expanding f in terms of Legendre polynomials of the cosine of the angle between the directions of field and electron velocity, it is argued that only two terms need be retained since the velocity direction is expected to be quickly randomized to an isotropic one. After a short time, only the spherical component f_0 is extant; this is written as a product of the steady-state Maxwellian or Druvestynian function and an auxiliary function g to be determined. Evolution of g is governed by $\partial g/\partial t = -Lg$, where L is a differential operator in the velocity variable for which an explicit form can be given. L is parameterized by collision cross sections and the external field. In this manner, a Fokker-Planck type of equation is derived fron the Boltzmann equation. Its solution is given in terms of the eigenfunctions of L as

$$g(v, t) = \sum_0^\infty b_n \exp(-\lambda_n t)\, \phi_n(v),$$

where ϕ_n, and λ_n are respectively the eigenfunctions and eigenvalues. The time-dependent energy and mobility are now given by

$$E(t) = \sum_0^\infty e_n \exp(-\lambda_n t) \quad \text{and} \quad \mu(t) = \sum_0^\infty \mu_n \exp(-\lambda_n t).$$

respectively. The coefficients b_n, e_n, and μ_n are related to each other and to the initial velocity distribution. These can be evaluated from the collision cross sections and the strength of the external field. In the long-time limit, all electron parameters (energy, mobility, etc.) will decay exponentially, dominated by the smallest nonzero eigenvalue. In particular, the relaxation time for thermalization will be given by $(\lambda_1)^{-1}$. Unfortunately the different cross sections, experimental and compiled, do not always agree, resulting in different computed thermalization times.

Using the cross sections compiled by Mozumder (1980b), Shizgal et al. (1989) obtained rather good agreement between computed values and the experimental results of Suzuki and Hatano (1986a, b) in Ar and Kr around 290–300 K. In units of $\mu s \propto torr$, the calculated values are ~870 for Ar and ~200 for Kr as compared with the experimental values ~800 and ~170–210, respectively. Somewhat worse agreement is found for He and Ne; the worst discrepancy of roughly a factor of two is comparable to the accuracy of the DPM method.

An interesting phenomenon arises in the thermalization of electrons in a Ramsauer gas displaying a strong power dependence of collision frequency on electron energy. In such cases, an electron population, initially formed in a non-Maxwellian distribution, can actually suffer a net displacement in the reverse field direction, exhibiting transient negative mobility. The effect was predicted by McMahon and Shizgal (1985) for Xe and confirmed by the experiment of Warman et al. (1985). Excellent agreement has been found by Shizgal et al. (1989) between calculation and experiment for the time dependence of electron mobility in Xe gas over the entire scale following a pulse of irradiation.

The electron degradation spectrum is defined as the path length of all secondary electrons per unit energy loss at E corresponding to a primary electron of energy E_0. Originally formulated as a stationary distribution (Inokuti, 1975), it was later extended to include time dependence (Dillon et al., 1988). In some sense, it serves the purpose of a distribution function in statistical mechanics. Once known as a function of energy, it can be used to predict a wide variety of yields for which the product cross section as a function of energy is obtainable. It is conjectured that, viewed as a transport equation, it should be related to the Boltzmann equation, although the exact form of this relationship has not been derived (Shizgal et al., 1989). Usually, however, the paucity of relevant data and the difficulty of solving for the degradation spectrum restricts the use of the degradation spectrum to relatively simple cases. Douthat (1975; 1983) has applied this method to electron thermalization in He and H_2 by considering the slowing down of a single electron between its initial and final energies. The averaging is taken over the distribution of subexcitation electrons in energy, giving the mean rate of energy loss. Compared with experiment, the agreement is qualitative and moderate. In the continuous slowing down approximation (CSDA), the degradation spectrum is very simply given by the reciprocal of the stopping power, $-dE/dx$. The mean rate of energy loss is then $-\langle dE/dt \rangle = (-dE/dx)(2E/m)^{1/2}$ and the time for electron energy moderation from initial energy E_0 to an energy E is given by

$$t(E) = \int_E^{E_0} \left(-\left\langle \frac{dE}{dt} \right\rangle \right)^{-1} dE.$$

The thermalization time for an electron swarm starting with an initial energy E_0 is then given by

$$\tau_{\text{th}} = \int_{3k_{\text{B}}T}^{E_0} f_{\text{s}}(E) t(E) \, dE,$$

where $f_{\text{s}}(E)$ is the degradation spectrum normalized within the limits of integration. This procedure can give a simple measure of the thermalization time, but the CSDA approximation is often not a good one. The calculation of the stopping power at subexcitational energy is also an equally difficult task.

8.3   ELECTRON THERMALIZATION
IN THE CONDENSED PHASE

In the condensed phase, the thermalization distance distribution is the more important consideration; the required time ($\sim 10^{-13}$ s) is usually so short as to be of no great consequence. Further, the coulombic field of the geminate positive ion provides an effective that which is almost irrelevant in the gas phase. The thermalization distance distribution is of paramount importance for calculating free-ion yield and scavenging reaction probabilities. Therefore, it would be profitable to read this section together with Chapter 9. In Sect. 8.3.1, we will discuss electron thermalization in a medium of low mobility, such as liquid hydrocarbons, taking n-hexane as a paradigm. Some effects of epithermal trapping and scavenging will be analyzed in Sect. 8.3.2.

8.3.1   THERMALIZATION DISTANCE DISTRIBUTION
IN LIQUID HYDROCARBONS

First, we want to make a comment about possible local temperature rise due to energy absorption. An early theory of radiation effect's was based on the point-heat hypothesis (Dessauer, 1923). Later analysis showed that the temperature rise would be too feeble and too transient for low-LET radiation to cause any real change (see Mozumder, 1969). There is no experimental evidence for temperature rise for low-LET radiations. The case of high-LET radiations is still open, though.

In earlier work, the importance of *thermalization tail*—that is, the distance traveled between subexcitational and thermal energies of the electron—was not recognized. Freeman and his associates (Freeman and Fayadh, 1965; Freeman, 1967) used initial electron positive-ion separations that were characterized by the absence of thermalization tail and a deficiency in low-energy electrons. This vaguely resembled stopping distance determined by electronic stopping power only, but it was not consistent with a realistic secondary electron distribution. Although only a relative distribution was used, a normalization problem remained (Hummel, 1967; Burns, 1968). Later, Dodelet and Freeman, (1972) proposed empirical distribution functions by comparing the calculated and observed relative increase of free-ion yield with external field (see Chapter 9). They concluded that if the zero field free-ion yield is <0.2, a truncated power law distribution best described the experiments; otherwise, a gaussian distribution centered at the origin should be used. No physical basis for these distributions has been offered. Further, as pointed out by Burns (1972), different distribution functions, while broadly agreeing with experiments, gave widely different effective thermalization lengths that could not be reconciled.

The mechanism of thermalization of subvibrational electrons by elastic collisions, which is important in the gas phase (see Sect. 8.2), can be discounted for molecular liquids on experimental grounds. If such were the mechanism, the mean fractional energy loss p per collision would be approximately $2m/M$, where m, and M are the electron and molecular masses, respectively. The number of collisions N needed to thermalize an electron of subvibrational energy E_v is given by $N = \ln(3k_BT/2E_v)/\ln(1 - p)$. Taking $E_v = 0.4$ eV (vide infra), $T = 300$ K, and $p = 1.26\sim10^{-5}$ appropriate to n-hexane, one gets $N = 1.88 \times 10^5$. On a random scattering basis, the rms travel distance would be $r_0 = N^{1/2}L$, where L is scattering mean free path. Taking $L = 5$ Å, this gives $r_0 = 2168$ Å. For an Onsager length of ~300 Å, the probability of escaping geminate recombination would then be $\exp(-300/2168) = 0.871$, which is much too high compared with the experimental value of ~0.03, (see Chapter 9). Mozumder and Magee (1967) introduced a more efficient mechanism of energy loss for molecular liquids by invoking intermolecular vibrational excitation. The qualitative features of this process have been detailed in the first paragraph of Sect. 8.1. Figure 8.3 shows the geometric and energetic relationship for electron thermalization in liquid hexane in the presence of the coulombic field of the geminate positive ion. The electron becomes subvibrational at O, a distance R_v from the positive ion. It is thermalized at P, at distance R_T, suffering a random walk (r, θ) in N collisions of mean free path L. The geometric and energy relations give

$$R_T^2 = R_v^2 + 2rR_v \cos \theta + r^2 \tag{8.4}$$

and

$$E_v - \frac{e^2}{\varepsilon R_v} = \frac{3}{2} k_B T - \frac{e^2}{\varepsilon R_T} + pN\hbar\omega, \tag{8.5}$$

where ε is the dielectric constant, T is the absolute temperature, k_B is the Boltzmann constant, and p is the probability of exciting an intermolecular vibration of quantum $\hbar\omega$ per elastic scattering. With $N\gg1$, the probability density of arriving at P in N steps is given by

$$W(\vec{r}, N) = \left(\frac{2\pi NL^2}{3}\right)^{-3/2} \exp\left(-\frac{3r^2}{2NL^2}\right). \tag{8.6}$$

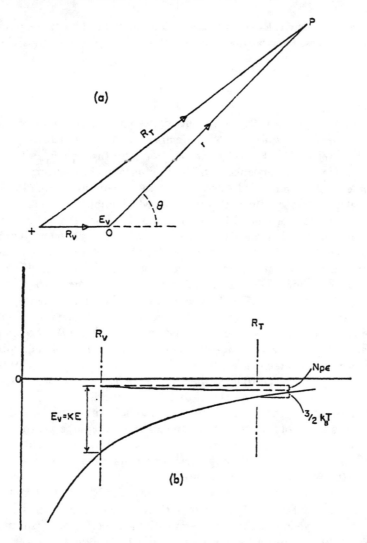

FIGURE 8.3 Geometric and energetic relationship for electron thermalization by random walk in liquid hexane in the presence of the geminate positive ion. Here $\varepsilon = \hbar\omega$. Reproduced from Mozumder and Magee (1967), with the permission of Am. Inst. Phys.©.

Given (E_v, R_v) the probability density of thermalization around P may now be computed by the following sequence of steps: (1) fix (r, θ) and compute R_T from Eq. (8.4); (2) evaluate N from Eq. (8.5) with assumed values of p and $\hbar\omega$; then (3) obtain W from Eq. (8.6). Finally, a consistency check should be made so that $r \leq NL$. Mozumder and Magee treated p as a parameter, taking $\hbar\omega = 0.01$ eV (*vide supra*), but the calculations depend only on the product $p\hbar\omega$, not on the individual factors. They calculated the weighted average escape probability $\rho(T)$ as follows:

$$\rho(T) = \int d\vec{r}\ W(\vec{r}, N)P(R_T, T). \tag{8.7}$$

In Eq. (8.7), $P = \exp(-r_c/R_T)$ is the Onsager escape probability for thermalization at distance R_T, with $r_c = e^2/\epsilon k_B T$. Temperature not only affects r_c, it influences R_T directly (see Eq. 8.5) and W indirectly through R_T, all of which contribute to the T-dependence of ρ. The authors found best agreement with the experiments of Hummel and Allen (1966) and of Hummel *et al.* (1966) in *n*-hexane at room temperature with $p = 0.055$, which implies, on the average, one excitation of intermolecular vibration in about 18 elastic scatterings. With no further adjustment, they calculated the variations of the free-ion yield in hexane with temperature and radiation quality. These calculations were in excellent agreement with observation; electrons from the decay of ^{37}Ar and ^{3}H were used to vary the radiation quality (Hummel *et al.*, 1966; Hummel, 1967). In all calculations, R_v was taken as 17 Å. This value is about the same as the stopping distance in water according to the Samuel-Magee theory. Although Mozumder and Magee employed a somewhat different procedure to obtain R_v, the underlying mechanism is probably the same—that is, excitation of *intramolecular* vibration. An *effective thermalization length* may be defined by $\langle R_{th} \rangle = r_c /\ln[1/\rho(T)]$. Mozumder and Magee (1967) give the variation of $\langle R_{th} \rangle$ with T for *n*-hexane with a comparison of calculated and measured free-ion yields.

The rationale for the value of $R_v \sim 17$ Å has been discussed before (see the second paragraph of Sect. 8.1). In hydrocarbon liquids such as *n*-hexane, it is expected that the vibration most effectively excited by electron impact is the CH stretch ($\sim$3000 cm^{-1}), although smaller quanta can also be excited. In any case, an effective upper limit in the subvibrational energy regime should be $\sim$0.5 eV, while the best agreement has been obtained for $E_v = 0.4$ eV. With thusly obtained values of E_v and R_v, Mozumder and Magee (1967) adjusted $p\hbar\omega$ to yield a free-ion yield in close agreement with the experiment of Hummel *et al.* (1966) at 293 K with an assumed total ionization yield of $\sim$4.0. The so-obtained value was 5.5×10^{-4} eV per collision, or a probability of 1 in 18 elastic collisions of exciting an intermolecular vibrational quantum of $\sim$0.01 eV. In all these calculations, the mean free path of elastic collision was taken equal to the intermolecular separation, $\sim$5 Å. The calculated values at other temperatures agreed well with experiment. In these calculations, and especially in the calculations for the free-ion yield with isotopic radiation ^{3}H and ^{37}Ar, allowance was made for nonisolated ion pairs in blobs and short tracks by first considering an inverse square dependence of generation of secondary electrons in energy, and then considering only the last ion-pair of such secondary (and higher-generation) tracks as contributing to the free-ion yield. Other intermediate ion-pairs were assumed to have quickly neutralized. The calculated G values of free-ion yields in n-hexane at 293 K and at 203 K for ^{37}Ar irradiation were 0.060 and 0.048 compared with the experimental value of 0.05 at 293 K (Hummel *et al.*, 1966).

Mozumder and Magee (1967) did not obtain the thermalization distance distribution directly, but evaluated an *effective thermalization length* (ETL) via the computed free-ion yield. Rassolov (1991) determined that distribution in n-hexane following essentially the same methodology, with two improvements: (1) true thermalization was invoked through both energy loss and energy gain collisions of the electron with intermolecular phonons; (2) the effect of an external field on the thermalization distribution was investigated. Considering phonons as harmonic oscillators, the ratio of energy gain and loss rates is given, with due consideration of state populations and degeneracies, by $\omega_g/\omega_1 = \exp(-\hbar\omega/k_BT)/(1 - \hbar\omega/E)$, where E is electron kinetic energy. Averaged with respect to Maxwellian distribution, these two rates would be equal, as required by detailed balance. Rassolov uses Metropolis *et al.*'s (1953) criterion for thermalization imposed on the terminal energy. The calculated thermalization distribution, is shown in Figure 8.4 for a subvibrational electron in n-hexane at 290 K starting an initial distance 23 Å from the geminate positive ion; other parameters used are the same as in Mozumder and Magee (1967). In the main, the distribution resembles a displaced gaussian with a peak at ~7 nm and a half-width ~2.5 nm. In the presence of an external field, the thermalization distribution is expected to be slightly skewed along the field direction. The calculated distribution $g(r, \theta)$ shows such an effect. When expanded in Legendre polynomials of cos θ, where θ is the angle between the radius vector and the field direction, it is mainly the spherically symmetrical zeroth-order polynomial that contributes to the free-ion yield after angular averaging over the field-dependent Onsager probability (see Chapter 9). The first-order Legendre term in the thermalization distribution function vanishes on integration. The second-order term, proportional to the square of the external field, makes a small contribution to the escape probability, but this is not perceptible up to the highest field yet experimentally studied. The net result is that the slope-to-intercept ratio of the of the free-ion yield as a function of the external field in the linear region remains the same as if the field had no effect on the thermalization distance (see Chapter 9). The calculated escape probability in n-hexane at 290 K agrees well with the free-ion yield measurement of Mathieu *et al.* (1967) up to the highest field, ~180 KeV/cm.

In hydrocarbon liquids other than n-hexane, the procedure for obtaining the thermalization distance distribution could conceivably be the same. However, in practice, a detailed theoretical analysis is rarely done. Instead, the free-ion yield extrapolated to zero external field (see Chapter 9) is fitted to a one-parameter distribution function weighted with the Onsager escape probability, and the mean thermalization length $\langle r_{th} \rangle$ is extracted therefrom (see Mozumder, 1974;

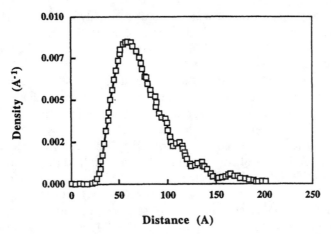

FIGURE 8.4   Electron thermalization distance distribution in n–hexane at 290K starting from an initial separation 23Å. See text for details. Reproduced from Rassolov (1991).

Dodelet and Freeman, 1975; Jay-Gerin *et al.*, 1993). The main outcome from such analysis is that the free-ion yield, and therefore by implication the $\langle r_{th} \rangle$ value, increases with electron mobility, which in turn increases with the sphericity of the molecule. The heuristic conclusion is that the probability of intermolecular energy losses decreases with the sphericity of the molecule, since there is no discernible difference between the various hydrocarbons for electronic or intramolecular vibrational energy losses. The $\langle r_{th} \rangle$ values depend somewhat on the assumed form of distribution and, of course, on the liquid itself. At room temperature, these values range from ~25 Å for a truncated power-law distribution in n-hexane to ~250 Å for an exponential distribution in neopentane.

Jay-Gerin *et al.* (1993) have sought empirical correlation between the zero-field free-ion yield G_{fi} and electron mobility μ by examining 52 nonpolar liquids, including liquefied rare gases. For low-mobility liquids ($\mu \leq 0.1 \text{ cm}^2\text{v}^{-1}\text{s}^{-1}$), they find $G_{fi} \sim 0.1$ per 100 eV, which is fairly independent of mobility. Above that mobility, the free-ion yield can be fitted to $G_{fi} = \alpha\mu^n$, with $n = 0.31 \pm 0.05$ and $\alpha = 0.21 \pm 0.02$. No special meaning has been attached to the fitting parameters. The most probable thermalization distance b obtained by fitting the free-ion yield to a presumably gaussian distribution can be expressed through the equation $G_{fi} = G_{\infty} \exp(-B/\varepsilon_s b)$, where ε_s is the static dielectric constant and B and G_{∞} are fitted constants. The thusly obtained value of $G_{\infty} = 2.1 \pm 0.3$, however, is too low to be taken as the total ionization yield.

While the variation of the mean thermalization length among different liquid hydrocarbons under high-energy irradiation has been well documented (see, e.g., Schmidt and Allen, 1968, 1970), the question of the dependence of thermalization length on electron energy became apparent only after laser photoionization with picosecond time resolution became available. With high-energy irradiation, the incident electron energies are so large ($\gg 100$ eV) that there is an inherent averaging with respect to a well-established subexcitation electron energy spectrum, sometimes called the "entry spectrum." The primary energy then becomes irrelevant. With photoionization, the initial electron energy can be controlled much better than 1 eV; direct observation of the e–ion recombination process can also be made with time resolution much better than 1 ns (see Choi et al., 1982; Schmidt et al., 1990; Hirata and Mataga, 1991, and references therein) Although the determination of mean thermalization length from these experiments is indirect, requiring a deconvolution connecting the kinetics of recombination with the thermalization distribution, there is convincing evidence that the mean thermalization length increases with electron energy in the few-eV regime.

Hirata and Mataga (1991) monitor the time-dependent absorption of a solution of BDATP [2,7-bis(dimethylamino)-4,5,9,10-tetrahydropyrene] in n-hexane at room temperature following picosecond laser pulse excitation at 280 nm and above. Absorption, monitored at various wavelengths ≥ 470 nm, is due to both the excitation of BDATP and to the BDATP cation radical. The delayed absorption, however, is only due to the cation, which can be separated and attributed to the e–ion recombination process. The observed decay of the geminate pair was fitted to delta function, gaussian, and exponential forms of the thermalization distance distribution using Bartczak and Hummel's (1987) Monte Carlo simulation. All three distributions gave acceptable evolutions of decay when compared with experiment with somewhat different mean thermalization distances. The variation of the b value of the gaussian thermalization distribution in n-hexane with excess electron energy is shown in Figure 8.5. It is seen that b starts to increase with E_{ex} at ~2 eV but tends to saturate beyond 3 eV, probably because of the existence of a low-lying excited state of the molecule at nearby energy. Similar results have been obtained by Choi et al. (1982) and by Schmidt et al. (1990), getting the b value in n-hexane ~5–6 nm at a shorter-wavelength excitation.

Guelfucci et al. (1997) have extended their earlier semiempirical method for determining the electron thermalization distance distribution in some nonpolar liquid hydrocarbons. The method relies on the fitting of experimental free-ion yield to the reciprocal of a linear function of the field. From fitting parameters, they compute moments of the distribution function with respect to the Onsager escape probability. The authors prefer a modified exponential distribution for electrons created by vacuum UV-irradiation, whereas a gaussian distribution could be used for high-energy irradiation.

It has been shown by Mozumder and Tachiya (1975) that, within the context of the diffusion model, the probability of generation of free ions is independent of *postthermal* electron scavenging, both in the absence and presence of an external field. Thus, the experimental finding—that the free-ion yield is reduced in neopentane (NP) by the addition of electron attaching solutes SF_6,

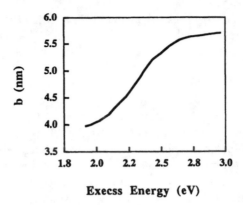

$$b \text{ (nm)}$$

Execss Energy (eV)

FIGURE 8.5 Variation of the b–value of the gaussian thermalization distribution in n–hexane with excess electron energy. While the b values were obtained by comparison with experimental results, other forms of initial distribution are also possible. Reproduced from Hirata and Mataga (1991).

CCl_4, and CS_2 (Schmidt, 1970; Schmidt and Allen, 1970), is interpreted on the basis of *epithermal* electron scavenging. Since the thermalization distance is long in NP, this possibility exists even at a relatively small concentration of an efficient scavenger. On a crooked path length of the epithermal electron between y and $y + dy$, the electron capture probability is given by $\phi_c \, dy = n\sigma \exp(-n\sigma y) \, dy$, where n is the scavenger concentration and σ is the capture cross section. If the electron executes N random walks of mean free path L to the capture point, then $y = NL$ and the probability of arriving at a vector distance between r and $r + dr$ is given by $4\pi r^2 W(r) \, dr$, where $W(r) = (2\pi NL^2/3)^{-3/2} \exp(-3r^2/2NL^2)$. Thus, the distribution of the negative ions formed by epithermal electron scavenging is given by $4\pi r^2 f(r) \, dr$, where

$$f(r) = \int_0^\infty \phi_c(y) W(r, y = NL) \, dy = \frac{3n\sigma}{2\pi L r} \exp\left[-\left(\frac{6n\sigma}{L}\right)^{1/2} r\right].$$

Replacement of the upper limit of integration y by ∞ causes little error since y is much greater than r. Similarly, replacing the lower limit r by zero causes correspondingly little error when used with the Onsager escape probability. The escape probability as a negative ion can now be expressed as

$$<P(n)> = \int_0^\infty x \exp\left[-\left(x + \frac{\gamma}{x}\right)\right] dx,$$

where $\gamma = r_c (6\pi\sigma/L)^{1/2}$.

This expression can be generalized in the presence of an external field, and the ratio of the escape probability as a negative ion to that as an electron in the absence of a scavenger computed as a function of the external field. From such an analysis and taking $L = 4$ Å, a typical intermolecular separation, Mozumder and Tachiya obtained electron attachment cross sections in NP as 4×10^{-16}, 5×10^{-17}, and 1×10^{-18} cm^2, respectively, for SF$_6$, CCl$_4$, and CS$_2$ with ~15% uncertainty.

8.3.2 ELECTRON THERMALIZATION IN POLAR MEDIA

Calculation of the electron thermalization distance in polar liquids is a difficult task for two reasons. First, the elastic and inelastic cross sections are not well known (however, see the later discussion for *solid* water). Second, it has never been clearly demonstrated that electrons in polar media thermalize first and then undergo trapping and solvation, although such a conjecture has been made on the basis of available theoretical and experimental evidence (Mozumder, 1988). An early attempt by Samuel and Magee (1953) used a classical random walk method for electron thermalization distance and time in liquid water with a starting energy ~15 eV. The procedure included the effect of the positive ion, which has generally been ignored in later work. They used a fixed scattering length and a fixed fractional energy loss per collision, both of which are suspect in view of vibrational and rotational processes. Kinetic energy loss in this model is partly due to inelastic collisions and partly working against the coulombic attraction. Taking the density-normalized scattering length from Bruche's (1929) gas phase work and a fractional energy loss in the interval 0.025–0.05, they computed $t_{th} \sim 2.8 \times 10^{-14}$ s and of thermalization distance ~12–18 Å.

Mozumder's (1988) conjecture on electron thermalization, trapping and solvation time scales in liquid water is based on combining the following theoretical and experimental information:

1. Migus *et al.*'s (1987) delineation of the formation of a primary species absorbing in the IR, which develops in ~110 fs and which transforms to the well-known spectrum of the hydrated electron in ~240 fs, which is consistent with the longitudinal dielectric relaxation time of water (Mozumder, 1969a, b).

2. Schnitker *et al.*'s (1986) finding, based on classical molecular dynamics simulation, of a large density (4.4 ml^{-1} at 10°C) of local potential minima qualifying as trapping sites.

3. The experimental demonstration by Knapp *et al.* (1987) that an electron can be bound to water clusters $(H_2O)_n$ under collision-free condition for $n > 10$.

4. Extrapolation (admittedly very approximate) of the thermalization time in humid air by microwave conductivity method (Warman et al., 1984) giving $t_{th} \sim 4.5 \times 10^{-15}$ s for unit water fraction.

5. Extrapolation, to liquid density, of thermalization time in gaseous water (also approximate) by Christophorou et al. (1975), based on drift velocity and transverse diffusion coefficient measurement, which gives $t_{th} \sim 2.0 \times 10^{-14}$ s.

6. Computation based on the Frohlich–Platzman (1953) equation, which gives $t_{th} = 5 \times 10^{-14}$ s for liquid water.

7. Calculation based on the stopping cross section implied by the experiments of Michaud and Sanche (1987) in solid water, giving $t_{th} \sim 2 \times 10^{-14}$ s.

Mozumder (1988) argues that a large fraction of incipient trapping sites found by Schnitker et al. do not bind the thermalized electron but merely scatter these because of insufficient potential strength. Imposing a quantum restriction on the volume of the phase space, the density of possible trapping sites in water, obtained from the work of Schnitker et al., turns out to be ~0.74 ml^{-1}. The model for electron traps in liquid water is then a trap of ~4 Å width and a potential depth of ~0.58–0.72 eV. Such traps would have a trapping cross section ~20 Å^2 and a binding energy ~$k_B T$. Thermalized electrons would then need to lose only a few $k_B T$ of energy to get trapped. On the basis of this kind of analysis, Mozumder concludes that the 10^{-13} s time scale is very important in liquid water, being dominated by energy loss of epithermal electrons. Thermalization, trapping, and solvation can then follow in quick succession.

Frohlich and Platzman's (1953) model for the energy loss of subexcitation electrons in highly polar media has already been described in Sect. 8.1. In liquid water, the time scale of thermalization according to this model is ~10^{-13} s. Notice that their equation gives an actual time rate and the authors do not hazard a stopping power computation, presumably due to lack of knowledge of the elastic scattering cross section. The calculation is for a straight electron trajectory and, furthermore, ignores direct vibrational excitation at close encounters. The energy loss is due to dipolar rotation of the Debye type. Since the rate of energy loss is independent of electron energy, the stopping time is proportional to the starting energy. Magee and Helman (1977) removed the linear trajectory limitation by using Monte Carlo simulation and a numerical fast Fourier transform technique. Their result for the time rate of electron energy loss on the diffusive trajectory is essentially similar to that of Frohlich and Platzman for the straight trajectory, except possibly for some pathological cases. In either scheme, vibrational excitations are ignored.

Rips and Silbey (1991) have reexamined the thermalization of photoelectrons (of a few eV in energy) with a master equation approach for the time rate of energy loss. Their method is quite general, and it includes both direct (energy loss) and inverse (energy gain) collisions according to the principle of detailed balance. As in the Frohlich-Platzman method, they first calculate the time rate of energy loss, thereby getting the thermalization time distribution. From this, an approximate thermalization distance distribution is obtained by invoking the time (t)–distance (r) relationship according to the chosen form of trajectory. We will discuss here only the diffusive trajectory, which is more realistic. In this, time and distance are related by $\langle r^2 \rangle = 6Dt$ and the authors effectively take a time-independent diffusion coefficient of the form $D = (1/6)va$, where v is the electron velocity and a is the intermolecular separation. In application, Rips and Silbey follow the Frohlich–Platzman model closely, also ignoring excitation of molecular vibration.

Considering an initial electron energy much larger than $k_B T$, Rips and Silbey show that the distribution of thermalization time is given by the first two moments of the energy loss function $\omega(\varepsilon)$ per unit time,

$$\langle \varepsilon \rangle = \int_{-\infty}^{\infty} d\varepsilon \; \varepsilon \omega(\varepsilon)$$

$$\langle \varepsilon^2 \rangle = \int_{-\infty}^{\infty} d\varepsilon \; \varepsilon^2 \omega(\varepsilon).$$

They derive $f(\varepsilon_{in}, t) = (\langle \varepsilon \rangle^2 / 2\pi t \langle \varepsilon^2 \rangle)^{1/2} \exp[-(\varepsilon_{in} - t\langle \varepsilon \rangle)^2 / 2t\langle \varepsilon^2 \rangle]$, where $f(t) \, dt$ is the probability that thermalization will occur between t and $t + dt$ and ε_{in} is the initial energy in units of $k_B T$. From this an asymptotic gaussian distribution of thermalization distance follows. This distribution can be written as

$$\phi(r) = C \exp\left(-\frac{r^2}{a^2 u_2 \varepsilon_{in}^{1/2}} - \frac{a^2 u_1^2 \varepsilon_{in}^{5/2}}{u_2 r^2} \right)$$

with

$$u_1 = \left| \langle \varepsilon \rangle \right|^{-1} \left(\frac{2k_B T}{ma^2} \right)^{1/2}$$

and

$$u_2 = \frac{2u_1 \langle \varepsilon^2 \rangle}{\langle \varepsilon \rangle},$$

where C is a normalization constant. Using the Frohlich-Platzman model, the authors find that the average thermalization distance varies as $\varepsilon_{in}^{3/4} \tau^{1/2}$, where τ is the dielectric relaxation time. As yet, this dependence has not been tested experimentally. The computed average thermalization distance in liquid water for an initial energy of 1 eV is ~12.7 nm, which is much greater than what is required by the spur diffusion model—that is, ~4 nm. Neglect of vibrational excitation could be the main reason for the discrepancy. Another contributing factor may be the use of a fixed diffusion coefficient.

Goulet *et al.* (1990; Goulet and Jay-Gerrin, 1988) have used Monte Carlo simulation for the electron thermalization time and the distance distribution in *solid* water using the elastic and inelastic cross sections obtained from the experiments of Michaud and Sanche (1987) on thin films irradiated by energetic electrons. In these calculations, thermalization is defined to be when the mean electron energy $\langle E \rangle$ falls below 0.205 eV, the smallest *intra*molecular vibrational quantum, for the first time. Electrons of smaller energy are assumed to be instantly trapped and solvated. The effect of the geminate positive ion on thermalization was found to be minor, accounting for ~5% recombination prior to thermalization. Similarly, prethermal dissociation was found to be insignificant, ~3%.

With a starting energy of 3 eV, the thermalization distance distribution showed a peak around 10 nm. Although the distribution resembles a displaced Maxwellian near the peak with a half-width ~7.5 nm, it has a long tail. The mean and median thermalization lengths are therefore very different. Calculated thermalization time and distance increase with energy; however both tend to plateau at ~8 eV. For example, the mean thermalization times for starting electron energies 0.5 and 7.2 eV are 18 and 198 fs, respectively. The corresponding mean thermalization distances are 2.5 and 38.4 nm, respectively. The authors calculate a total thermalization distance distribution using a Platzman-like subvibrational entry spectrum of the form $0.053/7.4 + 181.2/(E_0 + 8.3)^3$ where E_0 is electron energy and the first electronic excitation potential has been taken to be 7.4 eV. This total distribution is highly skewed. Although there is a sharp peak at ~1.5 nm, the mean thermalization distance is calculated to be 14 nm, again too large compared with the required spur size in *liquid* water radiolysis. On the other hand, the calculated mean distance may be correct for solid water, for which there is not enough experimental work to compare with. The computed mean thermalization time, averaged over the entry spectrum, is 62 fs; the results are claimed to be consistent with Mozumder's (1988) conjecture. The mean recombination and dissociation probabilities prior to thermalization are computed to be 0.047 and 0.033, respectively.

8.4   ELECTRON THERMALIZATION IN HIGH-MOBILITY LIQUIDS

In liquids where the thermal electron mobility is high, it is expected that the electron will encounter less resistance to motion on the way to thermalization. Therefore, the thermalization distance would be longer resulting in greater free-ion yield. In such liquids, one then expects a correlation between mobility and free-ion yield. Sano and Mozumder (1977) employed a Fokker-Planck method for thermalization in such liquids. They applied the model to three high-mobility liquids, tetramethylsilane (TMS), methane, and neopentane (NP). The method was considered unsuitable for low-mobility hydrocarbons because of excessive trapping. Very high mobility liquefied rare gases were also ruled out, because in these liquids the subexcitation electrons start with much higher energies in the absence of molecular vibrations, so that a constant momentum relaxation time, which is one of the model assumptions, cannot remain valid.

8.4.1 THE FOKKER-PLANCK APPROACH TO THERMALIZATION

The Fokker-Planck equation is essentially a diffusion equation in phase space. Sano and Mozumder (SM)'s model is phenomenological in the sense that they identify the energy-loss mechanism of the subvibrational electron with that of the quasi-free electron slightly heated by the external field, without delineating the physical cause of either. Here, we will briefly describe the physical aspects of this model. The reader is referred to the original article for mathematical and other details. SM start with the Fokker-Planck equation for the probability density W of the electron in the phase space written as follows:

$$\frac{\partial W}{\partial t} + \vec{v} \cdot \mathrm{grad}_{\vec{r}} W + \vec{\Gamma} \cdot \mathrm{grad}_{\vec{v}} W = \beta \, \mathrm{div}_{\vec{v}}(W\vec{v}) + q \nabla_{\vec{v}}^2 W.$$

Here $\vec{r}$ and $\vec{v}$ are respectively the electron position and velocity, $\Gamma = -(e^2/\varepsilon m)(\vec{r}/r^3)$ is the acceleration in the coulombic field of the positive ion and $q = \beta k_B T/m$. The mobility of the quasi-free electron is related to β and the relaxation time τ by $\mu = e/m\beta = e\tau/m$, so that $\beta = \tau^{-1}$. In the spherically symmetrical situation, a density function $n(v_r, v_t, t)$ may be defined such that $n \, dr \, dv_r \, dv_t = W \, d\vec{r} \, d\vec{v}$; here, v_r and v_t and are respectively the radical and normal velocities. Expectation values of all dynamical variables are obtained from integration over n. Since the electron experiences only radical force (other than random interactions), it is reasonable to expect that its motion in the v_t-space is basically a free Brownian motion only weakly coupled to r and v_r by the centrifugal force. The correlations[1], $\kappa(r, v_t^2)$ and $\kappa(v_r, v_t^2)$ are then neglected. Another condition, $\sigma(r)^2 \ll \langle r \rangle^2$, implying that the electron distribution is not too much delocalized on r, is verified a posteriori. Following Chandrasekhar (1943), the density function may now be written as an uncoupled product, $n = gh$, where

$$g(v_\psi, t) = \frac{v_\psi}{\langle v_\psi^2 \rangle} \exp\left(-\frac{v_\psi^2}{2\langle v_\psi^2 \rangle}\right)$$

---

[1] The correlation $\kappa(A, B)$ of two dynamical variables A and B is defined by $\kappa(A, B) = \langle (A - \langle A \rangle)(B - \langle B \rangle) \rangle$, where $\langle \, \rangle$ refers to the expectation value. The standard deviation $\sigma(A)$ is defined by $\sigma(A)^2 = \kappa(A, A)$.

and

$$h(r, v_r, t) = (2\pi C)^{-1}(FG - H^2)^{-1/2} \exp\{-[G(r - \langle r \rangle)^2 - 2H(r - \langle r \rangle)$$
$$(v_r - \langle v_r \rangle) + F(v_r - \langle v_r \rangle)^2]/2(FG - H^2)\}.$$

In the preceding $F = \kappa(r, r)$, $H = \kappa(r, v_r)G = \kappa(v_r, v_r)$ and the normalization constant C is fixed by equating the volume integral of n to unity. For further tractability, Sano and Mozumder expand $\langle r^{-p} \rangle$ in a Taylor's series and retain the first two terms only. The validity of this procedure can be established *a posteriori* in a given situation. At first, the authors obtain equations for the time derivatives of the expectation values and the correlations of dynamical variables. Then, for convenience of closure and computer calculation, these are transformed into a set of six equations, which are solved numerically. The first of these computes lapse time through the relation

$$t = \int \frac{d\langle r \rangle}{\langle v_r \rangle}.$$

The rest are first-order simultaneous equations of $\langle v_r \rangle$, $\langle v_\psi^2 \rangle$, $\kappa(r, r)$, $\kappa(r, v_r)$, and $\kappa(v_r, v_r)$ against the variable $\langle r \rangle$. To solve these equations, it is supposed that the electron starts from a given position with a given radial velocity—that is, at $t = 0$, $\langle r \rangle = r_0$, $v_r = v_0$, $\langle v_\psi^2 \rangle = 0$, and all correlation functions equal zero. The interpretation here is that v_0 is the subvibrational velocity (3.75×10^7 cm/s) and r_0 is the distance where it first occurs. The model treats r_0 as a parameter. However, agreement with experimental free-ion yield in molecular liquids such as tetramethylsilane (TMS), methane, and neopentane (NP), where the electron may be taken as reasonably quasi-free, is obtainable only for a narrow range of r_0. Thus, $r_0 = 4.0$ nm was always used. With the initial conditions, solution of the final set of equations asymptotically give a Maxwellian velocity distribution with $\langle K \rangle \to (3/2)k_B T$, where K is kinetic energy. An operational thermalization time t_{th} is defined so that $\langle K \rangle(t_{th}) = 1.05(3/2)k_B T$, the numerical factor 1.05 signifying ~5% error in numerical calculations. The mean thermalization distance $\langle r \rangle_{th}$ and its dispersion $\sigma(r)_{th}$ are now evaluated at $t = t_{th}$ and the distribution of thermalization distance is given as follows:

$$f(r) = C^{-1}(2\pi)^{-1/2}\sigma(r)_{th}^{-1} \exp\left[-\frac{\left(r - \langle r \rangle_{th}\right)^2}{2\sigma(r)_{th}^2}\right] \qquad (t = t_{th}).$$

With this distribution, the escape probability is calculated using the Onsager (1938) formula (see Chapter 9),

$$P(\text{cal}) = \int_0^\infty f(r) \exp\left(-\frac{r_c}{r}\right) dr$$

and compared with experiment. In the SM model, the connection between mobility (μ) and free-ion yield (G_{fi}) is made through the relaxation time τ. Analysis of temperature dependence of mobility gives τ through the quasi-free mobility μ_{qf} (see Chapter 10). With estimated r_0 and V_0, τ generates the thermalization distance distribution, which in turn gives the free-ion yield through the Onsager formula. Thus, unlike in most other models, this relationship here is deduced, not conjectured. The quasi-free electron mobilities in TMS, methane, and NP, obtained from measured mobilities and activation energies at 296 K, 118 K, and 296 K, are 100, 510, and 156 cm²v⁻¹s⁻¹ respectively. The momentum relaxation time is τ are then computed to be 5.7, 29.0, and 8.9 × 10⁻¹⁴, respectively. Using these values, SM calculate electron thermalization times in these liquids as 9.4, 64.0, and 14.8 × 10⁻¹⁴ s, respectively. The mean thermalization distance and its standard deviation are given (in nm) respectively by (17.6, 4.2) for TMS, (70.0, 18.0) for methane, and (24.7, 6.7) for NP. The calculated escape probabilities for these liquids, 0.17, 0.28, and 0.27, compare reasonably well with the respective experimental values, 0.17, 0.25, and 0.20.

Silinsh and Jurgis (1985) extended the SM model for geminate charge-pair separation in pentacene crystals to include the presence of an external electric field. Silinsh *et al.* (1989) further applied the model to polyacene crystals such as naphthalene and anthracene. In these comprehensive calculations, the effect of the external field was investigated in all of its aspects. First, the force exerted by the field appears as an additive term in Γ in the Fokker-Planck equation (*vide supra*). Then, it modifies the relaxation process and introduces a drift during the thermalization process. Silinsh and his associates utilize the mathematical structure of the SM model, however, modifying the physical basis to suit the situation. The thermalizing entity is now not a quasi-free electron but an adiabatic nearly small molecular polaron (MP), generated by photoionization and having initial kinetic energy ~1 eV. The effective mass m_{eff} has been shown to increase exponentially with temperature in pentacene, reaching ~100–1000m in the 165–300 K range, where m is the bare electron mass. The energy loss mechanism is phonon emission, and the thermalization time scale is determined by the inverse of the relaxation rate β. Thus, for a one-phonon process, $t_{th} \approx \beta^{-1} = n/v_{ph}$; $n = E_k/hv_{ph}$, where v_{ph} is phonon frequency, E_k is the initial energy, and n is the number of phonons required for thermalization. Considering optical phonons with energy quantum ~0.015 eV and $E_k = 1$ eV, the authors compute $t_{th} = 2 × 10^{-11}$ s, or $\beta = 5.0 × 10^{10}$ s⁻¹, which is considered a lower limit. An upper limit ~10¹³ s⁻¹ is urged on the basis of the characteristic hopping rate of a charge carrier in the *ab* plane. Since multiphonon scattering is the more likely process for thermalization, the authors settle for a value of $\beta = 10^{12}$ s⁻¹. The rest of the procedure is as in the SM model—that is, the calculation of expectation values and correlations of dynamical variables through the moment equations.

Over the temperature interval 165 K to 300 K, the calculations of Silinsh and Jurgis (1985) indicate that the thermalization rate in pentacene decreases from 3×10^{12} to 0.8×10^{12} s^{-1}. The trend is opposite to what would be expected in liquid hydrocarbons and may be attributed to the rapid increase of m_{eff} with temperature. The calculated mean thermalization distance increases with incident photon energy fairly rapidly, from ~3 nm at 2.3 eV to ~10 nm at 2.9 eV, both at 204 K. With increasing temperature, $\langle r \rangle_{th}$ decreases somewhat. These thermalization distances have been found to be consistent with the experimental photogeneration quantum efficiency when Onsager's formula for the escape probability is used.

In the presence of an external electric field, the calculated escape probability, *including the effect of the field on thermalization distance distribution*, increases uniformly with the field. It also increases with the incident photon energy, rapidly at first and then saturating around 3 eV. Over a field strength of 0–8×10^4 V·cm^{-1}, Silinsh and Jurgis found good agreement between the calculated escape probability and the experimental photogeneration quantum efficiency in pentacene. The progress of thermalization found by the authors is similar to that in the SM model for hydrocarbons. Starting with $E_k = 57k_B T$ at $r_0 = 0.79$ nm, the authors find little distributional width when about half the kinetic energy is lost and the mean radial distance has increased to 3 nm. At $E_k = 9k_B T$, the distributional width is significant at 0.42 nm and the mean radial distance is 6 nm. At thermalization, $\langle r \rangle = 9.3$ nm and the width of the radial distribution has increased to 1.3 nm, still much smaller than $\langle r \rangle$. Because of the field-induced drift, the thermalization distance distribution is asymmetrical. For example, with a field of 1.2×10^5 V·cm^{-1} in pentacene, the calculated mean thermalization lengths along and opposite to the field direction are 12.5 and 8.0 nm, respectively. Such "maximum" thermalization lengths increase both with incident photon energy and the field strength.

Silinsh *et al.* (1989) applied their thermalization procedure to naphthalene and anthracene at low temperatures, ~35 K or less. A stationary state was envisaged in the presence of an external field. Calculations have been performed for the saturation drift velocity, friction coefficients, and effective mass as functions of the external field. The conclusions are almost the same as for pentacene.

8.4.2   ELECTRON THERMALIZATION
IN LIQUEFIED RARE GASES

Liquefied rare gases(LRGs) are very important both from the fundamental point of view and in application to ionization chambers. In these media, epithermal electrons are characterized by a very large mean free path for momentum transfer ~10–15 nm, whereas the mean free path for energy loss by elastic collision is only ~0.5 nm. This is caused by coherence in momentum transfer scattering exhibited by a small value of the structure factor at low momentum transfers

(Lekner, 1967; Lekner and Cohen, 1967). From the experimental viewpoint, LRGs are excellent materials for the operation of ionization chambers, scintillation counters, and proportional counters on account of their high density, high electron mobility, and large free-ion yield (Kubota *et al.*, 1978; Doke, 1981). Since the probability of free-ion formation is intimately related to the thermalization distance in any model (see Chapter 9), at least a qualitative understanding of electron thermalization process is necessary in the LRG.

Early determination of thermalization length in LRG by Robinson and Freeman (1973) gave 133, 88, and 72 nm respectively in LAr, LKr, and LXe. These are not independent valves, but were obtained by fitting the Onsager escape probability, averaged over a gaussian distribution of thermalization distances, to the experimental ratio of free-ion yield at zero external field to the total ionization yield. The values quoted are the fitted *b parameters* (standard deviation) of the gaussian distribution. In these experiments, both the free-ion yield and the total ionization yield are unacceptably large, which, in part, may be due to imprecise dosimetry (Aprile *et al.*, 1993). Since the only available mechanism of energy loss of subexcitation electrons in LRG is elastic collision, the decrease of thermalization length with the mass of rare gas atom is not consistent. A later measurement by Huang and Freeman (1977) produced total ionization yields that are smaller by a factor of about 2. The evaluated thermalization lengths for the LRG were in the interval 100–200 nm, a typical value being 154 nm in LAr. These are *b* values of the gaussian distribution to which a power tail was added; however, no trend could be noticed among the liquefied rare gases.

Taking LAr as an example, more recent measurements give the zero-field escape probability as 0.35 (see Doke *et al.*, 1985). The Onsager length in LAr at 84 K is 127 nm, implying a mean thermalization length ~121 nm, which is almost the same as the inter-positive-ion separation (~124 nm) on a low-LET track (Burns and Mozumder, 1987). This shows that even at the minimum LET the ionizations produced in LRG are not isolated. In view of the nongeminate character of the problem, the Onsager model should not apply. Therefore, thermalization lengths extracted by using the Onsager formula in LRG are suspect. A similar problem has been addressed by Engler *et al.* (1993) in tetramethylsilane (TMS), where, under experimental conditions, the extracted initial electron–ion separation based on the Onsager equation is comparable to the mean inter-positive-ion separation. Engler *et al.* use a correction to the Onsager formula by incorporating a Debye screening factor due to neighboring ions. The ansatz is questionable for mixing spherical and cylindrical symmetries and for attaching the screening factor to the probability rather than to the potential. Nonetheless, the so-evaluated thermalization length in TMS, ~20 nm, compares favorably with that obtained by the Fokker-Planck method, ~18 nm (see Sect. 8.4.1). In another set of experiments, Thomas and Imel (1987) carefully measured the free-ion yield in LAr and LXe going down to very small fields. They rejected the Onsager model on the basis that the measured slope to intercept ratio near zero external field disagreed with the prediction of the Onsager theory (see Chapter 9). In their analysis, they disregarded the electron-ion interaction and also the diffusion term in the Jaffe' theory. Although the calculated escape probability as a function of the external field agrees well experiment with only one adjusted parameter, namely, the critical field E_c = 0.84 KV$\propto$cm^{-1}—no meaning can be given to it. Consequently, no thermalization length can be derived from such an analysis.

Subexcitation electrons in LRG have considerable energy, since the first excited state is close to the lowest ionized state. Thus, in LAr, with an ionization potential 14 eV, electrons can become subexcitational at 12 eV. In the absence of vibrational excitation and in view of the great disparity between the masses of the electron and the rare gas atoms, the calculated thermalization distance is expected to be so large that the effect of the electrostatic field of the ions on the thermalization distribution should be negligible. On the other hand, the motion of epithermal electrons in LRG is governed by two distinct transport mean free paths; one for momentum transfer, called Λ_1, and another for energy loss, called Λ_0. Although both are derived from the same differential cross section for elastic collision, only the first involves coherence through the liquid structure (Lekner, 1967; Lekner and Cohen, 1967). An approximate calculation by Mozumder (1982) replaces the structure factor at a given energy with that corresponding to mean momentum transfer at that energy. With this approximation, one gets Λ_1 = 15 nm in LAr at near thermal energies. It gradually falls with energy, the value at 6 eV being ~2 nm. The energy loss mean free path Λ_0 at near thermal energies is 0.7 nm, reaches a peak of 1.1 nm at ~1.5 eV, and then gradually falls to 0.3 nm at 6 eV. Since epithermal processes dominate the thermalization distribution, it can be assumed that $\Lambda_1 \gg \Lambda_0$ over the entire time scale. The thermalization can therefore be conceived as a special kind of random walk in which there is a sequence of several tens of very small energy-loss elastic collisions without any appreciable change in the direction of electron motion, followed by a large angle momentum transfer collision. The sequence then repeats itself until thermal equilibrium is established. This process of thermalization is quite different from what prevails in molecular liquids (see Sect. 8.3.1).

The mean energy loss in an elastic collision may be taken as $\delta(m/M)[\langle\varepsilon\rangle - (3/2)k_BT]$ where $\langle\varepsilon\rangle$ is the mean electron energy, m/M is the ratio of electron mass to that of the rare gas atom, and δ is a numerical parameter. The collision rate may be approximated by $\Lambda_0^{-1}(2\langle\varepsilon\rangle/m)^{1/2}$. The equation for the rate of energy loss may now be given as follows:

$$-\frac{d\langle\varepsilon\rangle}{dt} = \delta\left(\frac{m}{M}\right)\left(\langle\varepsilon\rangle - \frac{3}{2k_BT}\right)\Lambda_0^{-1}\left(\frac{2\langle\varepsilon\rangle}{m}\right)^{1/2}.$$

Ignoring the effect of the initial energy, since that is much greater than $k_B T$, the solution of the above equation gives the thermalization time t_{th} as

$$t_{th} = \frac{t_0}{\sqrt{6}} \ln \frac{\sqrt{\varepsilon_f / k_B T} + \sqrt{3/2}}{\sqrt{\varepsilon_f / k_B T} - \sqrt{3/2}}$$

where ε_f is the terminal energy and the time scale of thermalization is given by $t_0 = \Lambda_0 (2m/k_B T)^{1/2}(M/m\delta)$. Taking $\varepsilon_f = (1.1)(3/2)k_B T$ as usual (see Sect. 8.2), one gets $t_{th} = 1.44 t_0$. Further taking $M/m = 7.34 \times 10^4$ for Ar, $k_B T$ at 84 K, and $\delta = 2$, $\Lambda_0 = 0.6$ nm, $m = 0.4 m_e$, where m_e is the electron mass, as in Lekner (1967), one gets $t_0 = 0.99$ ns and $t_{th} = 1.4$ ns, which is comparable to the determination of Warman and Sauer (1975) in the gas phase. A slightly smaller thermalization time has been obtained by Mozumder (1982) by using the displaced pseudo-Maxwellian (DPM) approximation (see Sect. 8.2). This method, however, gives the variation of the electron diffusion coefficient with time during thermalization shown in Figure 8.6. The DPM procedure, including the time dependence of D, gives a gaussian distribution of thermalization distance with a standard deviation ~1000 nm. Since below ~1 eV, the most important energy regime, the momentum transfer mean free path Λ_1 varies little (Lekner, 1967), a very simple thermalization distance distribution may be obtained by taking it as a constant. The average fractional energy loss in an elastic collision $2m/M$ is very small, requiring a large number of such collisions for thermalization:

$$n_{th} = \frac{M}{2m} \ln \frac{\varepsilon_1 - 3k_B T/2}{\varepsilon_f - 3k_B T/2}$$

Here ε_1 and $\varepsilon_f = (1.1)(3/2)k_B T$ are respectively the initial and final electron energies. The special random walk with different mean free paths for energy and momentum transfer still generates a gaussian position distribution at thermalization, since $n_{th} \gg 1$. The initial energy is relevant but not too important in the subexcitation regime, since n_{th} depends logarithmically on it. The crooked path length to thermalization is given approximately by $n_{th}\Lambda_0$, within which there is, on average, a large angle momentum transfer scattering at every Λ_1. Therefore, the mean square deviation b^2 of the thermalization length is given by

$$b^2 = \frac{n_{th}\Lambda_0}{\Lambda_1} \Lambda_1^2 = \frac{M}{2m} \Lambda_0 \Lambda_1 \ln \frac{\varepsilon_1 - 3k_B T/2}{\varepsilon_f - 3k_B T/2}$$

Taking $\Lambda_1 = 15$ nm, $\varepsilon_1 = 5$ eV, and other values as before, the b value for LAr is evaluated as 1400 nm, which is much larger than 133 nm, obtained by fitting the free-ion yield to the Onsager formula (*vide supra*). Similar calculations for LKr and LXe give b values of the gaussian thermalization distribution

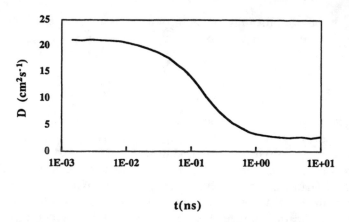

FIGURE 8.6   Evolution of the electron diffusion coefficient in LAr starting with an initial velocity 4.8 times the thermal velocity. Reproduced from Mozumder (1982).

as ~3500 and ~4500 nm, respectively, showing the expected trend. These large thermalization distances have been shown to be consistent with the electron mobility and electron-ion recombination rate constants in these liquids within a modified Jaffe model of recombination, to be described in the next chapter.

REFERENCES

Aprile, E., Bolotnikov, A., Chen, D., and Mukherjee, R. (1993), Phys. Rev. *A48*, 1313.
Bartczak, W.M., and Hummel, A. (1987), *87*, 5222.
Blaunstein, R.P., and Christophorou, L.G. (1968), J. Chem. Phys. *49*, 1526.
Braglia, G.L., Caraffini, G.L., and Diligenti, M. (1981), Nouvo Cimento *B62*, 139.
Bruche, E (1929), Ann. Physik *1*, 93.
Burns, W.G. (1968), J. Chem. Phys. *48*, 1876.
Burns, W.G. (1972), Int. J. Radiat. Phys. Chem. *4*, 249.
Burns, W.G., and Mozumder, A. (1987), Chem. Phys. Lett. *142*, 381.
Chandrasekhar, S. (1943), Rev. Mod. Phys. *15*, 1.
Chen, J.C.Y. (1964), J. Chem. Phys. *40*, 3507, 3513.
Chen, J.C.Y., and Magee, J.L. (1962), J. Chem. Phys. *36*, 1407.
Choi, H.T., Sethi, D.S., and Braun, C.L. (1982), J. Chem. Phys. *77*, 6027.
Christophorou, L.G., McCorkle, D.L., and Carter, J.G. (1971), J. Chem. Phys. *54*, 253.
Christophorou, L.G., Gant, K.S., and Baird, J.K. (1975), Chem. Phys. Lett. *30*, 104.
Dean, A.G., Smith, D., and Adams, N.G. (1974), J. Phys. *B7*, 644.
Dessauer, F. (1923), Z. Physik *20*, 288.
Dillon, M.A., Inokuti, M., and Kimura, M. (1988), Radiat. Phys. Chem. *32*, 43.
Dodelet, J.-P., and Freeman, G.R. (1972), Can. J. Chem. *50*, 2667.

Dodelet, J.-P., and Freeman, G.R. (1975), Int. J. Radiat. Phys. Chem. 7, 183.

Doke, T. (1981), Portugal. Phys. 12, 9.

Doke, T., Hitachi, A., Kikuchi, J., Masuda, K., Tamada, S., Mozumder, A., Shibamura, E., and Takahashi, T. (1985), Chem. Phys. Lett. 115, 164.

Douthat, D. (1975), Radiat. Res. 61, 1.

Douthat, D. (1983), J. Chem. Phys. 79, 4599.

Engelhardt, A. G., and Phelps, A. V. (1963), Phys. Rev. 131, 2115.

Engler, J., Gils, H. J., Knapp, J., Rebel, H., and Supper, R. (1993), Nucl. Instrum. Methods A327, 128.

Freeman, G.R. (1967), J. Chem. Phys. 46, 2822.

Freeman, G.R., and Fayadh, J.M. (1965), J. Chem. Phys. 43, 86.

Frohlich, H., and Platzman, R.L. (1953), Phys. Rev. 92, 1152.

Gerjuoy, E., and Stein, S. (1955), Phys. Rev. 97, 1671.

Gilardini, A. (1972), Low Energy Collisions in Gases, ch. 1, Wiley, New York.

Goulet, T., and Jay-Gerin, J.-P. (1988), J. Phys. Chem. 92, 6871.

Goulet, T., Patau, J.-P., and Jay-Gerin, J.-P. (1990), Radiat. Prot. Dosim. 31, 33.

Guelfucci, J.P., Filey-Rey, J., Cassanovas, J., and Baird, J.K. (1997), J. Chem. Phys. 106, 9497.

Herzenberg, A., and Mandl, F. (1962), Proc. Roy. Soc. A270, 48.

Hirata, Y., and Mataga, N. (1991), J. Phys. Chem. 95, 1640.

Huang, S.S.-S., and Freeman, G.R. (1977), Can. J. Chem. 55, 1838.

Hummel, A. (1967), Ph.D. Thesis, Free University, Amsterdam.

Hummel, A., and Allen, A.O. (1966), J. Chem. Phys. 44, 3426.

Hummel, A., Allen, A.O., and Watson, F.H., Jr. (1966), J. Chem. Phys. 44, 3431.

Huxley, L.G., and Crompton, R.W. (1974), The Diffusion and Drift of Electrons in Gases, ch. 3, John Wiley, New York.

Inokuti, M. (1975), Radiat. Res. 41, 6.

Jay-Gerin, J.-P., Goulet, T., and Billard, I. (1993), Can. J. Chem. 71, 287.

Klots, C.E., and Reinhardt, P.W. (1970), J. Phys. Chem. 74, 2848.

Knapp, M., Echt, D., Kreisel, D., and Recknagel, E. (1987), J. Phys. Chem. 91, 2601.

Knierim, K.D., Lin, S. L., and Mason, E.A. (1981), J. Chem. Phys. 75, 1159.

Knierim, K.D., Waldman, M., and Mason, E.A. (1982), J. Chem. Phys. 77, 943.

Koura, K. (1982), J. Chem. Phys. 76, 390.

Koura, K. (1989), Radiat. Phys. Chem. 34, 51.

Kowari, K., Demeio, L., and Shizgal, B. (1992), J. Chem. Phys. 97, 2061.

Kubota, S., Nakamoto, A., Takahashi, T., Hamada, T., Shibamura, E., Miyajima, M., Masuda, K., and Doke, T. (1978), Phys. Rev. B17, 2762.

LaVerne, J.A., and Mozumder, A. (1984), Radiat. Phys. Chem. 23, 637.

Lekner, J. (1967), Phys. Rev. 158, 130.

Lekner, J., and Cohen, M.H. (1967), Phys. Rev. 158, 305.

Magee, J.L., and Helman, W.P. (1977), J. Chem. Phys. 66, 310.

Mathieu, J., Blanc, D., Caminade, P., and Patau, J.-P. (1967), J. Chim. Phys. 64, 1979.

McMahon, D.R.A., and Shizgal, B. (1985), Phys. Rev. A31, 1894.

McMahon, D.R.A., Shizgal, B., and Ness, K. (1986), J. Phys. B19, 2759.

Mentzoni, M.H., and Rao, K.V.N. (1965), Phys. Rev. Lett. 14, 779.

Mentzoni, M.H., and Row, R.V. (1963), Phys. Rev. 130, 2312.

Metropolis, N., Rosenbluth, A.W., Rosenbluth, N.M., and Teller, A.H. (1953), J. Chem. Phys. 21, 1087.

Michaud, M., and Sanche, L. (1987), Phys. Rev. A36, 4684.

Migus, A., Gauduel, Y., Martin, J.L., and Antonetti, A. (1987), Phys. Rev. Lett. 58, 1559.

Mozumder, A. (1969a), J. Chem. Phys. 50, 3153.

Mozumder, A. (1969b), Advances in Radiat. Chem. 1, 1.

Mozumder, A. (1974), J. Chem. Phys. *60*, 4305.

Mozumder, A. (1980a), J. Chem. Phys. *72*, 1657.

Mozumder, A. (1980b), J. Chem. Phys. *72*, 6289.

Mozumder, A. (1982), J. Electrostatics. *12*, 45.

Mozumder, A. (1988), Radiat. Phys. Chem. *32*, 287.

Mozumder, A., and Magee, J. L. (1967), J. Chem. Phys. *47*, 939.

Mozumder, A., and Tachiya, M. (1975), J. Chem. Phys. *62*, 979.

Nishigori, T., and Shizgal, B. (1988), J. Chem. Phys. *89*, 3275.

Onsager, L. (1938), Phys. Rev. *54*, 554.

Pimblott, S.M., and Mozumder, A. (1991), J. Phys. Chem. *95*, 7291.

Platzman, R.L. (1955), Radiat. Res. *2*, 1.

Platzman, R.L. (1967), in *Radiation Research* (Silini, G., ed.), p. 20, North-Holland, Amsterdam.

Rassolov, V.A. (1991), M.S. dissertation, University of Notre Dame.

Rips, I., and Silbey, R.J. (1991), J. Chem. Phys. *94*, 4495.

Robinson, M.G., and Freeman, G.R. (1973), Can. J. Chem. *51*, 641.

Samuel, A.H., and Magee, J.L. (1953), J. Chem. Phys. *21*, 1080.

Sano, H., and Mozumder, A. (1977), J. Chem. Phys. *66*, 689.

Schmidt, K.H., Sauer, M.C., Jr., Lu, Y., and Liu, A. (1990), J. Phys. Chem. *94*, 244.

Schmidt, W.F. (1970), Radiat. Res. *42*, 73.

Schmidt, W.F. (1977), Can. J. Chem. *55*, 2197.

Schmidt, W.F., and Allen, A.O. (1968), J. Phys. Chem. *72*, 3730.

Schmidt, W.F., and Allen, A.O. (1970), J. Chem. Phys. *52*, 2345.

Schnitker, J., Rossky, P.J., and Kenney-Wallace, G.A. (1986), J. Chem. Phys. *85*, 2986.

Shimamori, H., and Hatano, Y. (1976), Chem. Phys. Lett. *38*, 242.

Shizgal, B. (1981), J. Chem. Phys. *74*, 1408.

Shizgal, B. (1983), Chem. Phys. Lett. *100*, 41.

Shizgal, B., McMahon, D.R.A., and Viehland, L.A. (1989), Radiat. Phys. Chem. *35*, 34.

Silinsh, E.A., and Jurgis, A.J. (1985), Chem. Phys. *94*, 77.

Silinsh, E.A., Shlichta, G.A., and Jurgis, A.J. (1989), Chem. Phys. *138*, 347.

Slater, J. (1946), Rev. Mod. Phys. *18*, 659.

Sowada, U., and Warman, J.M. (1982), J. Electrostatics *12*, 37.

Sowada, U., Warman, J.M., and deHaas, M.P. (1982), Phys. Rev. *B25*, 3434.

Suzuki, E., and Hatano, Y. (1986a), J. Chem. Phys. *84*, 4915.

Suzuki, E., and Hatano, Y. (1986b), J. Chem. Phys. *85*, 5341.

Takasaki, T., Ruan (Gen), J., Kubota, S., and Shiraishi, F. (1982a), Nucl. Inst. Meth. *196*, 83.

Takasaki, T., Ruan (Gen), J., Kubota, S., and Shiraishi, F. (1982b), Phys. Rev. *A25*, 2820.

Tembe, B.L., and Mozumder, A. (1983a), J. Chem. Phys. *78*, 2030.

Tembe, B.L., and Mozumder, A. (1983b), Phys. Rev. *A27*, 3274.

Tembe, B.L., and Mozumder, A. (1984), J. Chem. Phys. *81*, 2492.

Thomas, J., and Imel, D. A. (1987), Phys. Rev. *A36*, 614.

Warman, J.M. (1981), Radiat. Phys. Chem. *17*, 21.

Warman, J.M., and deHaas, M.P. (1975), J. Chem. Phys. *63*, 2094.

Warman, J.M., and Sauer, M.C., Jr. (1970), J. Chem. Phys. *52*, 6428.

Warman, J.M., and Sauer, M.C., Jr. (1975), J. Chem. Phys. *62*, 1971.

Warman, J.M., Zhou-Lei, M., and Van Lith, D. (1984), J. Chem. Phys. *81*, 3908.

Warman, J.M., Sowada, U., and deHaas, M.P. (1985), Phys. Rev. *A31*, 1974.

Electron Escape:
The Free-Ion Yield

When ionization is produced in liquids of low dielectric constant, most of the ions and electrons do not survive long—they undergo *initial or geminate* recombination. A small fraction, however, escapes initial recombination and can be collected as *free-ions* by a modest external field. The yield of such free ions per 100 eV of energy deposition is indicated by G_{fi}; the total ionization yield may be denoted by G_{tot} or G_i. The free-ions can undergo *general or homogeneous* recombination, which is a second-order process. Alternatively the free ions can form negative ions or be collected by an electric field. The experimental importance of free-ion yield is underscored by the nature of chemical changes that occur in irradiated liquids. The fundamental significance lies in the behavior of electron motion in noncrystalline condensed media. There is general agreement that the electron executes diffuse motion, but the theoretical details are still being worked out.

Measured free-ion yields depend strongly on molecular structure and the LET of radiation. G_{fi} increases with molecular sphericity (*vide infra*) and decreases sharply with the LET. Unless otherwise specified, the free-ion yield in this chapter will refer to near-minimum ionizing low-LET radiations, such as γ-rays or X-rays or electrons of a few MeV energy. Many experiments have been performed with alkane liquids, and certain other liquids have also been studied. Liquefied rare gases (LRGs) form a separate, interesting group in which G_{fi} is quite large and G_{tot} can be directly measured by collecting all the ions with a moderate external field. Onsager's (1938) theory is the main framework by which experimental free-ion yields are analyzed (see Sect. 9.2). Since this theory uses the Nernst-Einstein relationship between the mobility (μ) and the diffusion coefficient (D)—namely, $\mu = eDk_BT$, valid under thermal equilibrium—the implication is that the electrons are thermalized before any recombination begins. It would therefore be profitable to read this chapter along with Chapter 8. In any case, there is little evidence of epithermal recombination, although with very efficient scavengers in high-mobility liquids there is some evidence of epithermal scavenging (see Chapter 8).

9.1   SUMMARY OF EXPERIMENTAL RESULTS AT LOW LET

There are three methods of measuring G_{fi}: (1) from the steady conductivity induced by irradiation, (2) collecting the free charges by the clearing field technique (Schmidt and Allen, 1968), and (3) from extrapolated scavenging yield at low concentrations. Early measurements in n-hexane were made by Freeman (1963a) and by Allen and Hummel (1963) using the steady conductivity method, which requires the determination of ion mobilities. In lieu of a direct measurement, Freeman used the empirical equation of Chang and Wilkie (1955) that contains, $inter$ $alia$, viscosity and solute molecular volume as principal variables. It is probably inapplicable to electrons in hydrocarbon liquids, which exhibit widely different mobilities for nearly the same viscosity. Nevertheless, Freeman obtained $G_{fi} \sim 0.2$, which is not in error by more than 40%. He also observed that addition of O_2 did not alter the conductance and concluded that the electron moves as a massive ion. A similar observation was also made by Hummel et al. (1966). However, this conclusion was premature. It is more likely that in these experiments the electron was scavenged $after$ thermalization. It has been shown rigorously that such scavenging does not change the free-ion yield (Mozumder and Tachiya, 1975), although it alters the nature of the carrier species and the rate of recombination. Because of this saving feature, the earlier free-ion yield measurements are still meaningful.

The conductivity κ induced by radiation absorption at dose rate I ($eV\propto cm^{-3}$ s^{-1}) is given by $\kappa = uc$, where c the is free ion concentration and u is the sum of mobilities of positive and negative carriers. The establishment of steady state requires equal rates of generation and recombination, or $IG_{fi}/100 = kc^2$ where k is the second-order recombination rate constant. Eliminating c between these equations gives $G_{\text{fi}} = 100(k/u)(\kappa^2/Iu)$. The method then entails four measurements: (1) dosimetry (I), (2) conductivity (κ), (3) mobilities (u), and (4) k/u.

Allen and his associates measured u and k/u by separate experiments, whereas Freeman and Fayadh (1965) continued to use the empirical formula of Chang and Wilkie. The mobilities were given by drift-time measurement in a known field when a wedge of ionization was created at a well-defined position in the liquid. If the steady irradiation is suddenly interrupted at $t = 0$ the current density J in the cell will fall according to the second-order neutralization process as $J^{-1} = J_0^{-1} + (k/u)Et$ where J_0 is the initial current density and E is the field. Thus, k/u is determined from the decay of the current. However its value is so close to Debye's (1942) theoretical value (9.6×10^{-7} V$\propto$cm for hexane) that the latter was always preferred (see Hummel and Allen, 1966). This procedure was followed by Hummel et al. (1966) for obtaining the temperature dependence of G_{fi} in hexane over the 200–300 K interval. They also measured G_{fi} with ^{37}Ar-irradiation, which consists of nearly monoenergetic electrons at 2400 eV. Hummel and Allen (1967) extended the procedure in hexane for measurement of radiation-induced conductivity in the presence of an external field E at different temperatures. For relatively low $E(< 40$ KV/cm) the free-ion yield obtained from conductivity data was proportional to E with a slope-to-intercept ratio given accurately by the Onsager (1938) theory. Since that theoretical ratio $e^3/2\varepsilon k_{\text{B}}^2 T^2$, where ε is the dielectric constant, is independent of the unknown variables such as the distribution of initial separation and is precisely calculable the authors concluded that the agreement between theory and experiment was good. Incidentally, the same value of slope-to-intercept ratio has been found to be valid for multiple ion-pair spurs (see Sect. 7.5).

In earlier experiments, Freeman (1963a–c) continued to use Chang and Wilkie's formula for ionic mobilities and even took $u_+ \sim u_-$, which in general is not justified. He determined the free-ion yields in hexane, cyclohexane, and cyclohexane solutions of O_2, I_2, CCl_4, naphthalene, anthracene, and so forth, showing minor solute effects. In any case, the relatively low-field determination of G_{fi} by Freeman is consistent with the Onsager model. He determined the activation energies for *conductance* at low and high fields (46.5 KV$\propto$cm^{-1}) in cyclohexane and obtained the values 2.3 and 0.3 kcal/mole, respectively. From this, one can say that the high-field current is relatively temperature-independent, but no definite statement can be made about $G_{\text{fi}}(E)$, as the temperature dependence of mobility was not obtained.

In later measurements, Tewari and Freeman (1968, 1969) measured the ion mobilities from drift-time measurement and obtained k/u values from the current decay following a pulse of X-rays of ~ 1 ms duration. The purpose was to find the dependence of G_{fi} on molecular structure. It was found that G_{fi} increased with the sphericity of the molecule. In liquid argon $G_{\text{fi}} \sim 5$ was measured, which indicated that all ionized electrons in argon are free. However, this conclusion was later retracted (Robinson and Freeman, 1973). One characteristic of Tewari and Freeman's experiment is the observation of a sudden rise in the cell current following an irradiation pulse in certain high-mobility liquids; the authors called this the *k/u-overshoot*.

More recent determinations of G_{fi} by Freeman and his associates use the clearing field technique first introduced by Schmidt and Allen (1968). In this method, one collects the total free charge irrespective of mobility (*vide infra*). Robinson, *et. al.* (1971a,b) made measurements at low temperatures and confirmed the increase of G_{fi} with sphericity of the molecule (see Table 9.1). Fuochi and Freeman (1972) measured the yields in propane, methyl-substituted propane, and liquid Ar, obtaining a still lower value (2.0) in the last liquid.

Dodelet *et al.* (1972) studied the effect of an external field F at different temperatures in various molecular liquids. In 2-methylpropane, they found somewhat higher yield for $F > 8$ kV/cm at the lower temperature (148 K) although at a higher temperature (183 K) the yield was greater at zero or low values of F. The phenomenon was interpreted by Mozumder (1974b) in terms of the Onsager theory and termed the *temperature inversion effect*. It has been further studied by Freeman and Dodelet (1973) in 2-methylpropane and by Dodelet and Freeman (1975a,b) in diethyl ether and in carbonyl sulfide. Robinson and Freeman (1973) investigated the external field effect on free-ion yield in atomic liquids. Extrapolating these yields to $F \rightarrow \infty$ they obtained very high total ionization yields, which are probably due to experimental artifacts (see Aprile *et al.*, 1993).

In the clearing field technique of Schmidt and Allen (1968), a field ~3 kV/cm is imposed on the conductivity cells a few microseconds after irradiation by a pulse of X-rays ~0.5 ms duration. This field simply sweeps all available free charges and deposits them across a capacitor, where the charge can be suitably measured. To prevent loss of ions by volume recombination, the dose must be small, yet a high *dose rate* increases measurement sensitivity. The combination therefore requires relatively short irradiation pulse. The external field must be strong enough to ensure maximum ion collection, yet small enough so that the (zero-field) free-ion yield is not significantly increased. Finally, the gap between end of irradiation and start of high voltage must be large enough so that *geminate* neutralization is not perturbed, yet short enough that there will effectively be no volume recombination.

Schmidt and Allen (1968, 1970) describe procedure by which ~ 99% of all free charges are collected. Needless to say, there are stringent purification requirements including, in the final stage, subjection of the liquid to high voltage for an extended period. Preliminary investigations of the authors, in agreement with the observations of Freeman and his associates, showed that G_{fi} depends on molecular structure.

In their first series of research, Schmidt and Allen (1968) studied 19 pure liquids at room temperature, 3 liquids over a range of temperature, and mixtures

TABLE 9.1 Free-Ion Yield in Selected Nonpolar Liquids in the Limit of Zero External Field

Liquid	T (K)	$G_{fi}{}^a$	ETL[b] (A)	$E_a{}^c$	Reference
Cyclohexane	296	0.2	100	2.3	Freeman (1963b)
Cyclohexane[d]	296	0.7	100	1.4	Freeman (1963c)
CH_4	120	0.8			Robinson et al. (1971b)
Ethane	183	0.13		0.6	Robinson et al. (1971b)
Propane	183	0.08		0.8	Robinson et al. (1971b)
Ethylene	183	0.02		0.5	Robinson et al. (1971b)
Propylene	183	0.04		1.0	Robinson et al. (1971b)
n-Butane	296	0.19	83		Schmidt and Allen (1970)
n-Pentane	296	0.15	72	1.5	Schmidt and Allen (1968, 1970)
3-Methyl Pentane	296	0.15	70		Schmidt and Allen (1970)
Isopentane	296	0.17	76		Schmidt and Allen (1970)
NP	296	0.86	178		Schmidt and Allen (1970)
n-Hexane	296	0.13	67	1.9	Schmidt and Allen (1968, 1970)
2,2-Dimethylbutane	296	0.30	92		Schmidt and Allen (1970)
Cyclohexane	296	0.15	66	1.8	Schmidt and Allen (1968, 1970)
Cyclohexene	296	0.15	62		Schmidt and Allen (1970) Shinsaka et al. (1975)
Squalane	296	0.12	59		Schmidt and Allen (1970)
Benzene	296	0.05	42		Schmidt and Allen (1970)
Toluene	298	0.05	42		Schmidt and Allen (1970) Capellos and Allen (1970)
C_3H_8	233	0.14	87	0.9	Dodelet et al. (1972)
$(CH_3)_3CH$	294	0.31	112	0.6	Dodelet et al. (1972)
NP	294	1.1	237		Dodelet et al. (1972)
Naphthalene (l)	357	0.09	33		Shinsaka and Freeman (1974)
Antracene (l)	500	0.1	20		Shinsaka and Freeman (1974)
TMS	296	0.74	159		Schmidt and Allen (1970)
Argon	87	2.0	~1200		Fuochi and Freeman (1972)
Krypton	148	5.8	880		Robinson and Freeman (1973)
Xenon	183	7.0	720		Robinson and Freeman (1973)

[a]Per 100 eV of energy absorption.

[b]Effective thermalization length, the b value for origin-centered gaussian distribution (see text): only when $G_{fi} < 0.2$ is a truncated power law distribution used by Freeman and his associates.

[c]For the free-ion yield in kcal/mole.

[d]With an external field of 46.5 KV/cm.

of neopentane with 3-methylpentane or cyclohexane over a complete composition range. The authors used the Onsager formula for escape and fitted their results to a gaussian initial distribution centered at the origin of the form $\sim (-r^2/b^2)$. The b values obtained were reported, and it was shown that the product bd, where d is liquid density, is nearly temperature-independent for a given alkane. The value of bd does not change too much from liquid to liquid as long as the molecular structure is nearly the same. Going to nearly spherical molecules (neopentane, tetramethyl-silane, etc.) increases bd considerably.

In a later report, Schmidt and Allen (1970) extended their measurement to 38 pure liquids and mixtures at room temperature and to 5 liquids as a function of temperature. The free-ion yields are arranged by the alkanes and their isomeric and cyclic counterparts, which show considerable differences in the results. Thus, the free-ion yield in neopentane (NP) is about seven times that in n-pentane. Some of the results are shown in Table 9.1. In mixtures of NP with CCl_4 or CS_2, the observed decrease of G_{fi} with the additive concentration has been interpreted by Mozumder and Tachiya (1975) as due to epithermal electron scavenging (*vide infra*).

Table 9.1 lists free-ion yields in a few representative nonpolar liquids according to Allen and his associates and Freeman and his associates. However, there are also measurements by other investigators, which have been summarized in a review (Allen, 1976). (Also see Tabata et al. (1991) for a compilation of b values of thermalization distance distributions in nonpolar liquids obtained by fitting free-ion yields to the Onsager formula.) In summary, it can be said that the temperature and external field effect on G_{fi} are described rather well by the Onsager theory, but the dependence of the free-ion yield on molecular structure is not easily explained. Certain conjectures, some of which are listed here, have been made; these however, are observational correlations rather than rational explanations. The correlations are (1) between G_{fi} and ε (Freeman and Fayadh, 1965; Dodelet and Freeman, 1975a,b); (2) between G_{fi} and μ (Dodelet and Freeman, 1972; Schiller and Vass, 1975; Jay-Gerin et al., 1993); (3) between G_{fi} and molecular shape (Robinson, et al., 1971b); (4) between G_{fi} and phonon emission rate in atomic liquids (Dodelet, et al., 1972); (5) between G_{fi} and bond structure (Schmidt and Allen, 1968; Robinson, et al., 1971a, b); (6) between G_{fi} and the presence of tertiary and quaternary C atoms (Schmidt and Allen, 1970); (7) between μ and V_0 (lowest electron energy in extended state—Kestner and Jortner, 1973); the correlation between V_0 and G_{fi} is already well established; (8) between G_{fi} and anisotropy of molecular polarizability (Dodelet and Freeman, 1972); and (9) between fluctuations of polarization and transfer energies and molecular shape, giving simultaneously higher μ and G_{fi} for spherical molecules. Some serious attempts have been made to correlate free-ion yield and mobility (see Jay-Gerin et al., 1993). We have already discussed the Sano-Mozumder model in Sect. 8.4.1, where the correlation is through the energy loss of low energy electrons (for high-mobility liquids), but this model does notgive the physical mechanism of energy loss. Schiller and Vass (1975) have also sought to derive the desired correlation through statistical fluctuation, but it is not easy to relate that to real systems.

9.2   ONSAGER'S THEORY OF GEMINATE-ION RECOMBINATION

The fundamental theory of electron escape, owing to Onsager (1938), follows Smoluchowski's (1906) equation of Brownian motion in the presence of a field F. Using the Nernst-Einstein relation $\mu = eD/k_B T$ between the mobility and the diffusion coefficient, Onsager writes the diffusion equation as

$$\frac{\partial f}{\partial t} = D \operatorname{div}\left\{\exp\left(\frac{-V}{k_B T}\right) \operatorname{grad}\left[f \exp\left(\frac{V}{k_B T}\right)\right]\right\},$$

where f is the probability density of finding the electron around r at time t, and V is the potential acting on the electron. He notes that the escape probability ϕ calculated from the stationary state treatment of the diffusion equation—with a source at r_0, the initial separation, and sinks at origin and infinity—satisfies the equation

$$\operatorname{div}\left[\exp\left(-\frac{V}{k_B T}\right) \operatorname{grad} \phi\right] = 0.$$

Under the joint influence of the external field F and the field of the geminate positive ion, V, is given by

$$V = -\frac{e^2}{\varepsilon r} - eFr \cos \theta,$$

where e is the magnitude of electronic charge, ε, is the dielectric constant of the medium, and θ is the angle between the radius vector and the "downstream" field direction. When $F = 0$, (no external field), Onsager obtained the solution for the escape probability simply as

$$\phi = \exp\left(-\frac{r_c}{r_0}\right), \tag{9.1}$$

where $r_c \equiv e^2/\varepsilon k_B T$ is a critical distance at which the electron–positive-ion potential energy equals numerically to $k_B T$. This distance is designated as the *Onsager length*. The initial separation r_0 has been interpreted by Mozumder and Magee (1967) as the thermalization length. In this sense, the treatment of Chapter 8 and that of the present chapter are complementary. Ultimately, the integral of the escape probability over the distribution of thermalization length is the quantity to be compared with experiment.

9.2.1   STATIONARY CASE: SIMPLIFIED TREATMENT

Mozumder (1969a) offers a simplified derivation of Eq. (9.1) following a suggestion by Magee regarding the equivalence of time-dependent and steady-flow treatments (see Magee and Taylor, 1972). Consider, as in figure 9.1, that an electron is being added at unit rate at separation r_0 from the positive ion. This unit current partitions between the inner, neutralization current $I_<$ and the outer, escape current $I_>$. The current density is given by the sum of diffusion and conduction currents as

$$\frac{I}{4\pi r^2} = -D\frac{\partial n}{\partial r} + \mu n \frac{\partial V}{\partial r},$$

where n is the stationary electron density and $V = e/\varepsilon r$ is the electrostatic potential for unit charge. Taking $\mu = eD/k_BT$ and using the definition of r_c, the right-hand side of the preceding equation is reduced to $-D(\partial n/\partial r + nr_c/r^2)$. The inner solution $(r = r_0)$ with Smoluchowski's boundary condition at the reaction radius r_1,—with $n(r_1) = 0$, may be given as follows:

$$n = \frac{I_<}{4\pi Dr_c}\left\{\exp[r_c(r^{-1} - r_1^{-1})] - 1\right\}.$$

Similarly the outer solution $r = r_0$ with a sink at infinity $n(\infty) = 0$) is given by

$$n = \frac{I_>}{4\pi Dr_c}\left[\exp\left(\frac{r_c}{r}\right) - 1\right].$$

Demanding continuity of n at $r = r_0$, using the Kirchhoff relation $I_> - I_< = 1$, and identifying $I_>$ (for unit input current) as the escape probability ϕ, we get

$$\phi = \exp\left(-\frac{r_c}{r_0}\right)\left\{\frac{1 - \exp[r_c(r_0^{-1} - r_1^{-1})]}{1 - \exp(-r_c/r_1)}\right\}. \tag{9.2}$$

Equation (9.2) shows the effect of the reaction radius r_1 on the escape probability, which, remarkably, is free of the diffusion coefficient. Normally $r_c \gg r_1$, which reduces Eq. (9.2) to the celebrated Onsager formula $\phi = \exp(-r_c/r_0)$ as given by Eq. (9.1).

The foregoing treatment can be extended to cases where the electron-ion recombination is only partially diffusion-controlled and where the electron scattering mean free path is greater than the intermolecular separation. Both modifications are necessary when the electron mobility is ~ 100 cm^{2}v^{-1}s^{-1} or greater (Mozumder, 1990). It has been shown that the complicated random trajectory of a diffusing particle with a finite mean free path can have a simple representation in fractal diffusivity (Takayasu, 1982). In practice, this means the diffusion coefficient becomes distance-dependent of the form

$$D(r)^{-1} = D_0^{-1}\left(1 + \frac{d}{r}\right), \tag{9.3}$$

where D_0 is the diffusion coefficient at interparticle separation (r) much greater than the mean free path of scattering, taken proportional to the parameter d. A value of d that is ~3.7 times the mean free path has been found suitable for electron-ion recombination in high-mobility liquids (Mozumder, 1990).

Equation (9.3) has been derived for one-dimensional diffusion and supported by molecular dynamics simulation in the three-dimensional case (Powles, 1985; Tsurumi and Takayasu, 1986; Rappaport, 1984). For the partially diffusion-controlled recombination reaction we again refer to Figure 9.1, where the inner (Collins-Kimball) boundary condition is now given as

$$\kappa n(r_1) = \frac{I_<}{4\pi r_1^2}, \tag{9.4}$$

where κ is the reaction velocity of the final chemical step. Using Eqs. (9.3) and (9.4), the solution of the diffusion equation in the inner region ($r_1 \le r \le r_0$) may be given as follows (Mozumder, 1990):

$$n(r) = \frac{I_<}{4\pi D_0 r_c}\left\{1 + \frac{d}{r} + \frac{d}{r_c} + \exp\left(\frac{r_c}{r} - \frac{r_c}{r_1}\right)\left[\frac{D_0 r_c}{\kappa r_1^2} - \left(1 + \frac{d}{r_1} + \frac{d}{r_c}\right)\right]\right\}. \tag{9.5}$$

For the outer region ($r \ge r_0$), the boundary condition at infinity should be changed for a finite dose to that at an outer radius given by $R = (3W/4\pi\rho\Gamma)^{1/3}$, where W is the mean energy needed to create a geminate pair, ρ is the medium density, and Γ is the dose (Mozumder, 1982). Using the outer boundary condition $n(R) = 0$, the solution of the diffusion equation in the outer region ($r_0 \le r \le R$) is given by

$$n(r) = \frac{I_>}{4\pi D_0 r_c}\left\{\left[\exp\left(\frac{r_c}{r} - \frac{r_c}{R}\right)\right]\left(1 + \frac{d}{r_c} + \frac{d}{R}\right) - \left(1 + \frac{d}{r_c} + \frac{d}{r}\right)\right\}. \tag{9.6}$$

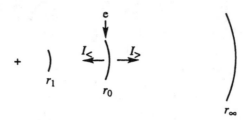

FIGURE 9.1 Simplified derivation of Onsager's escape probability formula. In the stationary state a unit electron current at r_o partitions as $I_<$ toward the reaction radius and as $I_>$ toward the sink at infinity; the latter is the escape probability. Reproduced from Mozumder (1969a), with the permission of John Wiley & Sons, Inc.©

Joining the solutions given by Eqs. (9.5) and (9.6) at the source ($r = r_0$) and recognizing that the escape probability $P_{esc} = 1 - I_< = I_>$, one gets

$$P_{esc} = \tag{9.7}$$

$$\frac{\exp(-r_c/r_0)(1 + d/r_0 + d/r_c) - \exp(-r_c/r_1)(1 + d/r_1 + d/r_c - D_0 r_c/\kappa r_1^2)}{\exp(-r_c/R)(1 + d/r_c + d/R) - \exp(-r_c/r_1)(1 + d/r_1 + d/r_c - D_0 r_c/\kappa r_1^2)}$$

The standard Onsager formula $P_{esc} = \exp(-r_c/r_0)$ can be retrieved from Eq. (9.7) in the brownian limit of vanishing reaction radius and zero dose,—that is $r_1 = d = 0$ and $R = \kappa = \infty$. The solution for the partially diffusion-controlled case within the context of brownian diffusion ($d = 0$) was given earlier by Monchick (1956) and by Sano (1983); it is identical to Eq. (9.7) for $d = 0$. Various other limits of Eq. (9.7) are possible, including the case of geminate neutrals, for which $r_c = 0$. Equation (9.7) shows that the reaction radius has little effect on the escape probability if that is one-tenth the Onsager length or less. On the other hand, P_{esc} increases significantly with the mean free path in all cases. Taking a typical case in liquid alkanes with $r_0/r_c = 0.25$ and $r_1/r_c < 0.1$, the *fractional* increase of escape probability is linear with the ratio d/r_0 in the interval 0 to 1, being about 0.25 at $d/r_0 = 0.2$. A similar increase of escape probability with electron mean free path has been obtained by Tachiya (1988) and by Tachiya and Schmidt (1989). In the earlier paper, a Monte Carlo procedure was used with a particular model of velocity randomization following a scattering. The conclusion remained unchanged in the later work, the probability bounded only by energy diffusion and the Onsager limit.

There is clear experimental evidence that the free-ion yield increases with temperature (Schmidt and Allen, 1968; Dodelet and Freeman, 1972). Schmidt and Allen investigated the temperature dependence of G_{fi} in n-pentane, n-hexane, and cyclohexane over the interval ~0° C to 95° C and found in some cases an increase by a factor of 2. Dodelet and Freeman studied G_{fi} in cyclohexane over a similar range of temperature and reported as high a free-ion yield as 0.24. In certain saturated hydrocarbon liquids—such as methane, ethane, propane, n-butane, isobutane, cyclopropane, and propylene—G_{fi} has been measured over the interval 100–300 K (Dodelet *et al.*, 1972; Fuochi and Freeman, 1972; Robinson and Freeman, 1974). In methane, G_{fi} increases sharply over 100–120 K from ~0.75 to ~1.12. In other cases, G_{fi} increases more moderately and in not such a clear-cut fashion.

The theoretical dependence of free-ion yield on temperature is complex. The main contributing factors are the Onsager length (r_c) and the effective thermalization length (ETL). The distance r_c is inversely proportional to T if we ignore the temperature dependence of the dielectric constant for the time being (see Eq. 9.1 et seq.). However, the temperature dependence of G_{fi} in n-hexane cannot be accurately explained by the temperature dependence of r_c alone (Mozumder and Magee, 1967). It stands to reason that the thermalization distance should be longer at lower temperatures, since the electron has to lose more energy at a lower temperature to become thermalized. This increase of ETL partially counteracts the effect of temperature on r_c in the Onsager formula. There is also a mild variation of the dielectric constant with temperature. All these effects have been incorporated by Mozumder and Magee (1967) in determining ETL in n-hexane over the temperature range 203–293 K following the procedure outlined in Ch. 8. The results are shown in Table 9.1 for the ETL and computed G_{fi}; these results compare well with the experimental results of Hummel *et al.* (1966).

9.2.2   KINETICS

Salient theoretical features of the kinetics of geminate electron-ion recombination have been presented in Sect. 7.5. Here we will discuss some experimental results and their theoretical implications.

With the advent of picosecond-pulse radiolysis and laser technologies, it has been possible to study geminate-ion recombination (Jonah *et al.*, 1979; Sauer and Jonah, 1980; Tagawa *et al.* 1982a, b) and subsequently electron-ion recombination (Katsumura *et al.*, 1982; Tagawa *et al.*, 1983; Jonah, 1983) in hydrocarbon liquids. Using cyclohexane solutions of 9,10-diphenylanthracene (DPA) and p-terphenyl (PT), Jonah *et al.* (1979) observed light emission from the first excited state of the solutes, interpreted in terms of solute cation–anion recombination. In the early work of Sauer and Jonah (1980), the kinetics of solute excited state formation was studied in cyclohexane solutions of DPA and PT, and some inconsistency with respect to the solution of the diffusion equation was noted.[1]

Tagawa *et al.* (1982a) studied excited solute state formation in solutions of cyclohexane, methylcyclohexane, and isooctane. The lifetime of the excited state

---

[1] It would be profitable to read this section along with Section 7.5.

of cyclohexane was evaluated to be 1.1 ns with 10% uncertainty. For a 0.1 ml^{-1} solution of biphenyl in cyclohexane, Tagawa et al. (1982b) observed a t$^{-1/2}$ decay of the biphenylide ion as is required by the diffusion theory. Katsumura et al. (1982) observed fluorescence from the first excited stated of neat cycloalkanes. The inferred electron-ion geminate recombination times in cyclohexane, methyl-cyclohexane, bicyclohexyl, and cis-decalin are reported as ≤5, ≤ 5,10–15, and 10–15 ps, respectively. Tagawa et al. (1983) determined the electron-ion recombination time in cyclohexane to be ~3 ps by absorption spectroscopy. In n-hexane and cyclohexane, Jonah (1983) studied electron-ion recombination by absorption spectroscopy. The kinetics was consistent with an exponential initial distribution. In n-hexane, a t$^{-1/2}$ law of decay was observed consistent with diffusive recombination, from which a half-life <100 ps was inferred.

Braun and Scott (1983) investigated electron-ion recombination in hexanes by IR-Stimulated dissociation of (trapped) electrons produced by UV photoionization of dissolved anthracene. This technique was first introduced by Lukin et al. (1980); in it, the trapped electrons are rethermalized following the IR absorption, probably resulting in a slightly longer thermalization length. Braun and Scott determined the half-life of e-ion recombination in n-hexane in this manner and obtained the values ~9 and 70 ± 20 ps at 296 and 214 K, respectively. These values are consistent with Hong and Noolandi's (1978a, b) theory and also consistent with the mobilities at the respective temperatures. The Hong-Noolandi theory gives the half-life of the e-ion recombination process as $t_{1/2} = 1.2 \tau$, where $\tau = \varepsilon r_0^3/3\mu e$ and r_0 is the initial separation. Using the dielectric constant and the electron mobility in hexane at 296 Ktr, and taking $r_0 = 4.5$ nm, one gets $\tau = 6.4$. The calculated half-life is then consistent with the measured value of ≤9 ps.

Scott and Braun (1985) used the same technique in hexane over the temperature interval 214–296 K and determined that the so-inferred median thermalization length is insensitive to temperature and has a value $r_{th} = 4.2$–5.2 nm. Scott and Braun (1986) studied the UV-wavelength dependence of the half-life of e-ion geminate recombination in the same system and found the values 108 ps for 355 nm and 930 ps for 266 nm excitation at 191 K. These results were analyzed via the Hong-Noolandi theory, and the extended half-life was attributed to longer thermalization length arsing from higher excitation energy. The inferred mean thermalization lengths are 3.0 and 7.5 nm, at 355 and 266 nm respectively, using the exponential initial distribution.

Braun and Scott (1987) used two-photon ionization of benzene and azulene in n-hexane and followed the e-ion recombination process by monitoring the transient absorption of the electron. The results are not very different from those obtained by the IR stimulation technique. A mean thermalization length of 5.0 nm was inferred at 223 K using a two-photon excitation at 266 nm. Hong and Noolandi's theory was used for the analysis. The absorption technique was

extended by Braun *et al.* (1991) to UV photoionization of a durene solution of
n-hexane with a 35-ps pulse at 266 nm. Transient absorption due to geminate
electrons at 1064 nm was monitored at 208 K. Excellent fit to the Hong-
Noolandi theory has been claimed for an initial r^2-exp distribution of the form
$g(r) = (r^2/2L^3) \exp(-r/L)$, where $g(r)$ dr is the probability of thermalization
length between r and $r + dr$. Taking the electron diffusion coefficient
$D = 9.7 \times 10^{-5}$ cm^{2}s^{-1} consistent with measured mobility, comparison with
experiment gives $\langle r \rangle = 5.7$ nm.

In conclusion we may state the following:

1. The kinetics of electron-ion recombination is well described by the diffusion model both for photoionization and for ionization induced by
   high-energy irradiation.
2. In photoionization experiments, there is evidence that the thermalization length increases with the photon energy.
3. The mean thermalization length in high-energy irradiation tends to be
   somewhat longer than in the photoionization case, possibly due to
   higher initial energy.

9.3   CASE OF MULTIPLE ION-PAIRS

Onsager's theory of geminate-ion recombination is well supported by experiment in low-mobility hydrocarbon liquids. In particular, the slope-to-intercept
ratio of the variation of G_{fi} with the external electric field depends only on the
liquid temperature and the dielectric constant of the medium; this has been
experimentally verified (see Sect. 9.5). Nevertheless, the model lacks a certain
realism in that the probability of multiple ion-pair formation is significant; in
such a situation, the Onsager theory is inapplicable (Allen, 1976). The probability of multiple ion-pair formation has been estimated somewhat differently
by various authors, but there is a general agreement that single ion-pairs a are
poor representation of the total ionization even at the lowest LET track.
According to Mozumder and Magee (1967) ~50% of all ionizations produced
by MeV electrons in *n*-hexane appear as single ion-pairs. Of the multiple ion-
pair spurs, those having initially two and three ion-pairs are the most important, accounting for ~22% and ~7% of total ionizations respectively. Thus, a
consideration of recombination processes in multiple ion-pair cases with a possible reconciliation with the Onsager theory seems necessary.

An early theory of ionic recombination in liquids was developed by Jaffe'
(1913) for application at a relatively high LET. However, in Jaffe's theory,
coulombic interactions are ignored and the positive and negative ions are
assigned the same mobilities and distribution functions. Therefore, its use in a

real situation is doubtful, although various modifications have been proposed to suit particular cases (see Sect. 9.6). In an extreme variation of the theory proposed by Kramers (1952), the diffusion terms are thrown out; this is equally objectionable. In the first studies of free-ion yield induced by low-LET irradiation, it was tacitly assumed that all ionizations in a multiple-ion spur would quickly neutralize leaving the *last ion-pair*, to which the Onsager model could be applied (Hummel and Allen, 1966). This concept was extended by Mozumder and Magee (1967) to the free-ion yield in n-hexane induced by ^3H and ^{37}Ar irradiations in which the ionizations on the secondary tracks were treated separately and similarly. The final result was roughly in agreement with experiment (Hummel, 1967), but that alone cannot be taken as a sufficient justification of the theoretical model. In any case, this hypothesis would imply a smaller total ionization yield than actual.

The next step employed by Mozumder (1971) and by Dodelet and Freeman (1975a,b) consists of reducing the multiple ion-pair problem to that of a (collection of) isolated ion-pairs. For spherical geometry (<500 eV energy deposition), Mozumder solves the electron diffusion equation by the prescribed diffusion method with a screened electrostatic potential due to the electron spatial distribution, while the positive ions are assumed to remain close to the origin—their mobility was neglected. This gives a certain time scale at which all but the farthest electron recombine. The kinetics of the entire recombination process can thus be calculated, and the *number* of escaping electrons is given by $\exp(-r_c/\bar{p}_n)$ where $\bar{p}_n$ is the average distance from the origin of the *last electron* with respect to the initial distribution that originally contained n ion-pairs. The initial distribution is taken to be gaussian, with variance the sum of the variances due to the electronic range and the thermalization tail. Calculated values of $\bar{p}_n$ in hexane vary from 12.1 nm for $n = 2$ to 24.9 nm for $n = 20$. The corresponding *numbers* of escaped electrons at 296 K are 0.080 and 0.292, respectively. Note that in this model, as long as the spherical symmetry can be maintained, the number of escaped electrons cannot exceed unity.

For energy deposition >500 eV, the short-track geometry is better described in cylindrical symmetry (Mozumder and Magee, 1966). In a cylinder of infinite axial length, no electron survives recombination. But in a short track of finite extent, a stage is reached when the mean radial distance of a surviving electron from the track axis, due to diffusional broadening and gradual recombination, becomes equal to the mean inter-positive-ion separation on the track axis. Mozumder (1971) calculates this time scale, the number of surviving electrons, and their mean radial distance self-consistently using prescribed diffusion and then applies the Onsager formula of escape for each of these. Establishing a time scale $t' = 4.7 \times 10^{-11}$ s in n-hexane consistent with electron mobility in that liquid, Mozumder (1971) computes the reduced time t_1/t' for the degeneration of the short track to the collection of isolated ion-pairs to be 0.594 at 500 eV and 0.382 at 5 KeV energy deposition respectively. The corresponding numbers of surviving electrons at those times are 1.54 and 62.5 respectively, and the number of escaping electrons are respectively 0.288 and 1.84.

Figure 4 of Mozumder (1971) compares the kinetics of neutralization in an isolated ion-pair, a spherical blob, and a short track. For the isolated ion-pair there is vey little recombination until ~5 × 10^{-11} s, but most recombination is over by ~1ns. The main difference in the recombination kinetics is between the isolated and multiple ion-pair cases. There is not such a great difference among the different multiple ion-pair blobs or short tracks. In the multiple-ion-pair cases, the neutralization is gradual and much faster than that in the isolated ion-pair case at short times. However, this could be an artifact of the model predicated by close proximity of the positive ions having essentially zero mobility (*vide infra*).

Dodelet and Freeman (1975) divide the geometry of a multiple-ion-pair spur into spherical concentric shells such that exactly one electron is contained in each shell with respect to the initial distribution $f(r)$ that is,

$$\int_{r_{i-1}}^{r_i} f(r)\, dr = n^{-1}; \; r_0 = 0 \,; r_n = \infty,$$

where n is the initial number of ion pairs and $f(r)$ is normalized in the sense that

$$\int_0^\infty f(r)\, dr = 1.$$

The innermost electron has an escape probability taken as $\exp(-nr_c/r)$, weighted with $f(r)$, and integrated between 0 and r_1; this is denoted by $\phi\,(1)$. The second electron sees a recombining positive hole of charge n if the first electron has escaped, but only sees a charge of $n - 1$ if the first electron recombines. Thus, the probability $\phi(2)$ that the second electron will escape recombination is given by the sum of two terms, weighted by $\phi(1)$ and $1 - \phi(1)$ respectively, of the integrated escape probabilities with respect to $f(r)$ between r_1 and r_2 calculated respectively for holes of charge n and $n - 1$. Analogously $\phi(n)$ is given by a sum of n terms conditioned respectively on 0, 1, 2, ...$n - 1$ prior neutralizations. The overall escape probability for the entire spur is thus given by

$$\Phi(n) = n^{-1}\left[\phi\!\left(1\right) + \phi\!\left(2\right) + \ldots \phi\!\left(n\right)\right].$$

If an external field is present, the procedure would be the same except that now in place of $\phi(i)$ and $\Phi(n)$ one would use the corresponding probabilities of electron escape as given by the Onsager equation in the presence of an external field (see Sect. 9.5).

Green and Pimblott (1991) have criticized the truncation procedures used by Mozumder (1971) and Dodelet and Freeman (1975a,b) to obtain, respectively, the probabilities of the farthest distance and the ordered distances of electrons in a multiple-ion spur. According to these authors, the statistically correct way of deriving these probabilities relate to the corresponding *marginal distributions* of indistinguishable electrons, all of which have an *identical and independent distribution* $f(r)$. Thus, the distance distribution of the farthest electron in an n-ion-pair spur would be given by $g_n(r) = n[F(r)]^{n-1}f(r)$, where

$$F(r) = \int_0^r f(r')\,dr'.$$

Considering a gaussian distribution $f(r) = [4\pi r^2/(2\pi\sigma^2)^{3/2}] \exp(-r^2/\sigma^2)$, this means that, for a two ion-pair spur, the average distance of the farther electron should be 1.975σ instead of the 2.134σ obtained by Mozumder's original prescription. The relative error of the calculated escape probability decreases with the width of the distribution, but it can be ~25% for $\sigma = 8$ nm, which is typical for n-hexane.

For application to the model of Dodelet and Freeman (1975a,b), the joint density of n ordered distances should be

$$n!\prod_i f(r_i),$$

where $r_1 < r_2 < \ldots < r_n$. The probability density for the kth distance is now given by $g_k(r) = [n!/(k-1)!(n-k)!][F(r)]^{k-1}[1-F(r)]^{n-k}f(r)$, which allows of an intuitive interpretation. Using a gaussian distribution $f(r)$ Green and Pimblott compute a slightly smaller probability of escape for a two ion-pair spur than that obtained by the original procedure of Dodelet and Freeman. The relative error decreases both at small and large values of the gaussian width parameter. Note that both the procedures of Mozumder and of Dodelet and Freeman overestimate the escape probability somewhat.

In Sect. 7.4.6, we discussed various stochastic simulation techniques that include the kinetics of recombination and free-ion yield in multiple ion-pair spurs. No further details will be presented here, but the results will be compared with available experiments. In so doing, we should remember that in the more comprehensive Monte Carlo simulations of Bartczak and Hummel (1986, 1987, 1993, 1997); Hummel and Bartczak, (1988) the recombination reaction is taken to be fully diffusion-controlled and that the diffusive free path distribution is frequently assumed to be rectangular, consistent with the diffusion coefficient, instead of a more realistic distribution. While the latter assumption can be justified on the basis of the *central limit theorem*, which guarantees a gaussian distribution for a large number of scatterings, the first assumption is only valid for low-mobility liquids.

In any case, Bartzack and Hummel have gradually extended their computations from a two ion-pair spur to a full track section up to 50 or 100 KeV in energy. Figure 9.2, shows the variation of the free-ion yield in n-hexane with incident electron energy according to Bartczak and Hummel (1997). For very low electron energies generating only a few ionizations, the escape probability decreases with the initial number of ion-pairs, because the inter-positive-ion distance remains small compared with the mean electron thermalization length. At high electron energy, the inter-positive-ion distance increases greatly, and the ionization track somewhat resembles a collection of isolated geminate pairs. In this regime, the escape probability increases with electron energy. Consequently, Bartczak and Hummel (1997) find a pronounced minimum of G_{esc} at a few KeV electron energy, which is in reasonable agreement with the experiment of Holroyd and Sham (1985) in 2,2,4-trimethylpentane. The computations are also in agreement with the earlier experiment of Hummel et al. (1966) using [37]Ar internal coversion electrons ~2 KeV energy in n-hexane. In these calculations, total ionization yield was taken to be 4–5 and a gaussian distribution was used for electron thermalization distance relative to its parent ion.

The importance of the simulation work of Bartczak and Hummel lies in the following findings:

1. The slope-to-intercept ratio of the variation of free-ion yield with the external field is about the same for multiple ion-pairs as for the isolated geminate pair, to within ~12%.
2. For the few ion-pair spur, the escape probability is nearly the same if the inter-positive-ion separation remains ~1 nm or less.
3. The long-time survival probability for the entire track approaches the limit $P(\tau)/P_{esc} = 1 + 0.6\tau^{-0.6}$, where τ is a normalized time, much earlier than the $P(\tau)/P_{esc} = 1 + (\pi\tau)^{-0.5}$ predicted by the free diffusion theory (Bartczak and Hummel,1997). Notice that the $~\tau^{-0.6}$ dependence of the existence probability had been established earlier in the experiment of van den Ende et al.(1984).

These results imply that the use of the representative single ion-pair distribution in the ionization produced by low-LET irradiation in liquid hydrocarbons can be approximately justified even though the track itself has considerable contribution from multiple-ion-pair spurs and short tracks. It also means that even in the case of an isolated ion-pair, the long-time limit of the existence probability is perturbed by the long-range coulombic field.

In a series of experiments, Brocklehurst and his associates have investigated time-resolved fluorescence (single photon counting) due to solute ion recombination, a model system being squalane solution of p-terphenyl (PT). Especially significant is the "tail" region. Baker et al. (1987) found that the relative intensity of the tail (200–300 ns) increases with photon energy (15–150 eV) under UV-synchrotron irradiation, while the magnetic field enhancement decreases.

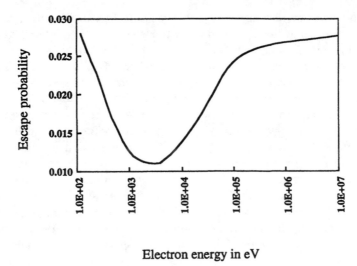

Electron energy in eV

FIGURE 9.2   Variation of the ion escape probability in a nonpolar liquid with incident electron energy according to the simulations of Bartczak and Hummel. The curve shown is for exponential intra-ionic separation with a *b* value of 5.12 nm. (See the original reference for other parametric values.) Agreement with various limited experiments in *n*-hexane is only approximate. Reproduced from Bartczak and Hummel (1997), with the permission of Am. Chem. Soc ©.

This has been attributed to cross-recombination in multiple ion-pair spurs. Baker *et al.* (1989) used UV-synchrotron radiation, in the 10–40 eV range, in squalane solution of PT and determined that in the long-time scale (>75 ns) the magnetic field enhancement of the recombination luminescence decreases with photon energy. Significantly, ^{90}Sr-β irradiation gives values in between those for 15- and 40 eV synchrotron photons, corresponding to one and two ionizations respectively. Brocklehurst (1989) determined the fractional enhancement of fluorescence in the same system between 100 and 300 ns and found that this factor decreases with photon energy in the low-energy region but increases with energy in the high-energy region. The value for ^{90}Sr-β irradiation hints that the asymptote at high energy is similar to the value for a low-energy UV photon (see Figure 9.2). These experiments can be taken as demonstrations of multiple ion-pair recombination in spurs.

In conclusion we may state that there is evidence for multiple ion-pair recombination in spurs; yet a theoretical analysis of free-ion yield and scavenging at low-LET based on the geminate ion-pair picture is meaningful in view of the similarity of the recombination process in the geminate and multiple ion-pair cases. However, if this analogy holds, the geminate ionization yield has to be somewhat less than the true ionization yield.

9.4 DEPENDENCE OF FREE-ION YIELD ON MOLECULAR STRUCTURE AND MOBILITY

After measurements were made of the free-ion yield, thermal electron mobility, and the energy of the lowest conduction state of the electron (V_0) in various hydrocarbon liquids, speculations appeared in the literature probing their intercorrelation as well as correlation between these and other physical properties of the molecule or of the liquid state. Some of these have been mentioned in Sect. 9.1 as conjectures. Noting that the nearly spherical molecules give higher free-ion yields and thermal electron mobilities (e.g., neopentane vs. n-hexane), we have alluded to a relationship between molecular structure on one hand and free-ion yield and mobility on the other. However, this suggestion is probably too simplistic to be true, since in the gas phase there is no such correlation. In fact, electron mobility in neopentane in the gas phase is lower than that of n-hexane. Undoubtedly, the liquid structure is important, although how it is related to the molecular shape and electron transport is not clear. Another important point is that, within the Onsager model, the *postthermal* geminate escape probability is independent of the mobility. Therefore, the dependence of the free-ion yield on electron mobility has to be indirect—for example, through the thermalization distance distribution. If this view is accepted, then most of the effective thermalization distance must be accumulated in the epithermal regime, and the epithermal mobility should vary among the liquids in the same manner as thermal mobility.

Schiller and Vass (1975) attempted a *theoretical* correlation between free-ion yield and electron mobility in a trapping model, and thereby further correlation with V_0. In this model, electrons are trapped due to local energy fluctuations with a probability

$$P = (2\pi\sigma^2)^{-1/2} \int_{-\infty}^{V_0} \exp\left[-\frac{(E - E_t)^2}{2\sigma^2} \right] dE$$

where E_t is the trapped electron energy and σ is the parameter of equilibrium gaussian energy fluctuation. The mobility is then given by $\mu = \mu_{qf}(1 - P)$ where μ_{qf} is the quasi-free electron mobility.

The authors assume different and statistically independent mechanisms of electron-ion recombination in the quasi-free and trapped states. Thus $P_{esc} = w_{qf}w_t$, where w_{qf} and w_t are respectively the probabilities of escaping recombination in the quasi-free and trapped states. Based on some heuristic and not entirely plausible arguments, w_{qf} is approximately equated to 1/2. The probability of finding a trapped electron at a distance between r and $r + dr$ from the positive ion is given by $(\sigma_t P \, dr/v) \exp(-\sigma_t Pr/v)$, where P, is again the probability of finding an electron in the trapped state at equilibrium, v is the trap volume, and σ_t is the trapping cross section. The trapped electrons are considered classical entities governed by the Onsager escape probability. Thus, the overall escape probability is given by

$$P_{esc} = \frac{\alpha^3}{4} \int_0^\infty x^2 \exp\left(-\frac{1}{x} - \alpha x\right) dx$$

with $\alpha = \sigma_t r_c P/v$, $x = r/r_c$ and r_c is the Onsager length.

Assuming the quasi-free mobility to be the same in different liquid hydrocarbons, the measured electron mobility gives P, the equilibrium localization probability. The escape probability can then be evaluated through the parameter α. Schiller and Vass find most reasonable values of parameters as $\mu_{qf} = 65$ $cm^2v^{-1}s^{-1}$, $E_t = -0.28$ eV, and $\sigma_t r_c/v = 10$. Taking v as the molecular volume, this implies $\sigma_t = 0.06$ nm^2 or a trapping radius of 0.15 nm. The computed free-ion yield, with a fixed total ionization yield $G_{ion} = 4.0$, is nearly constant at 0.2 up to $\mu \sim 4$ cm^{2}v^{-1}s^{-1}, beyond which it increases with the mobility, slowly at first and steeply for $\mu \geq 20$ cm^{2}v^{-1}s^{-1}. The dependence of the free-ion yield on the quasi-free electron energy V_0 can be demonstrated in the same manner, since the latter is strongly related to the mobility by experiment as well as by the theory of Schiller and Vass (*vide supra*). Comparison with experiment certainly verifies the trend of increase of G_{fi} with the electron mobility (or with the decrease of V_0), but the quantitative predictions are far from satisfactory.

Jay-Gerin et al. (1993) have made an extensive compilation of mobilities and free-ion yields in 52 nonpolar liquids including the liquefied rare gases (LRGs). The most probable thermalization lengths (b) are also reported using the Onsager formula and taking the total ionization yield ~4. The authors note that G_{fi} remains nearly constant at ~0.1 for $\mu \leq 0.1$ cm^{2}v^{-1}s^{-1}. For 35 liquids in which $\mu > 0.1$ $cm^2v^{-1}s^{-1}$, the free-ion yield has been *empirically* fitted to the equation

$$G_{fi} = \alpha\mu^n, \tag{9.8}$$

where $\alpha = 0.21\pm0.02$ and $n = 0.31 \pm 0.05$ have been obtained by a log-log fit. As yet, no particular theoretical significance has been found for these parameters. The actual fit is qualitatively reasonable but quantitatively inaccurate. Part of the problem is that the experimental free-ion yields show local variations against what is implied by Eq. (9.8). At times, the free-ion yield varies considerably at or near the same value of electron mobility. Sometimes, the free-ion yield actually decreases *locally* with the mobility. The extension of Eq. (9.8) to LRGs does not seem justified, because in these liquids the zero-field free-ion yield cannot be established (see Sect. 9.6). Despite these criticisms, Eq. (9.8) is useful in condensing a large volume of experimental data upon which later theories could be based.

Jay-Gerin *et al.* (1993) also found an *empirical* correlation between G_{fi} and the most probable thermalization length b as follows:

$$G_{fi} = G_{\infty} \exp\left(-\frac{B}{\varepsilon_s b}\right), \tag{9.9}$$

Here ε_s is the static dielectric constant, and the fitted parameters have the values $G_{\infty} = 2.1\pm0.3$ and $B = 324 \pm 22$ Å. However, the value of G_{∞} cannot be interpreted as the total ionization yield, since it is too low. Furthermore, the procedure is inconsistent, since the values of b were already determined with a much greater value of the total ionization yield. The involvement of the static dielectric constant is not evident, since it does not vary a great deal among these nonpolar liquids and also because the determined value of b depends on the form of the thermalization distribution function. For high-mobility liquids, the extraction of b by fitting into the simple Onsager equation is not quite correct, because in such cases the free-ion yield also depends on the electron mean free path (see Sect. 9.2.1). In spite of these shortcomings, Eq. (9.9) serves the useful purpose of correlating a considerable amount of free-ion yield and thermalization length data.

9.5   DEPENDENCE OF FREE-ION YIELD ON EXTERNAL FIELD

In all liquids, the free-ion yield increases with the external electric field E. An important feature of the Onsager (1938) theory is that the slope-to-intercept ratio (S/I) of the linear increase of free-ion yield with the field at small values of E is given by $e^3/2\varepsilon k_B^2 T^2$, where ε is the dielectric constant of the medium, T is its absolute temperature, and e is the magnitude of electronic charge. Remarkably S/I is independent of the electron thermalization distance distribution or other features of electron dynamics; in fact, it is free of adjustable parameters. The theoretical value of S/I can be calculated accurately with a known value of the dielectric constant; it has been well verified experimentally in a number of liquids, some at different temperatures (Hummel and Allen, 1967; Dodelet *et al.*, 1972; Terlecki and Fiutak, 1972).

With an increase of E beyond a certain value specific to the liquid, the free-ion yield increases sublinearly with the field, eventually showing a saturation trend at very high fields (see Mathieu *et al.*,1967). Freeman and Dodelet (1973) have shown that a fixed electron-ion initial separation underestimates the free-ion yield at high fields, and that a distribution of thermalization distance must be used to explain the entire dependence of P_{esc} on E. Therefore, the theoretical problem of the variation of free-ion yield with external field is inextricably mixed with that of the initial distribution of electron-cation separation.

At this stage, it should be recalled that although the S/I ratio has been derived in the Onsager model for a geminate pair, there is extensive Monte Carlo simulation work to extend the same value for multiple-ion-pair cases and, therefore, to the entire track as well (see Sect. 9.3). Hence, the comparison of the experimental value of S/I with the theoretical value of Onsager is meaningful. For example, Hummel and Allen (1967) measure $S/I = 6.0 \times 10^{-5}$ cm/V in n-hexane at 298 K, whereas the theoretical value is 5.81×10^{-5} cm/V.

Onsager's (1938) formula for the probability of escaping geminate-ion recombination in the presence of an external field E may be written as

$$P_{esc}(r_0, E, \theta) = \exp\left[-\frac{r_c}{r_0} - \beta r_0(1 + \cos \theta) \right]$$

$$\sum_{m,n=0}^{\infty} \frac{[\beta(1 + \cos \theta)]^{n+m} r_c^m r_0^n}{m!\,(m + n)!}, \qquad (9.10)$$

where r_0 is the initial electron-ion separation and θ is the angle between the initial separation vector and the downstream direction of E. For a small external field E, Eq. (9.10), on random averaging over $\cos \theta$, reduces to a linear dependence on E—namely, $P_{esc}(r_0, E) = \exp(-r_c/r_0)(1 + e^3E/2\varepsilon k_B^2 T^2) + O(E^2)$, where the slope-to-intercept ratio S/I is $e^3/2\varepsilon k_B^2 T^2$ (*vide supra*). The highest field E_1 at which P_{esc} is proportional to E depends on the criterion of linearity and the particular liquid, since the approximation used to reduce Eq. (9.10) to a linear form is not uniform on r_0. Mozumder (1974a, b) has calculated E_1 for various compounds. Generally, E_1 decreases with the free-ion yield, typical values for n-hexane and neopentane being 40 and 7 kV/cm, respectively.

As it stands, the Onsager equation (9.10) is not convenient for computer calculation, because the series is *formally* doubly infinite and r_0, θ, E, and so on, are coupled. Various authors have rearranged the series to suit specific purposes (Freeman, 1963a–c; Terlecki and Fiutak, 1972; Abell and Funabashi, 1973; Mozumder, 1974a, b). Freeman's treatment contains a discrepancy in an improper averaging over θ, which overestimates the field coefficients by ~10–50% (Mozumder, 1974a), although accidentally the final effect is not so great because of the alternating signs of the coefficients. Mozumder's (1974a) rearrangement of the Onsager equation results in uniform approximation, which is easily computer adapted. He averages Eq. (9.10) over the solid angle and essentially expresses the various coefficients in terms of exponential functions, finally obtaining

$$P_{esc} = 1 - (2\zeta)^{-1} \sum_{k=0}^{\infty} A_k(\eta)A_k(2\zeta), \qquad (9.11)$$

where $\zeta = \beta r_0 = eEr_0/2k_BT$, $\eta = r_c/r_0$, and the A_k coefficients are given inductively by $A_0(x) = 1 - \exp(-x)$ and $A_{k+1}(x) = A_k(x) - \exp(-x)x^{k+1}/(k+1)!$.

Note that in Eq. (9.11) the coefficient in the external field is factored out. A form very similar to Eq. (9.11) has been employed by Abell and Funabashi (1973) in which the expansion coefficients are expressed as products of incomplete gamma functions of order k,—that is,

$$\Gamma(k, x) = \left[(k - 1)!\right]^{-1} \int_0^x \exp(-t)\, t^{k-1} dt.$$

Expansion of P_{esc} in powers of E is generally not recommended, because that expansion *truncated* to a finite number of terms results in large error as in a similar expansion of an exponential function of large argument.

For relatively low fields, though, the first 66 field coefficients have been tablulated by Mozumder (1974a). He has also shown that for very large fields $P_{esc} \to 1 - e < r^{-2}>/\varepsilon E$ where $\langle \cdots \rangle$ indicates averaging over the distribution of initial separation. The E^{-1} asymptotic approach of P_{esc} to unity has been seen in a few experiments.

One aspect of the analysis of Onsager's equation is the realization that, in presence of an external field, P_{esc} can actually decrease with T. This phenomenon, called the *temperature inversion effect* (TIE), can be explained as follows. At relatively low fields, the S/I ratio—namely, $e^3/2\varepsilon k_B^2 T^2$—is expected to be larger at a lower temperature. Therefore, at a sufficiently high field, the line at a lower temperature may cross over that at a higher temperature, yielding TIE. Whether it will actually occur depends mainly on the variation of εT^2 with T. Generally, upon cooling, ε increases but εT^2 may be expected to decrease. To observe TIE, the decrease of εT^2 upon cooling must be sufficiently large as to offset the zero-field decrease of P_{esc}. There is an experimental indication of TIE in 2-methylpropane between 148 and 183 K (Dodelet *et al.*, 1972).

Hummel and Allen (1967) have confirmed the validity of the theoretical S/I ratio in *n*-hexane over the temperature interval 219—298 K. In this liquid the field dependence of ionization current was extended beyond 180 KV/cm by Mathieu *et al.* (1967). Schmidt (1970) determined the free-ion yield in NP up to ~15 KV/cm. These and certain other experiments have been analyzed by Abell and Funabashi (1973) after averaging the field-dependent escape probability (see Eq. 9.11.) over an assumed initial distribution of electron-ion separations. These authors also calculated the electron scavenging probability as a function of scavenger concentration in these liquids for various initial distributions. They demonstrated that in most cases the exponential distribution gives the best agreement with experiments for both scavenging and field dependence of escape probability, although in some cases the results for the gaussian distribution are also acceptable. Using the exponential distribution, the distance parameter (*b*-value) for thermalization found by Abell and Funabashi is 5.6 nm in both *n*-hexane and cyclohexane as determined by scavenging analysis, and 5.0 and 25.0 nm respectively in *n*-hexane and neopentane as determined by analyzing the field effect on the free-ion yield. A similar conclusion has been reached by Friauf *et al.* (1979) by a scavenging study in cyclohexane.

Based on the validity of the exponential distribution, Abell and Funabashi proceed to justify it on the basis of quasi–quantum mechanical arguments. Dodelet *et al.* (1972) measured the free-ion yield, as a function of temperature and external field, in propane (123–233 K), 2-methylpropane (148–294 K), neopentane (295 K), and also in LAr and O_2. In neopentane, the G_{fi} determination agreed with Schmidt (1970) up to the highest field. In all hydrocarbons, the S/I ratio agreed with the Onsager value at all temperatures. The analysis of Dodelet *et al.* (1972) and that of Freeman and Dodelet (1973), based on the random angular averaging of the Onsager escape probability (see Eq. 9.10), further averaged over the initial distribution, gives a fairly consistent total ionization yield of ~4.1–4.3 in the different hydrocarbons. However, to obtain a global agreement of G_{fi} over the entire range of E, the authors find it necessary to add a power tail over a truncated gaussian core of the form $(4r_0^2/\pi^{1/2}b^3)\exp(-r_0^2/b^2)$.

Mozumder (1974b) used Eq. (9.11) and five different assumed forms of the initial distribution to fit the experimental G_{fi} values in 19 hydrocarbon liquids and certain other organic liquids. These five forms are (1) delta function, (2) gaussian centered at origin, (3) exponential, (4) gaussian centered away from the origin, and (5) a truncated power law of the type $f(r_0) = (2.7/4\pi b^3)(r_0/b)^{-5.7}$ $(r_0 \geq b)$ (see Leone and Hamill, 1968). For any particular liquid, the range parameter (b value) varied over a factor of 1.6 to 1.8 for the five distributions. Thus, the smallest b value, obtained for the power-law distribution, was 28 Å for 1,4-dioxane and 133 Å for neopentane. In every case, these fitted b values, with any assumed form of the initial distribution, were much greater than the electronic range of a low-energy (~100 eV or less) secondary electron. This gives some credence to the extraordinary importance of the subvibrational tail of the thermalization length (see Chapter 8).

In Sect. 8.3, we discussed the effect of an external field on the electron thermalization distance according to Rassolov's (1991) calculation. Briefly, that distribution is slightly skewed in the presence of external field. However, when expanded in a Legendre series in the cosine of the angle between the radius vector and the field direction, only the spherically symmetric zeroth-order term makes a substantial contribution to the free-ion yield in a low-mobility liquid. The first-order term vanishes upon angular integration. The second-order term makes a small contribution. *This means that in a low-mobility* (and, therefore, in a low free-ion yield) *liquid, it is safe to use a field-independent thermalization distance distribution in computing the effect of the field on the free-ion yield.* However, the situation is not clear for a high-mobility liquid or in computing the scavenging probability in the presence of an external field.

9.6   FREE-ION YIELD IN LIQUEFIED
RARE GASES

An understanding of free-ion yield in liquefied rare gases (LRGs) is important both from the theoretical viewpoint (Cohen and Lekner, 1967; Gruhn and Loveman, 1979) and for application to ionization chambers (Kubota et al., 1978; Doke, 1981; Aprile et al., 1987). Electron mobility (see Chapter 10) and free-ion yield in LRGs are among the highest in the liquid phase. Because of their relatively high density and small Fano factor, approaching ~0.05 in LXe, LRG proportional counters have good energy resolution. The Fano factor represents the ratio of the mean square fluctuation of ionization to its average number. Furthermore, high saturation drift velocity, large photon yield, and relatively short decay times in these liquids are important to detector physics.

Establishing the zero-field free-ion yield in LRGs has proven to be extraordinarily difficult experimentally. The early measurements of free-ion yields and total ionization yields in LRGs gave rather high values, probably due to experimental artifacts (Robinson and Freeman, 1973). Later measurements gave more reasonable values (Huang and Freeman, 1977; Takahashi et al., 1973, 1974; Aprile et al., 1993). Charge collection as a function of applied field by Scalettar et al. (1982) and by Thomas and Imel (1987) can also be taken as proportional to field-dependent free-ion yield. However, the free-ion yield seems to fall continuously with the decrease of field at ~100 V/cm or less, even though a value of P_{esc} = 0.35 has been argued for LAr at zero field (Doke et al., 1985). It appears that the fall of P_{esc} at low fields is not the result of volume recombination alone or the result of impurities. The collected charge is usually low (~fC), for which a simple correction can be made due to volume recombination (vide infra); the impurity level is also low (~ppb or less). It is likely that this fall of P_{esc} is in the nature of track recombination in LRGs and a zero-field free-ion yield predicted by the Onsager model should not be expected.

Taking P_{esc} = 0.35 in LAr at zero field and the Onsager length as 127 nm implies a mean initial electron-ion separation ~127 nm, which is about an order of magnitude smaller than the thermalization distance (see Chapter 8), but almost the same as the inter-positive-ion separation on a low-LET rack (Burns and Mozumder, 1987). Thus, the Onsager model should not apply in view of the non geminate character of the problem. The situation is the same in the other LRG; therefore, the initial distance obtained by fitting the escape probability to the Onsager equation is suspect in all these cases. Engler et al. (1993) have combined Onsager's escape probability with Debye's srceening factor due to neighboring ions. The ansatz is questionable for mixing spherical and cylindrical symmetries and for attaching the screening factor to the probability rather than to the potential. It is important to realize that in LRG the track geometry is more nearly cylindrical even at the minimum LET.

Thus, a Jaffe'-type theory treats the track geometry correctly (Jaffe', 1913; Kramers, 1952; Burns and Mozumder, 1987). However, these models suffer from incorrect electron and ion distributions arising from their vastly different mobilities and from incorrect electron-ion recombination rate. The data of Scalettar *et al.* (1982) shows that the Onsager slope-to-intercept ratio is not obeyed in LAr at low fields. Further, Thomas and Imel (1987) objected to the ~r⁻¹ potential of the Onsager theory. They ignored electron-ion interaction and, further the diffusion terms in the Jaffe' model and obtained $P_{esc} = \xi^{-1}$ ln$(1 + \xi)$, where $\xi = E_c/E_1$; E_c is a critical field, defined in terms of electron mobility, initial ion density, and the recombination coefficient; and E is the external field. Although $E_c = 0.84$ KV/cm fits the experimental data in LAr, no meaning can be given to it. Also, in certain limits, vastly different theretical models can give the same functional dependence of P_{esc} on E (Mozumder, 1974a), which therefore cannot be taken to justify the model.

Mozumder (1995a, b, 1996) has recently introduced a reencounter model of free-ion yield in LRGs. The key concept in this model is the gradual recombination of electrons drawn at the track axis by the cylindrical field of the line of positive charges R_2^+ (R = Ar, Kr, orXe) while the external field is separating the opposite charge distributions. Following an unsuccessful encounter, the electrons travel to the classical turning point (CTP) determined by the linear density of remaining positive charges. The dynamics is repeated until the charge distributions are completely separated by the external field. Referring to Figure 9.3 and considering an external field E parallel to the track axis, the electrons start from an initial distance r_0 and are drawn by the cylindrical field of the positive ions to a radial distance ~$(n_+^0)^{-1}$, where n_+^0 is the initial linear density of positive charges determined by the LET and the W-value for ionization. Quickly thereafter, electrons recombine with positive ions with a probability κ, which is related to the specific rate k_r through the fractal diffusion parameter and the recollision probability following an unsuccessful encounter (Lo'pez-Quintella and Buja'n-Nu'nez, 1991; Mozumder, 1994). Successive values of r_1, the CTP, increase due to gradual recombination; eventually the sum of return times just exceeds the track separation time $T = L_0/\mu E$, where μ is the electron mobility and L_0 is the initial track length. At this stage, recombination stops and all residual charges are collected by the external field.

The total number of recombinations is given by

$$N_R = \kappa \sum_{s=1}^{m} n_+^s L_s,$$ (9.12)

where L_s is the overlapping track length, n_+^s is the linear positive-ion density after s return passes and m is the maximum value of s determined by the track separation time T. Obviously,

$$n_+^{s+1} = \left(1 - \kappa\right)n_+^s \quad \text{and} \quad L_{s+1} = L_s - \mu E\tau_s,$$ (9.13)

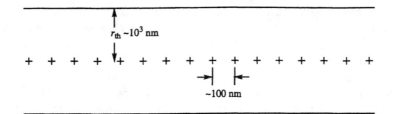

FIGURE 9.3   A low-LET track in a liquefied rare gas (LRG). Even at a minimum LET, the electron thermalization length (~10^3 nm) greatly exceeds the inter-positive-ion separation of R_2^+ (~10^2 nm). Thus, the geometry approximates cylindrical symmetry rather than a collection of isolated ionizations. Reproduced from Mozumder (1995a), with the permission of Elsevier©.

where τ_s is the sth electron return time. Since the electron remains epithermal due to forward and inverse elastic collisions, we have $(2e^2 n_+ /\varepsilon)\ln(r_1 \sqrt{r_2}) = (2e^2 n_+^s + 1/\varepsilon)\ln(r_1^s + 1/r_2)$, where r_1^s is the sth CTP and $r_2 = (n_+^0)^{-1}$ is the lower limit of the cylindrical field. Combining this relation with Eq. (9.13) gives $r_1^s + 1/r_2 = (r_1^s/r_2)^{1/(1-\kappa)}$ connecting successive CTPs. The radial velocity of the electron is $dr/dt = -2\mu e n_+^s/\varepsilon r$, giving the sth return time as $\tau_s = \varepsilon(r_1^s)^2/2\mu e n_+^s$, whereupon m is determined by

$$\sum_{s=1}^{m} \tau_s \approx L_0/\mu E,$$

but not greater than this quotient. Once m is thusly determined, N_R is obtained from (9.12) and (9.13) as follows:

$$N_R = n_+^0 L_0 [1 - (1 - \kappa)^{m+1}] - V_d n_+^0 \left[\sum_{s=1}^{m} \tau_s (1 - \kappa)^s - (1 - \kappa)^{m+1} \sum_{m=1}^{s} \tau_s \right], \quad (9.14)$$

Here, $V_d = \mu E$ is the drift velocity. The recombination and escape probabilities are now given by $P_R = N_R/n_+^0 L_0$ and $P_{esc} = 1 - P_R$. Since $V_d \propto \mu$, but $\tau_s \propto \mu^{-1}$ these probabilities are independent of mobility. However, the initial separation r_0 is expected to depend (increase) with electron mobility, thus making the escape probability indirectly dependent on the mobility. These effects are quite similar to those in the Onsager theory.

Detailed comparison of calculated and experimental results for the variation of the escape probability with the external field in Lar, LKr, and LXe has been made by Mozumder (1995a, b, 1996) using the data on LET, W value, mobility, and so forth. Experiments are with ~MeV electrons or beta-emitters having minimum LET in these liquids. The external field generally does not have any preferred direction relative to the track axis. Mozumder (1995a) argues that in such a case E should be replaced by $E/2$ and κ by $\kappa/2$ to represent isotropic averaging. The initial distance r_0 should be given by thermalization, which occurs in the LRG by a large number ($\sim 10^5$) of elastic collisions in the absence of intra- or inter molecular vibrations. However, because of liquid structure, there is a great difference between energy and momentum transfer mean free paths, denoted by Λ_0 and Λ_1, respectively. The mean square thermalization length is then given by $r_{th}^2 = \Lambda_0 \Lambda_1 (M/2m) \ln(E_i/E_{th})$, where M, and m are the respective masses of the rare gas atom and the electron, E_i is the initial electron energy, and E_{th} is the thermal energy at the liquid temperature (see Sect. 8.4.2).

Values of r_{th} in LAr, LKr, and LXe at their respective temperatures are 1568, 3600, and 4600 nm. Wide variation of thermalization length around the rms value is of course expected. However, the escape probability has been found to be insensitive to r_0 in these cases around their rms values. Therefore, averaging P_{esc} over the distribution of thermalization length has been deemed unnecessary when r_0 is taken equal to r_{th}.

The encounter reaction probability κ, which is related to the electron-ion recombination rate k_r, has the value 0.01, 0.1 and 0.01 for LAr, LKr, and Lxe, respectively. The corresponding values of k_r/k_D are 0.11, 0.22, and $\sim 5 \times 10^{-3}$ respectively, where k_D is the Debye rate constant. The relationship between κ and k_r has been consistently derived in the fractal diffusion theory of reencounters (Mozumder, 1994). Using these input data and Eq. (9.14), Mozumder (1995a, b, 1996) finds good agreement between model calculation and experimental results of Scalettar et al. (1982) and Takahashi et al. (1980) for LAr, of Aprile et al. (1993) and Kubota et al. (1979) for LKr, and of Takahashi et al. (1980) for LXe. The quality of agreement for LAr is shown in Figure 9.4 where absolutely calculated P_{esc} is plotted along with the escape probability relative to that at a 22-kV/cm external field. The significance of the relative escape probability lies in the fact that even at this maximum field, the computed escape probability is 0.95. Below ~ 1 kV/cm, the experimental values fall below the calculated ones for loss of charge collection due to volume recombination. A simple correction is indicated by $P_{fi}/P_{esc} = (1 + k_r c_0 t_c)^{-1}$ where P_{fi} is the probability of free-ion formation including volume recombination, c_0 is the concentration of homogeneous free ions, and t_c is the charge collection time at the electrode. Using parametric values appropriate to the experiment of Takahashi et al. (1980), Mozumder calculates $P_{fi}/P_{esc} = 0.35, 0.43$, and 0.70 in LAr at $E = 140, 200$, and 600 V/cm respectively, compared with the respective experimental values of 0.30, 0.42, and 0.78. Similar corrections have also been made for LKr and LXe, with comparable agreement with experiment.

9.7   POLAR MEDIA

Although free-ion yield has been measured in a number of polar liquids (see Allen, 1976, and Tabata et al., 1991, for tables), and in some as a function of temperature, neither the free-ion yield nor the total ionization yield is understood

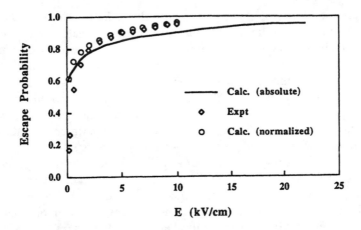

FIGURE 9.4 Variation of the escape probability with the external field in LAr for a 1-MeV incident electron. Full curve, absolute calculation; experimental points and calculated values normalized to 22 KV/cm are denoted by diamonds and circles, respectively. See text for explanation of parameter values used in the calculation. Reproduced from Mozumder, (1995a), with the permission of Elsevier©.

theoretically. The yield of the hydrated electron at the shortest time scale is believed to be around 4.6. Since this is already much greater than the total ionization yield in the gas phase for high-energy electrons (3.3 ± 0.1), there is an unresolved problem of ionization yield in the liquid phase. Tentatively, one may assume that all ionized electrons in water hydrate. Their subsequent decay is due not only to geminate neutralization, but also to various spur reactions, of which the reaction with the OH radical plays a dominant role. The situation is probably similar in liquid ammonia, where a contributing factor is the slowness of the reaction of the electron with the conjugate acid, NH_4^+. In liquids where the dielectric constant is between 3 and 20, the measured free-ion yield, usually below 1.0, generally increases with the dielectric constant, but there are many exceptions (see Allen, 1976).

Mozumder (1969b) pointed out that in the presence of freshly created charges due to ionization, the dielectric relaxes faster—with the longitudinal relaxation time τ_L, rather than with the usual Debye relaxation time τ applicable for weak external fields. The evolution of the medium dielectric constant is then given by

$$\varepsilon^{-1} = \varepsilon_s^{-1} + (\varepsilon_\infty^{-1} - \varepsilon_s^{-1}) \exp\left(\frac{-t}{\tau_L}\right), \tag{9.15}$$

where ε_s and ε_∞ are respectively the static and high-frequency dielectric constants and $\tau_L = (\varepsilon_\infty/\varepsilon_s)\tau$. Mozumder (1969b) used Eq. (9.15) in a Smoluchowski equation to predict the probability of escaping geminate recombination as solvated ions. Different diffusion coefficients were introduced for unsolvated and solvated electrons. In retrospect, it is doubtful if these calculated yields are applicable to highly polar liquids such as water, ammonia, or alcohols, partly because of spur reactions (which violate the geminate model) and partly because of the uncertainty of the total ionization yield. Nevertheless, relations like Eq. (9.15) have frequently been attempted to explain the evolution of the absorption spectrum of the solvated electron. In the case of liquid water, the longitudinal relaxation time is computed to be ~250 fs, which agrees with experiment, but the anticipated mechanism may be different (see Sects. 6.5.2 and 6.5.3).

There is greatly renewed interest in electron solvation, due to improved laser technology. However it is apparent that a simple theoretical description such as implied by Eq. (9.15) would be inadequate. That equation assumes a continuum dielectric with a unique relaxation mechanism, such as molecular dipole rotation. There is evidence that structural effects are important, and there could be different mechanisms of relaxation operating simultaneously (Bagchi, 1989). Despite a great deal of theoretical work, there is as yet no good understanding of the evolution of free-ion yield in polar media.

REFERENCES

Abell, G.C., and Funabashi, K. (1973), J. Chem. Phys. *58*, 1079.

Allen, A.O. (1976), *Yields of Free Ions Formed in Liquids by Radiation*, NSRDS-NBS57, National Bureau of Standards, Washington, D.C.

Allen, A.O., and Hummel, A. (1963), Disc. Faraday Soc. *36*, 95.

Aprile, E., Ku, W.H.-M., Park, J., and Schwartz, H. (1987), Nucl. Instrum. Methods *A261*, 519.

Aprile, E., Bolotnikov, A., Chen, D., and Mukherjee, R. (1993), Phys. Rev. *A48*, 1313.

Bagchi, B. (1989), Annu. Rev. Phys. Chem. *40*, 115.

Baker, G.J., Brocklehurst, B., and Holton, I. R. (1987), Chem. Phys. Lett. *134*, 83.

Baker, G.J., Brocklehurst, B., Hayes, M., Hopkirk, A., Holland, D.H.P., Munro, I.H., and Shaw, D.A. (1989), Chem. Phys. Lett. *161*, 327.

Bartczak, W.M., and Hummel, A. (1986), Radiat. Phys. Chem. *27*, 71.

Bartczak, W.M., and Hummel, A. (1987), J. Chem. Phys. *87*, 5222.

Bartczak, W.M., and Hummel, A. (1993), J. Phys. Chem. *97*, 1253.

Bartczak, W.M., and Hummel, A. (1997), Radiat. Phys. Chem. *49*, 675.

Braun, C.L., and Scott, T.W. (1983), J. Phys. Chem. *87*, 4776.

Braun, C.L., and Scott, T.W. (1987), J. Phys. Chem. *91*, 4436.

Braun, C.L., Smirnov, S.N., Brown, S.S., and Scott, T.W. (1991), J. Phys. Chem. *95*, 5529.

Brocklehurst, B. (1989), Radiat. Phys. Chem. *34*, 513.

Burns, W. G., and Mozumder, A. (1987), Chem. Phys. Lett. *142*, 381.

Capellos, C., and Allen, A. O. (1970), J. Phys. Chem. *74*, 840.

Chang, P., and Wilkie, C.R. (1955), J. Phys. Chem. *59*, 592.

Cohen, M.H., and Lekner, J. (1967), Phys. Rev. *158*, 305.

Debye, P. (1942), Trans. Electrochem. Soc. *82*, 265.

Dodelet, J.-P., and Freeman, G.R. (1972), Can. J. Chem. *50*, 2667.

Dodelet, J.-P., and Freeman, G.R. (1975a), Can. J. Chem., *53*, 1263.

Dodelet, J.-P., and Freeman, G.R. (1975b), Int. J. Radiat. Phys. Chem. *7*, 183.

Dodelet, J.-P., Fuochi, P.G. and Freeman, G.R. (1972), Can. J. Chem. 50, 1617 and references therein.

Doke, T. (1981), Portugal Phys. 12, 9.

Doke, T., Hitachi, A., Kikuchi, J., Masuda, K., Tamada, S., Mozumder, A., Shibamura, E., and Takahashi, T. (1985), Chem. Phys. Lett. 115, 164.

Engler, J , Gils, H.J., Knapp, J., Rebel, H., and Supper, R. (1993), Nucl Instrum. Methods A327, 128.

Freeman, G.R. (1963a), J. Chem. Phys. 38, 1022.

Freeman, G.R. (1963b), J. Chem. Phys. 39, 988.

Freeman, G.R. (1963c), J. Chem. Phys. 39, 1580.

Freeman, G.R., and Dodelet, J.P. (1973), Int. J. Radiat.Phys. Chem. 5, 371.

Freeman, G.R., and Fayadh, J.M. (1965), J. Chem. Phys. 43, 86.

Friauf, R.J., Noolandi, J., and Hong, K.M. (1979), J. Chem. Phys. 71, 143.

Fuochi, P.G., and Freeman, G.R. (1972), J. Chem. Phys. 56, 2333.

Green, N.J.B., and Pimblott. S.M. (1991), Radiat. Phys. Chem. 37, 161.

Gruhn, C.R., and Loveman, R. (1979), IEEE Trans. Nucl. Sci. NS-26, 110.

Holroyd, R.A., and Sham, T. (1985), J. Phys. Chem. 89, 2909.

Hong, K.M., and Noolandi, J. (1978a), J. Chem. Phys. 68, 5168.

Hong, K.M., and Noolandi, J. (1978b), J. Chem. Phys. 69, 5026.

Huang, S.S.-S., and Freeman, G.R. (1977), Can. J. Chem. 55, 1838.

Hummel, A. (1967), Ph.D. Thesis, Free University, Amsterdam.

Hummel, A., and Allen, A.O. (1966), J. Chem. Phys. 44, 3426.

Hummel, A., and Allen, A.O. (1967), J. Chem. Phys. 46, 1602.

Hummel, A., and Bartczak, W.M. (1988), Radiat. Phys. Chem. 32, 137.

Hummel, A., Allen, A.O. and Watson, F.H., Jr. (1966), J. Chem. Phys. 44, 3431.

Jaffe', G. (1913), Ann. Physik 42, 303.

Jay-Gerin, J.-P., Goulet, T., and Billard, T. (1993), Can. J. Chem. 71, 287.

Jonah, C.D. (1983), Radiat. Phys. Chem. 21, 53.

Jonah, C.D., Sauer, M.C., Jr., Cooper, M.C., and Trifunac, a.d. (1979), Chem. Phys. Lett. 63, 53.

Katsumura, Y., Tabata, Y., and Tagawa, S. (1982), Radiat. Phys. Chem. 19, 267.

Kestner, N.R., and Jortner, J. (1973), J. Chem. Phys. 59, 26.

Kramers, H.A. (1952), Physica 18, 665.

Kubota, S., Nakamoto, A., Takahashi, T., Hamada, T., Shibamura, E., Miyajima, M., Masuda, K., and Doke, T. (1978), Phys. Rev. B17, 2762.

Kubota, S., Hishida, M., Suzuki, M., and Ruan(Gen), J. (1979), Phys. Rev. B20, 3486.

Leone, J.A., and Hamill, W.H. (1968), J. Chem. Phys. 49, 5304.

Lo'pez-Quintella, M.A., and Buja'n-Nu'nez, M.C. (1991), Chem. Phys. 157, 307.

Lukin, L.V., Tolmachev, A.V., and Yakovlev, B.S. (1980), Chem. Phys. Lett. 81, 595.

Magee, J.L., and Taylor, A.B. (1972), J. Chem. Phys., 56, 3061.

Mathieu, J., Blanc, D., Camade, P., and Patau, J.-P. (1967), J. Chim. Phys. 64, 1674.

Monchick, L. (1956), J. Chem. Phys. 24, 381.

Mozumder, A. (1969a), Advances in Radiat. Chem. 1, 1.

Mozumder, A. (1969b), J. Chem. Phys. 50, 3153.

Mozumder, A. (1971), J. Chem. Phys. 55, 3020.

Mozumder, A. (1974a), J Chem Phys. 60, 4300.

Mozumder, A. (1974b), J. Chem. Phys. 60, 4305.

Mozumder, A. (1982), J. Chem. Phys. 76, 5107.

Mozumder, A. (1990), J. Chem. Phys. 92, 1015.

Mozumder, A. (1994), J. Chem. Phys. 101, 10388.

Mozumder, A. (1995a), Chem. Phys. Lett. 238, 143.

Mozumder, A. (1995b), Chem. Phys. Lett. 245, 359.

Mozumder, A. (1996), Chem. Phys. Lett. 253, 438.

Mozumder, A., and Magee, J.L. (1966), Radiat. Res. 28, 203.

316 Chapter 9 Electron Escape: The Free-Ion Yield

Mozumder, A., and Magee, J.L. (1967), J. Chem. Phys. 47, 939.
Mozumder, A., and Tachiya, M. (1975), J. Chem. Phys. 62, 979.
Onsager, L. (1938), Phys. Rev., 54, 554.
Powles, J.G. (1985), Phys. Lett. 107A, 403.
Rappaport, D.C. (1984), Phys. Rev. Lett. 53, 1965.
Rassolov, V.A. (1991), M.S. Dissertation, University of Notre Dame.
Robinson, M.G., and Freeman, G.R. (1973), Can. J. Chem. 51, 641.
Robinson, M.G., and Freeman, G.R. (1974), Can. J. Chem. 52, 440.
Robinson, M.G., Fuochi, P.G., and Freeman, G.R. (1971a), Can. J. Chem. 49, 984.
Robinson, M.G., Fuochi, P.G., and Freeman, G.R. (1971b), Can. J. Chem. 49, 3657.
Sano, H. (1983), J. Chem. Phys. 78, 4423.
Sauer, M.C., Jr., and Jonah, C.D. (1980), J. Phys. Chem. 84, 2539.
Scalettar, R.T., Doe, P.J., Mahler, H.J., and Chen, H.H. (1982), Phys. Rev. A25, 2419.
Schiller, R., and Vass, S. (1975), Int. J. Radiat. Phys. Chem. 7, 193.
Schmidt, W.F. (1970), Radiat. Res. 42, 73.
Schmidt, W.F., and Allen, A.O. (1968), J. Phys. Chem. 72, 3730.
Schmidt, W.F., and Allen, A.O. (1970), J. Chem. Phys. 52, 2345.
Scott, T.W., and Braun, C.L. (1985), Can. J. Chem. 63, 228.
Scott, T.W., and Braun, C.L. (1986), Chem. Phys. Lett. 127, 501.
Shinsaka, K., and Freeman, G.R. (1974), Can. J. Chem. 52, 3495.
Shinsaka, K., Dodelet, J.-P., and Freeman, G. R. (1975), Can. J. Chem. 53, 2714.
Smoluchowski, M. Von (1906), Ann. Physik 21, 756.
Tabata, Y., Ito, Y., and Tagawa, S., eds. (1991), CRC Handbook of Radiation Chemistry, Table V.6, pp. 310–312 and 408–411. CRC Press, Boca Raton, Florida.
Tachiya, M. (1988), J. Chem. Phys. 89, 6929.
Tachiya, M., and Schmidt, W.F. (1989), J. Chem. Phys. 90, 2471.
Tagawa, S., Katsumura, Y., and Tabata, Y. (1982a), Radiat. Phys. Chem. 19, 125.
Tagawa, S., Washio, M., Tabata, Y., and Kobayashi, H. (1982b), Radiat. Phys. Chem. 19, 277.
Tagawa, S., Washio, M., Kobayashi, H., Katsumura, Y., and Tabata, Y.(1983), Radiat. Phys. Chem. 21, 45.
Takahashi, T., Miyajima, M., Konno, S., Hamada, T., Shibamura, E., and Doke, T. (1973), Phys. Lett. 44A, 23.
Takahashi, T., Konno, S., and Doke, T. (1974), J. Phys. C7, 230.
Takahashi, T., Konno, S., Hitachi, A., Hamada, T., Nakamoto, A., Miyajima, M., Shibamura, E., Hoshi, Y., Masuda, K., and Doke, T. (1980), Sci. Papers Inst. Phys. Chem. Res. (Riken, Saitama 351, Japan) 74, 65.
Takayasu, H. (1982), J. Phys. Soc. Jpn. 51, 3057.
Terlecki, J., and Fiutak, J. (1972), Int. J. Radiat. Phys. Chem. 4, 469.
Tewari, P.H., and Freeman, G.R. (1968), J. Chem. Phys. 49, 4394.
Tewari, P.H., and Freeman, G.R. (1969), J. Chem. Phys. 51, 1276.
Thomas, J., and Imel, D.A. (1987), Phys. Rev. A36, 614.
Tsurumi, S., and Takayasu, H. (1986), Phys. Lett. 113A, 449.
Van den Ende, C. A.M., Warman, J.M., and Hummel, A. (1984), Radiat. Phys. Chem. 23, 55.

Electron Mobility In Liquid Hydrocarbons

In the presence of an external electric field E an electron, freed from its geminate cation, quickly attains a stationary drift velocity v. For relatively low fields v is proportional to E:

$$v = \mu E, \tag{10.1}$$

and the mobility μ is *defined* by Eq. (10.1). This *zero-field drift mobility* is related to the diffusion coefficient D by the Nernst-Townsend relation, also known as the Nernst-Einstein equation: $\mu = eD/k_B T$. The radiation-induced conductivity, or the current density per unit electric field, is given by

$$\sigma = e(n_e \mu + n_p \mu_+) \equiv e n_e \mu, \tag{10.2}$$

where n_e and n_p are the concentrations of the electron and the positive ion, respectively, and μ_+ is the mobility of the positive ion. The extreme right hand side of Eq. (10.2) applies when, as is often the case, the mobility of the positive ion is much less than that of the electron. Soon after the discovery of radiation-induced conductance in liquid hydrocarbons (Tewari and Freeman, 1968; Schmidt and Allen, 1969; Minday et al., 1971), it was realized that electron mobilities in these substances fall between those of anions and of quasi-free electrons in liquefied rare gases (LRGs). Even so, there is a wide variation of mobility among different hydrocarbons at the same temperature. Another remarkable point is that these liquids are normally insulating, not because electrons are less mobile in them, but because of the dearth of free electrons, except those produced by an external agency such as irradiation.

The mobility is one of the more significant and readily measurable properties of the electron in a medium. Since it is sensitive to the energy and the state of the electron, a detailed study of mobility can be used as a tool for their evaluation. Knowledge of mobility can be applied to dosimeters and ionization chambers togther with a measurement of induced conductivity and free-ion yield (see Eq. 10.2). The subject of electron mobility is vast. In this chapter, we will limit ourselves to electron mobility in liquid hydrocarbons with occasional reference to other liquids. Also, the treatment will be mainly descriptive. The subject matter has been reviewed by Munoz (1991).

10.1   SUMMARY OF EXPERIMENTS ON ELECTRON MOBILITY

Experimental techniques for mobility measurement have been reviewed by Allen (1976) and by Munoz (1991). Broadly speaking there are two ways to measure electron mobility: (1) by measuring radiation-induced conductance, or (2) by measuring electron drift time over a fixed distance in a cell (the time-of-flight method). In the first method, current i is measured in the cell following a known dose of pulsed irradiation (Fuochi and Freeman, 1972; Dodelet et al., 1973). Normally, the electron is the dominant carrier and the current decays by e-ion recombination. Extrapolation of the current to the pulse end gives the initial current i_0 carried by the freed electrons. With known dose and free-ion yield (G_{fi}) the electron concentration n_e is calculated. Equation (10.2) may now be written as $i_0 = e n_e \mu V A / d$, where V is the voltage across plane parallel electrodes of area A separated by distance d in the cell. The cell constant A/d can be determined accurately by measuring the conductance of an ionic liquid of known conductivity. The mobility is then given by $\mu = i_0 / [e n_e V(A/d)]$.

In the time-of-flight method the drift time t_D over a fixed distance d is measured directly. The drift velocity is then $v = d/t_D$. If the potential difference across d is V then the mobility is given from the defining equation (10.1) as $\mu = d^2/(V t_D)$. The concept is simple; however, there are different experimental designs to measure the drift time. In one procedure (Schmidt and Allen, 1970), a thin layer of ionization is created by switching X-rays collimated close to the cathode. The time required for the current to reach its maximum is the drift time to the anode. Alternatively, the entire cell is exposed to a short X-ray pulse (Bakale and Schmidt, 1973a). The current decreases lineraly with time as the ions are extracted by the field. The drift time is that time at which the current is reduced to insignificance. In another variation, called the Hudson method (see Schmidt and Allen, 1970), X-rays are turned on at a known time, irradiating the cell uniformly. The current rises parabolically with time, and the time to reach the maximum is the electron drift time across the cell.

Shutters or other gating devices can also be used effectively to measure the drift time. In the double-shutter method, retarding potentials are applied periodically to two grids placed at a fixed distance in the drift space. The current is maximized when the drift time between the gates is a multiple of the period between the opening of the gates. Reportedly, this method is very accurate, with measurement errors of only ~1–2% in LAr (Jahnke et al., 1971). In the single-shutter variation of the method, only one grid near the cathode is pulsed, or an UV-ionization of the cathode is used and the light intensity is modulated with a shutter (Minday et al., 1972)

Generally speaking time-of-flight methods give more accurate mobility measurements. They are also free from separate dose and free-ion yield measurements. However, accurate drift time measurement requires a very low dose to avoid space charge effects and volume recombination. It also requires high sample purity. Drift times are usually of the order of several microseconds. Assuming electron-attaching impurities with typical specific rates ~10^{13} $M^{-1}s^{-1}$, a mean electron lifetime ~10 μs implies that the impurity concentration must be below 10^{-8} M, or roughly <1ppb. The conductance method of mobility measurement can tolerate at least an order of magnitude higher concentration of impurities. It also requires a much greater dose for reliable measurement of the dose and the free-ion yield. However, the combined errors of all these measurements result in errors of ±30% in mobility determination, whereas the time-of-flight method can routinely give mobility values with error bounds of ±10% (Allen,1976).

Table 10.1 lists electron mobility in some representative nonpolar liquids, most of which are hydrocarbons with some other liquids added for the sake of comparison. A more comprehensive table will be found in the (Tabata et al. (1991). It is convenient to classify the liquids as low, intermediate, or high mobility accordingly as the electron mobility is less than 1, between 1 and 10, or greater than 10 $cm^2v^{-1}s^{-1}$ respectively. There are some special cases where the same liquid, under different conditions of temperature and/or pressure, would be classified in different categories. In such cases the classification is mainly, but perhaps not exclusively, determined by density (Munoz, 1991). Usually, the liquids in each category are characterized by a specific transport mechanism and follow a specific form of field-dependent mobility (see Sect. 10.1.4). Presently, though we are concerned with the zero-field mobility. The activation energy is reported in cases where the mobility increases with temperature, and it can be approximately fitted into an Arrhenius equation.

From the measured mobilities, certain general systematics can be observed. Of these, the dependences of mobility on temperature and molecular srtucture, which are of obvious importance, will be discussed in the following subsections. In n-alkanes, at and around room temperatures, the electron mobility gradually falls with the carbon number, but it becomes nearly constant at n > 7. One interpretation attributes this to electron scattering by a finite part of the alkane

TABLE 10.1   Electron Mobility in Nonpolar Dielectric Liquids

Liquid	T (K)	Mobility $(cm^2v^{-1}s^{-1})$	$\varepsilon_{ac}{}^a$	Reference
Low-mobility liquids				
He	4.2	0.02		Meyer *et al.* (1962)
N$_2$ (l)	77	2.5×10^{-3}		Holroyd *et al.* (1972)
Ethane	111	1.3×10^{-3}	0.13	Schmidt (1977)
Ethylene	170	0.003		Robinson and Freeman (1974)
Propane	175	0.05	0.13	Robinson and Freeman (1974)
Propylene	234	0.008	0.17	Robinson and Freeman (1974)
Cyclopropane	234	0.004	0.17	Robinson and Freeman (1974)
n-Butane	293	0.27	0.17	Robinson and Freeman (1974)
Butene-1	293	0.006		Dodelet *et al.* (1973)
trans-Butene-2	293	0.029	0.23	Dodelet *et al.* (1973)
n-Pentane	295	0.15	0.2	Nyikos *et al.* (1977)
2-Methylpentane	293	0.29		Dodelet *et al.* (1976)
3-Methylpentane	295	0.22	0.2	Kalinowski *et al.* (1975)
n-Hexane	296	0.09	0.24	Schmidt and Allen (1970)
Cyclohexane	294	0.45	0.12	Dodelet and Freeman (1972)
Methylcyclohexane	296	0.07	0.2	Allen and Holroyd (1974)
n-Heptane[b]	295	0.05	0.17	Nyikos *et al.* (1976)
Benzene	300	0.6		Minday *et al.* (1971)
Toluene	300	0.54		Minday *et al.* (1972)
m-Xylene	292	0.057	0.19	Shinsaka and Freeman (1974)

(Continued)

chain that repeats itself for large *n*. In another model (see Sect. 10.2.3), this behavior arises from the similarity of quasi-free electron mobility and trapping potential. There is an empirical correlation between electron mobility and V_0, the electron energy at the bottom of the conduction band; the smaller the value of V_0, the larger the electron mobility. Since the absolute value of V_0 is generally small and there is considerable uncertainty in its determination, one relies on an empirical equation between μ and V_0 such as the one given by Wada et al. ((1977)—namely, $V_0 = (1/15)[\ln(0.3271/\mu - 0.002778)]$. A somewhat simpler form due to Hamill (1981b) is often useful; this is $\mu = 0.36 \exp(-0.35\,V_0/k_B T)$. Apart from conjectures, no systematic derivation of these equations is available. There is a strong correlation between electron mobility and free-ion yield, some aspects of which were discussed in Sect. 9.4. The increase of G_{fi} with the mobility is probably an indirect effect via the thermalization length distribution. If

TABLE 10.1 (Continued)

Liquid	T (K)	Mobility (cm²v⁻¹s⁻¹)	ε_{ac}[a]	Reference
Intermediate-mobility liquids				
Isobutane	294	5.0		Fuochi and Freeman (1972)
Isobutene	293	1.44		Dodelet *et al.* (1973)
2-methylbutene-2	300	3.6	0.11	Minday *et al.* (1971)
Cyclopentane	296	1.1		Schmidt and Allen (1970)
2,2-Dimethylbutane	296	10.0		Schmidt and Allen (1970)
2,3-Dimethylbutene-2	293	5.8		Dodelet *et al.* (1973)
Cyclohexene	293	1.0		Dodelet *et al.* (1973)
2,2,4-Trimethylpentane	296	7.0		Schmidt and Allen (1970)
2,2,3,3-Tetramethylpentane	295	5.2	0.06	Dodelet and Freeman (1972)
High mobility liquids				
Ar	85	475		Miller *et al.* (1968)
Kr	117	1800		Miller *et al.* (1968)
Xe	163	2200		Miller *et al.* (1968)
Methane	111	400		Bakale and Schmidt (1973a)
Neopentane	296	68	0.01	Minday *et al.* (1972)
Tetramethylsilane	223–292	100	~0	Allen *et al.* (1975)
2,2,4,4-Tetramethylpentane	295	24	0.06	Dodelet and Freeman (1972)
2,2,5,5-Tetramethylhexane	293	12	0.05	Dodelet and Freeman (1972)
Tetramethylgermanium	296	90		Sowada (1976)

[a]Activation energy of mobility in eV.

[b]For *n*-alkanes beyond heptane, the mobility and activation energy at room temperature remain nearly constant at ~0.04 cm²v⁻¹s⁻¹ and ~0.2 eV, respectively.

there is a similarity between the scattering mechanisms of thermal and epithermal electrons, then the effective thermalization length should increase with the thermal mobility, giving a positive correlation with the free-ion yield.

The dependences of electron mobility on medium density and on phase change are complex and poorly understood. In Ar, Kr and Xe, the mobility increases by a factor of about 2 in going from the liquid to the solid phase. This has generated speculation that long-range order is not *necessary* for high electron mobility. On the other hand, electron mobility in Ne increases from ~10^{-3} to 600 cm²v⁻¹s⁻¹ on solidification at 25.5 K (see Allen, 1976). In liquid He, the electron mobility above the λ-point (2.2 K) varies approximately inversely with the viscosity, consistent with the bubble model. Below the λ-point, the mobility remains finite even though the viscosity vanishes. The mobility increases from 0.033 cm^{2}v^{-1}s^{-1} at 2.2 K to 220 cm^{2}v^{-1}s-1 at 0.5 K. At still lower temperatures, the mobility varies as the inverse cube of absolute temperature (see Allen, 1976). In liquefied rare gases, electron mobility as a function of medium density shows a sharp peak, which is mainly determined by density on the high-density side, in apparent agreement with theory. On the low-density side, though, other details become important. When the temperature is continuously increased, several hydrocarbon liquids pass from low to intermediate mobility, or from intermediate to high mobility (see Tabata *et al.*, 1991). In this sense, the categorical divisions are somewhat circumstantial.

10.1.1   TEMPERATURE DEPENDENCE

Electron mobility is an activated process in low-mobility hydrocarbon liquids. The mobility in these liquids increases with temperature with activation energy exceeding 0.1 eV. In a few high-mobility liquids, the measured activation energy lies between ~0 to 0.05 eV (see Table 10.1); in many others, the mobility decreases with temperature. Intermediate mobility liquids show complex behavior, although some of these have activation energies ~0.1 to ~0.05 eV.

Since liquid density can change considerably over the temperature interval of mobility measurement, care must be taken in giving a meaning to the directly measured activation energy (Munoz,1991). On the other hand, density correction to mobility at constant temperature cannot be effected in liquids with the ease that is customary in low-density gases. In the latter case, the product of density and mobility remains essentially constant while density is changed by variation of pressure at constant temperature, implying the dominance of scattering by individual molecules. Clearly, such is rarely the case in the liquid phase. In any case, a positive activation energy usually signifies the existence of a trapped electron state. In the *two-state model,* this may be interpreted as the shift of the equilibrium population of electrons toward the quasi-free state, in which the mobility is orders of magnitude greater than that in the trapped state (see Sect. 10.2.2). However, the quasi-free mobility itself may be temperature-dependent. Usually, this mobility is expected to decrease with temperature due to increased thermal velocity, which shortens the relaxation time (see Sect. 10.2). This effectively reduces the activation effect from the trapped state somewhat. Therefore, the activation energy of mobility should not be simply equated to the electron binding energy in the trap. Nevertheless, in certain theoretical models, these two energies are seen to be correlated (see Sect. 10.2.3).

In an alternative model, quantum-mechanical tunneling of the electron is invoked from trap to trap without reference to the quasi-free state. The electron, held in the trap by a potential barrier, may leak through it if a state of matching energy can be found on the other side of the barrier. An incipient trap close to the trapped electron usually is not expected to have the right configuration for energy matching. But thermal fluctuation continually changes the trap configuration, occasionally providing the energy matching when the trap-to-trap transfer becomes possible. At a higher temperature, this possibility becomes more frequent, and the process is therefore called *phonon-assisted hopping* (see Sect. 10.2.1).

Where mobility data are available over a considerable range of temperature, the activation energy is often found to be temperature dependent. Thus, in *n*-pentane the activation energy increases with temperature whereas in ethane it decreases (Schmidt, 1977). Undoubtedly, part of the explanation lies in the temperature dependence of density, but detailed understanding is lacking. In very high mobility liquids, the mobility is expected to decrease with temperature as in the case of the quasi-free mobility. Here again, as pointed out by Munoz (1991), density is the main determinant, and similar results can be expected at the same density by different combinations of temperature and pressure. This is true for LAr, TMS, and NP, but methane seems to be an exception.

10.1.2 DEPENDENCE ON MOLECULAR STRUCTURE

Since the early days of electron mobility measurement, several researchers have noticed a strong relationship between mobility and molecular shape (Allen, 1976; Schmidt, 1977; Gee and Freeman, 1983, 1987; Gyorgy *et al.*, 1983). Simply stated, the more spherical the molecule is, the greater is the electron mobility. Zero-field mobility is largest in LAr, LKr, and LXe, the liquids that exhibit Ramsauer effect in the gas phase (see Table 10.1). Electron mobility in liquid He (also in liquid Ne) is low, interpreted in terms of a bubble state of the electron. The most surprising thing about electron mobility in liquid alkanes is the wide variability, by about three orders of magnitude, correlated by molecular sphericity, while other ordinary properties are quite similar. Molecules in the three liquids of highest mobility, methane, TMS, and NP (see Table 10.1), are all nearly spherical in shape. An apparent counterexample is the fact that the mobility in *cis*-butene (lower symmetry) is much greater than that in *trans*-utene (higher symmetry) (Dodelet *et al.*, 1973). This particular case has generated some discussion on the difference between molecular sphericity and liquid structure symmetry (Stephens, 1986; Freeman, 1986).

The correlation between mobility and sphericity has given rise to different speculations relating molecular shape and physical properties that could influence electron transport. However, it should be stressed that the liquid structure is important as well (Stephens, 1986). For example, although the electron mobility in *liquid* NP is several orders of magnitude larger than that in *liquid* n-hexane, the mobility in NP *gas* is actually somewhat less than that in *gaseous* n-hexane. Molecular shape and liquid structure are probably correlated in a manner that is not precisely understood, but some conjectures are available (Allen, 1976). More spherical molecules provide more uniformly varying potentials in the liquid. Anisotropy of molecular polarizability creates strong irregularities of the potential in the liquid. These irregularities scatter the electron randomly and also cause the electron to become localized at potential minima. Both these effects can reduce the mobility drastically. However, no quantitative explanation is as yet available.

10.1.3   DRIFT AND HALL MOBILITIES

In the joint presence of an electric field $\vec{E}$ and a magnetic field $\vec{B}$ in a medium, the stationary electron velocity can be written as

$$\vec{v} = \mu\vec{E} + \mu_H\mu\vec{E}x\vec{B}, \tag{10.3}$$

where μ_H is *defined* to be the Hall mobility and μ stands for the ordinary drift mobility. Since the magnetic field $\vec{B}/v$ has the dimension of $\vec{E}$, the dimension of the Hall mobility is the same as that of the drift mobility, and both mobilities may be expressed in the same unit.

Hall mobility is of special significance for *inhomogeneous* electron transport models—for example, the two-state, trapping and percolation models—because in such cases the Hall mobility is closely related to the mobility in the higher conducting state of the electron. Unfortunately, the experimental determination of Hall mobility requires that the drift mobility be sufficiently large that a reasonable Hall voltage may be expected (see Eq. 10.3). Hence, Hall mobility measurement is limited to high mobility liquids, whereas theoretically it would be even more important to have data in low-mobility liquids.

Experimental measurement of Hall mobility produces values of the same order of magnitude as the drift mobility; their ratio $r \equiv \mu_H/\mu$ may be called the Hall ratio. If we restrict ourselves to high-mobility electrons in conducting states in which they are occasionally scattered and if we adopt a relaxation time formulation, then it can be shown that (Smith, 1978; Dekker, 1957)

$$\mu = \frac{e}{m^*}\langle\tau\rangle, \quad \mu_H = \frac{e}{m^*}\frac{\langle\tau^2\rangle}{\langle\tau\rangle}, \quad \text{and} \quad r = \frac{\langle\tau^2\rangle}{\langle\tau\rangle^2},$$

where $\langle\cdots\rangle$ represents averaging over the distribution of momentum relaxation time τ and m^* is the electron effective mass.

Specific values of r are helpful for identifying the nature of the scattering process. For example, if the momentum relaxation time is proportional to a given power of electron energy—$\tau(\varepsilon)\propto\varepsilon^{\gamma}$—then $r = \Gamma(5/2)\Gamma(5/2 + 2\gamma)/[\Gamma(5/2 + \gamma)]^2$, where Γ is the gamma function. Thus, for scattering of low–energy electrons by neutral impurities and by optical phonons, $\gamma = 0$ and $r = 1.0$; for scattering by dislocations and acoustic phonons (or, when the mean free path of scattering is independent of electron energy), $\gamma = -0.5$ and $r = 1.18$; and for scattering by ionized inpurities, $\gamma = 1.5$ and $r = 1.93$. Still, the experimental value of r may not agree with any of the known scattering mechanisms which therefore cannot be traced from the r-value alone.

It is important to realize that even in the presence of traps, the measured Hall mobility refers to that in the higher conducting state (Munoz, 1991). Thus, a value of r significantly >1.0, and increasing with temperature in a certain interval, has been taken as an evidence in favor of traps in NP near the critical point (Munoz, 1988; Munoz and Ascarelli, 1983). Similarly, a nearly constant value of r near 1.0 in TMS over the temperature interval 22–164°C has been taken to indicate absence of trapping in that liquid. The scattering mechanism in TMS is consistent with that by optical phonons (Doldissen and Schmidt, 1979; Munoz and Holroyd, 1987).

Measurement of Hall mobility requires the ability to measure very small currents in a liquid of high purity. A charge-sensitive amplifier, rather than a current amplifier, is needed for measuring such small Hall currents as ~1 nA over a drift time ~100 μs, producing a collected charge ~10^{-13} C as compared with typical rms noise level of ~10^{-15} C. The purity requirement is about an order of magnitude stricter than that for drift mobility measurement. The latter, for a high mobility liquid like TMS, has been estimated as an oxygen equivalent of <3 ppb if an electron lifetime ~100 μs is to be desired.

Besides these, there are other technical requirements for good Hall mobility measurement, which have been detailed by Munoz (1991). In many Hall mobility experiments, a modification of the Redfield (1954) technique is used (Munoz, 1988). In the Redfield method, instead of measuring the Hall current directly, a counter electric field is applied to cancel it; or, more often, the transverse voltage needed to cancel the difference between the signals on switching the direction of the magnetic field is recorded. This is therefore a null method. Consider that two parallel resistive plates of length L are immersed in the liquid along the y direction of a right–handed rectangular coordinate system. A voltage V_L impressed along the length of each of these plates produces an electric field $E_L = V_L/L$ in the y direction and a corresponding longitudinal drift current, i_L. A voltage difference ΔV_t is applied across the plates (along the x direction) separated by distance t such that it would cancel the difference between the Hall signals obtained for magnetic fields +B to −B, applied along the z direction. The Hall mobility is then given by

$$\mu_H = \frac{1}{2B}\frac{\Delta V_t}{t}\frac{L}{V_L}.$$

Munoz and Holroyd (1987) have measured Hall mobility in TMS from 22 to 164° C. This measurement parallels very well the variation of drift mobility with temperature in this liquid, and the Hall ratio remains essentially constant at 1.0±0.1. Both the drift and Hall mobilities in TMS decrease with temperature beyond 100°C, becoming 50 cm²v⁻¹s⁻¹ at 164°C. The overall conclusion is that TMS is essentially trap-free in this temperature range, and the decrease of mobilities is due not to trapping, but to some other scattering mechanism that is more effective at higher temperatures.

Hall mobility in NP has been measured by Munoz and Ascarelli (1983, 1984) as a function of temperature up to the critical point (160°C). It falls relatively slowly from 220 cm²v⁻¹s⁻¹ at 140°C to 170 cm²v⁻¹s⁻¹ at the critical temperature. The drift mobility, however, falls precipitously over that temperature interval to ~30 cm²v⁻¹s⁻¹ at the critical temperature. Consequently the Hall ratio r increases sharply from 1.5 at 130° to 5.5 at 160°C. This has been taken as evidence for intrinsic trapping in this liquid.

Itoh et al. (1989) have measured Hall mobility in 2,2,4,4-tetramethylpentane (TMP),2,2,4-trimethylpentane (isooctane) and 2,2-dimethylbutane (DMB). In 2,2,4,4-TMP and in 2,2-DMB, the Hall ratio remains effectively constant, at slightly more than 1.0 over the temperature range 20 to 160°C. Therefore, it is believed that traps are relatively insignificant for electron transport in these liquids, even though the absolute value of the Hall mobility remains well below 100 cm²v⁻¹s⁻¹ over most of the temperature interval. In isooctane, the Hall ratio is ~3.5 over a significant temperature range, signifying the existence and importance of traps. It should be noted that in all these branched hydrocarbons, the absolute value of the Hall mobility, ~10–30 cm²v⁻¹s⁻¹, is much smaller than theoretically predicted (Davis and Brown, 1975; Berlin and Schiller, 1987).

Hall and drift mobilities have been measured in mixtures of n-pentane and NP by Itoh et al., (1991) between 20 and 150°C. They found both mobilities to decrease with the addition of n-pentane to the extent that the Hall mobility in a 30% solution was reduced by a factor of about 5 relative to pure NP. However the Hall ratio remained in the range 0.9 to 1.5. This indicates that, up to 30% n-pentane solution in NP, the incipient traps are not strong enough to bind an electron permanently. However, they are effective in providing additional scattering mechanism for electrons in the conducting state.

10.1.4   FIELD DEPENDENCE OF MOBILITY

The dependence of drift mobility on the external field must be interpreted by a theoretical model, and as such it can elucidate the transport mechanism in a given case. In this section, we will only describe the phenomenological and experimental aspects. Their theoretical significance will be taken up in Sect. 10.2.

At low enough fields, the electron drift velocity is proportional to the field, which is called *ohmic* behavior. With increasing field strength, the drift velocity is no longer proportional to the field, and the mobility, still defined by Eq. (10.1), becomes field-dependent. Basically, two kinds of phenomena are observed: *sublinear* and *supralinear*. In high-mobility liquids often the drift velocity increases less than proportionately with the field and the phenomenon is termed sublinear. At very high fields, a saturation drift velocity (v_s) is often observed in sublinear cases—that is, $\mu \propto E^{-1}$. Supralinear behavior is generally seen in low-mobility liquids where, at sufficiently high fields, the drift velocity increases more than proportionately with the field. In such cases, instead of drift velocity saturation, electron amplification and eventual avalanche and dielectric breakdown may occur at very high fields. These latter effects are important for insulation studies, but they are clearly outside the field of radiation chemistry. Sublinear drift velocity, characteristic of quasi-free electrons, is a signature of the hot electron effect. When the rate of energy input by the electric field into the electron population exceeds the rate of energy loss by *thermal* electrons, the average electron energy rises above thermal. This lowers the relaxation time, thus lowering the effective mobility.

For the *supralinear* case, the hopping rate, whether activated or due to tunneling, increases with the field; therefore, the mobility also increases with the field. Alternatively, in the trapping model, the field shifts the population equilibrium in favor of the quasi-free state, again causing the mobility to increase with the field. It should be remembered that sub- and supralinear classification is approximate and the same liquid can exhibit both phenomena at different temperatures, or at different field regimes at the same temperature. For example, drift velocity is supralinear with field in ethane at 170 K beyond 100 KV/cm, whereas at 303 K it is sublinear for fields exceeding 5 KV/cm (Schmidt, 1977). In LAr, LKr, and LXe near the respective critical temperatures, the drift velocity is *sublinear* for E exceeding a few KV/cm; *at the same temperature*, it is *supralinear* for fields of about a few hundreds of V/cm or less (Jahnke *et al.*, 1971; Huang and Freeman, 1978, 1981). In all cases, an experimental critical field E_c can be defined such that for $E < E_c$, the mobility is field-independent.

Figure 10.1a shows electron drift velocity as a function of electric field in methane, NP, and TMS (sublinear cases) according to the data of Schmidt and co-workers. These are contrasted in Figure 10.1b with supralinear drift velocity in neohexane, ethane, 2,2,2,4-TMP, and butane at the indicated temperatures In the case of neohexane, the drift velocity has been found to be proportional to the field up to 140 KV/cm (Bakale and Schmidt, 1973b).

Table 10.2 lists the critical field E_c in various nonpolar liquids along with the approximate nature of field dependence of mobility when $E > E_c$. It is remarkable that the higher the zero-field mobility is, the smaller is the value of E_c, indicating the role of field-induced heating. Also note that in the sublinear case, E_c is larger in the case of molecular liquids than for liquefied rare gases,

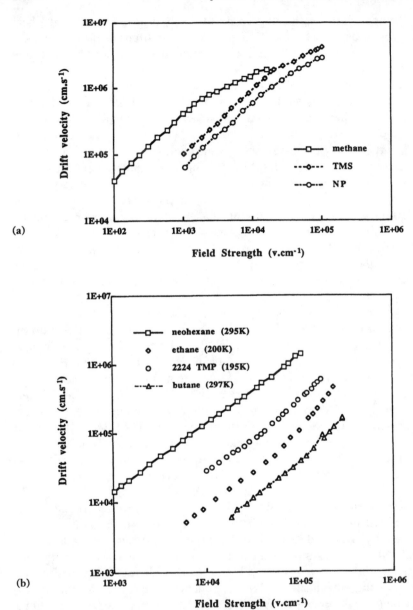

(a)

(b)

FIGURE 10.1 (a) Electron drift velocity versus external electric field in methane, NP, and TMS showing sublinear dependence. See text for details. (b) Electron drift velocity versus external electric field in neohexane, ethane, 2,2,2-TMP, and butane at the indicated temperatures. Generally, the field dependence is supralinear, whereas for neohexane, linear variation has been seen up to 140 kV/cm. See text for details. Reproduced from Schmidt (1977), with permission of National Research Council of Canada©.

which shows the importance of cooling modes provided by molecular vibrations and rotations. For sublinear molecular liquids, the power index for the variation of mobility with field has generally the asymptotic value of -0.5. However, this form of field dependence is not expected to be accurate close to the linear regime. Among the low-mobility liquids, n-hexane and cyclohexane might be expected to exhibit supralinear drift velocity at sufficiently high electric fields, but experimentally drift velocity has been found to be linear with the field up to 83 and 100 KV/cm, respectively, in these liquids (Schmidt and Allen, 1970). The values of Λ obtained by fitting the field dependence of mobility to the Bagley equation in the supralinear case should correspond to *some* kind of jump distance of the localized or trapped electron. However, its large value and temperature variation are not easy to explain with simple models of transport (see Sect. 10.2). Experimentally, Λ has been found to increase with temperature

TABLE 10.2 Critical Field E_c and Nature of Field Dependence of Mobility in Various Nonpolar Liquids

Liquid	T (K)	E_c (KV/cm)	Nature[a]	Reference
Sublinear cases				
LAr	85	0.2	0.63[b]	Miller et al. (1968)
LKr	117	0.02	0.48[c]	Jacobsen et al. (1986)
LXe	163	0.03	0.35[d]	Miller et al. (1968)
CH$_4$	111	1.5	0.5	Schmidt and Bakale (1972)
NP	296	20.0	0.5	Bakale and Schmidt (1973b)
TMS	296	20.0	0.5	Schmidt (1977)
Isooctane	296	90.0	0.5	Schmidt (1977)
Supralinear cases				
Ethane	111–216	9.2–40	1.0–4.5	Schmidt et al. (1974)
Propane	130–260	36–72	3.0	Sowada et al. (1976)
Butane	230–300	113–83	1.7–3.0	Schmidt (1977)
Pentane	296	200	1.2	Sowada et al. (1976)
Cyclopentane	200–300	83–55[e]	2.0–4.5	Schmidt (1977)

[a]In sublinear cases, a value of n is given when the mobility varies asymptotically as E^{-n}. In supralinear cases, the mobility is fitted to the Bagley equation, $\mu(E)/\mu(0) = \sinh(\eta)$ with $\eta = e\Lambda E/2k_BT$, and the value of Λ is reported in nm.

[b]For $E < 10$ KV/cm; for $E > 10$ KV/cm, $n = 1$ in LAr.

[c]For $E < 0.2$ KV/cm; for $0.2 < E < 4$ KV/cm, $n = 0.75$, and for $E > 4$ KV/cm, $n = 1$ in LKr.

[d]For $E < 0.12$ KV/cm; for $0.12 < E < 2.75$ KV/cm, $n = 0.81$, and for $E > 2.75$ $n = 1$ in LXe.

[e]E_c in cyclopentane is estimated from the Λ value obtained by fitting the temperature dependence of mobility by a generalized Bagley equation (see Sect. 10.2). Schmidt and Allen (1970) found drift velocity in this liquid to be linear with field up to at least 62 KV/cm.

TABLE 10.3   Saturation Drift Velocity in High-Mobility Liquids

Liquid	T (K)	μ (cm^{2}v^{-1}s^{-1})^a	u_s (cm/s)
LAr	85	400	6.4×10^5
LKr	120	1200	4.8×10^5
LXe	165	2000	2.6×10^5
CH$_4$	111	400	6.0×10^6
TMS	296	100	8.0×10^6

^{a}Zero field drift mobility.

After Schmidt (1988).

in all cases except for propane, in which it is temperature-independent. No obvious explanation has been offered for such variation of Λ with temperature in different liquids.

In certain liquids, the electron drift velocity shows peculiar behavior under special circumstances, some of which will now be described.

1. In liquid ethane, the electron mobility changes from a low to a high value (~4 cm^{2}v^{-1}s^{-1}) at T≥260 K, eventually reaching 44 cm^{2}v^{-1}s^{-1} at the critical temperature (305.3 K) (Doldissen et al., 1980). Not surprisingly, the field dependence changes from supralinear at 111 K to sublinear at 305.3 K with E_c < 10 KV/cm. At 239 K, the drift velocity is linear to the field up to the highest field investigated.

2. In liquid Ne, evidence has been found for a high-mobility species, which may be a delocalized electron, that converts to a low-mobility species in several tens of nanoseconds (Sakai et al., 1992). Field dependence of the low-mobility species is supralinear, but the lifetime of the high-mobility species increases with the field strength and decreases with temperature from ~2 to ~100 ns.

3. The saturation drift velocity in high-mobility molecular liquids can attain very high values (see Table 10.3), and in LRGs it can be increased by the addition of a small amount of a polyatomic compound (Yoshino et al., 1976). Both of these effects indicate that even at the highest field the electron can remain somewhat cooler by the excitation of intramolecular, modes which effectively increases the momentum relaxation time of the hot electron. Since Shockley's (1951) theory predicts a strong field dependence of mobility when the drift velocity approaches the speed of sound c, there have been attempts to correlate the saturation drift velocity u_s with c. However, no clear-cut theoretical picture has emerged. In many cases, u_s greatly exceeds c. According to Spear and LeComber (1969), u_s in LRGs should be given by $u_s/c = (64W/9\pi k_B T)^{1/2}$, where W

is an energy given through the energy dependence of the effective mass—namely, $m_0^*/m^* = 1 - \varepsilon/W$, where m^* and m_0^* are respectively the effective massses at electron energy ε and at the bottom of the conduction band. Holroyd and Schmidt (1989) note that the saturation drift velocity is not explained by the Cohen-Lekner theory (1967).

4. (IV) In high-mobility liquids, there are considerable data on the density dependence of mobility; this information has been reviewed by Munoz (1991). Density in LRGs, liquid methane, NP, and TMS can be varied by change of temperature and/or pressure, and the consequent variation of mobility is qualitatively similar in these liquids. Roughly speaking, the mobility increases sharply with density starting from the critical temperature, reaches a maximum at around 1.5–2.0 times the critical density, and then continues to fall until the density at the triple point is reached. On the high-density side, a mobility minimum has been found in LAr and TMS. In cases where the same density can be obtained at different pressures, as in LAr and NP, the mobility shows pressure dependence on the low-density side of the mobility maximum. On the high-density side, the mobility depends mainly on the density. In almost all cases, the dependence of mobility on density mimics the dependence of V_0, the electron energy at the bottom of the conduction band.

10.2   THEORETICAL MODELS
OF ELECTRON TRANSPORT

Theories of electron mobility are intimately related to the state of the electron in the fluid. The latter not only depends on molecular and liquid structure, it is also circumstantially influenced by temperature, density, pressure, and so forth. Moreover, the electron can simultaneously exist in multiple states of quite different quantum character, between which equilibrium transitions are possible. Therefore, there is no unique theory that will explain electron mobilities in different substances under different conditions. Conversely, given a set of experimental parameters, it is usually possible to construct a theoretical model that will be consistent with known experiments. Rather different physical pictures have thus emerged for high-, intermediate- and low-mobility liquids. In this section, we will first describe some general theoretical concepts. Following that, a detailed discussion will be presented in the subsequent subsections of specific theoretical models that have been found to be useful in low- and intermediate-mobility hydrocarbon liquids.

The electron in a condensed medium is never entirely free, being in constant interaction with the molecules. It is designated *quasi-free* when its wave function is *delocalized* and extends over the medium geometry. Such quasi-free electrons do not pertrurb the medium structure significantly and their mobility is high. By contrast, *localized* electrons perturb the medium structure significantly, and in the extreme case the perturbation is so great that the electrons becomes self-trapped. Trapping potentials may *preexist* in a liquid, or they may arise due to the electron-medium interaction. In some cases, as in liquid helium, hydrogen, and ammonia, the electron can create a *cavity* by repulsive interaction and get trapped in it.

Springett, *et al.*, (1968) have given the criterion for *energetic stability* of the trapped electron as $E_t < V_0$, where E_t is the trapped electron energy in its ground state and V_0 is the lowest energy of the quasi-free electron in the conduction band. The trapped electron energy $E_t = E_e + 4\pi R^2 \gamma + (4\pi/3)R^3 P$, where E_e is the ground state electronic energy, R is the cavity radius, P is the pressure, and γ is the surface tension of the liquid. The last two terms of this equation represent the work necessary to create a void of radius R in the liquid. Jortner (1970) simplifies the calculation by considering a very high trapping potential when the electron is entirely confined within the cavity, giving $E_e = h^2/8mR^2$; he also ignores the pressure-volume work in the limit of low P and gets $E_t = h^2/8mR^2 + 4\pi R^2 \gamma$. The cavity or bubble adjusts to a radius, giving a minimum value of E_t such that $\partial E_t/\partial R = 0$ and $E_t = h(2\pi\gamma/m)^{1/2}$. The stability criterion may now be stated as $4\pi\beta\gamma/V_0^2 < 1/4\pi^2 = 0.0253$ with $\beta = \hbar^2/2m$.

The more incisive calculation of Springett, *et al.*, (1968) allows the trapped electron wave function to penetrate into the liquid a little, which results in a somewhat modified criterion often quoted as $4\pi\beta\gamma/V_0^2 \leq 0.047$ for the stability of the trapped electron. It should be noted that this criterion is also approximate. It predicts correctly the stability of quasi-free electrons in LRGs and the stability of trapped electrons in liquid ^{3}He, ^{4}He, H_2, and D_2, but not so correctly the stability of delocalized electrons in liquid hydrocarbons (Jortner, 1970). The computed cavity radii are 1.7 nm in ^{4}He at 3 K, 1.1 nm in H_2 at 19 K, and 0.75 nm in Ne at 25 K (Davis and Brown, 1975). The calculated cavity radius in liquid He agrees well with the experimental value obtained from mobility measurements using the Stokes equation $\mu = e/4\pi R\eta$, with perfect slip condition, where η is liquid viscosity (see Jortner, 1970). Stokes equation is based on fluid dynamics. It predicts the constancy of the product $\mu\eta$, which apparently holds for liquid He but is not expected to be true in general.

Another way of classifying the states of an electron in a liquid is by the role of positional disorder. In a perfect crystalline lattice, the electrons propagate as Bloch waves without scattering. In a real crystal, they are ocassionally scattered due to disorder, generating a finite mean free path l. Modern research has shown that this picture holds in many solids on melting, and a *very* long range order is not necessary for a relatively long mean free path of scattering (Mott and Davis, 1979). According to Ioffe and Regel (1960), the crucial quantity is the product kl, where k is the *wave number* associated with the propagating electron. When $kl \gg 1$, band motion applies, the quasi-free mobility is large and given by $\mu = e\tau/m^*$, where m^* is the effective mass of the electron, and τ is the momentum relaxation time. When $kl \ll 1$, the electron-medium interaction is so strong that new bound states appear and the electron is trapped. Consequently, the mobility is low. When $kl \sim 1$, no simple model applies, but mobilities ~1–10 cm²v⁻¹s⁻¹ may be expected.

Cohen and Lekner (1967) developed a theory of electron mobility in LRGs that may be taken as a paradigm for quasi-free electrons. For this, they had to address two nontrivial theoretical problems: (1) calculation of screening of electron-atom potential in the presence of neighboring atoms *at liquid density*; and (2) incorporation of spatial correlation of atomic scattering centers. At higher external fields, the stationary electron energy distribution can significantly deviate from Maxwellian; its calculation poses another part of the theory. After obtaining all these ingredients self-consistently, Cohen and Lekner proceeded to calculate the drift velocity using the Boltzmann transport equation. Their expression for the zero-field mobility may be given as

$$\mu = \frac{2}{3N} \left(\frac{2}{\pi m k_B T} \right)^{1/2} \frac{e}{4\pi a^2 S(0)}, \tag{10.4}$$

where N is the atomic number density of the liquid at temperature T, a is the scattering length ($4\pi a^2$ is the zero-energy scattering cross section), and $S(0)$ is the limit of the liquid structure factor at zero momentum transfer ($K \to 0$). The structure factor is related to the pair correlation function $g(r)$ of the liquid as follows:

$$S(K) = 1 + 4\pi \int [g(r) - 1] \frac{\sin(Kr)}{Kr} r^2 \, dr.$$

Thermodynamics relates $S(0)$ to the isothermal compressibility by $S(0) = N k_B T \chi$, which is useful in calculating the low-field mobility using Eq. (10.4). The Cohen-Lekner theory was reasonably successful in predicting the low-field mobility in LAr and the sublinearity of the drift velocity at intermediate fields, although the saturation drift velocity was not well explained. The theory is based on a single-scattering model where each scattering is greatly weakened due to atomic correlation effect, since $S(0)$ is $\ll 1$. However, it cannot address the problem of density variation of mobility (see Holroyd and Schmidt, 1989) as found by Jahnke et al., (1971). For that purpose, Basak and Cohen (1979) apply a modification of deformation potential theory of solid state physics. That potential is due to long-wavelength fluctuation of the density given in terms of the derivatives of V_0 with respect to N. With an assumed dependence of V_0 on N, Basak and Cohen were able to explain the density dependence of mobility in LAr. However, the experimental determination of $V_0(N)$ in LAr and LXe by Reininger et al., (1983) has considerably weakened the agreement. Basically, theory predicts too low a mobility at low densities and too high at high densities. Nishikawa (1985) advocates the use of adiabatic compressibility

$\chi_s = (\rho c^2)^{-1}$ in place of its isothermal counterpart, where ρ is the mass density of the liquid and c is the speed of sound. With this modification, the agreement between theory and experiment improves somewhat; yet some fundamental difference remains unexplained.

In high-mobility liquids, the quasi-free electron is often visualized as having an effective mass m^* different fron the usual electron mass m. It arises due to multiple scattering of the electron while the mean free path remains long. The ratio of mean acceleration to an external force can be defined as the inverse effective mass. Often, the effective mass is equated to the electron mass m when its value is unknown and difficult to determine. In LRGs values of $m^*/m \sim 0.3$ to 0.5 have been estimated (Asaf and Steinberger,1974). Ascarelli (1986) uses $m^*/m = 0.27$ in LXe and a density–dependent value in LAr.

Schmidt (1976) has given a classical model for the field dependence of quasi-free electron mobility that predicts $\mu(E) \propto E^{-1/2}$ in the high-field limit. At any field E, the mobility is given in the relaxation time formulation as $\mu = (e/m)\Lambda/v_{el}$, where Λ is the mean free path and v_{el} is the electron random velocity. At low fields, $v_{ef} = V_{th}$, the thermal velocity, and a field-independent mobility ensues. At a high E, when the drift velocity v_d is no longer negligible compared with v_{th}, the energy gained by the electron between collisions is $(eE)(v_d\tau) = (e^2E^2/m)(\Lambda^2/v_{el}^2)$. For a stationary drift velocity this must equal the energy loss in a collision, which is $f(mv_{el}^2/2)$, where f denotes the fractional energy loss per collision. Equating these expressions one gets

$$v_{el} = \left(\frac{eE\Lambda}{m} \sqrt{\frac{2}{f}} \right)^{1/2}$$

and $\mu(E) \propto v_{el}^{-1} \propto E^{-1/2}$. This derivation assumes constancy of f and Λ, whereas in reality both of these would depend on electron energy. Nevertheless, in many high-mobility liquids, such a field dependence of mobility has been observed, at least over a limited range of E (see Table 10.2).

We remarked earlier that the saturation drift velocity in LRGs is not well understood in the original framework of the Cohen-Lekner theory. In a modified approach, Nakamura et al., (1986) introduced additional inelastic collision processes that reproduced the experimental field dependence of drift velocity in LAr with an adjusted variation of cross section with energy. This cross section increases rapidly beyond 1 eV, but its origin remains obscure. Atrazev and Dmitriev (1986) argued that the influence of liquid structure on scattering cross-section should diminish with energy, and beyond a few eV the cross section should approach that of an isolated atom. With this conjecture, they were able to explain the field dependent drift mobility in Lar. In yet another approach, Kaneko et al., (1988) employed a gas kinetic formulation and claimed good agreement with experiment for the field-dependent mobility with a constant scattering cross section (2.5 and 3.0 Å respectively in LAr and LCH$_4$). However, the scattering process is unexplained.

Jortner and Kestner (1973), using an effective medium theory, introduced a percolation model for *inhomogeneous* transport with the aim of expressing the electron mobility as a function of V_0 in all liquid hydrocarbons. The liquid is envisaged as a random composition of regions of high ($\mu_0 \approx 100$ cm^{2}v^{-1}s^{-1}) and low mobility ($\mu_1 \approx 0.01$ cm^{2}v^{-1}s^{-1}). Rotational fluctuation generates these regions of widely different mobility. With gaussian random cellular potential characterized by standard deviation ξ the fraction C of volume of high mobility is calculated respectively to be

$$\frac{1}{2}\left[1 + \operatorname{erf}\left(\frac{V_0 - E_t}{\sqrt{2}\,\xi}\right)\right] \quad \text{for } V_0 < E_t,$$

and

$$\frac{1}{2}\left[1 - \operatorname{erf}\left(\frac{V_0 - E_t}{\sqrt{2}\,\xi}\right)\right] \quad \text{for } V_0 > E_t.$$

Here E_t, an energy cut-off parameter, and ξ have been adjusted to be -0.27 and 0.26 eV respectively for all hydrocarbon liquids. Application of the semiclassical effective medium theory then gives the average mobility as

$$\mu = \mu_0\left[a + \sqrt{a^2 + \frac{x}{2}}\,\right]^{1/2},$$

where $2a = (3C/2 - 1/2)(1 - x) + x/2$ and μ_1/μ_0. For a given liquid, one first determines V_0, which gives C with the adjusted values of the energy parameters. One then calculates x and a, using the assumed values of μ_0 and μ_1. Finally, the effective mobility is obtained from the preceding equation. It should be noted that there is a minimum value of $C = C^*$, called the critical percolation limit, such that open extended channels can only exist for $C > C^*$. The mobility is extremely low for $C < C^*$. Jortner and Kestner choose $C^* = 0.2$ as appropriate for a three-dimensional disordered system. Also, for $x < 0.01$ and $C > 0.4$, f is independent of x and the effective mobility is simply $\mu = (\mu_0/2)(3C - 1)$. The computed value of μ falls precipitously for $C < 0.4$, eventually reaching μ_1 at $C = 0$. With many adjustable parameters, the percolation model of Jortner and Kestner is partially successful in explaining the variation of mobility with V_0 in liquid hydrocarbons. But a more serious objection has been raised with respect to Hall mobility (see Munoz, 1991). The Hall mobility should only refer to the high-mobility regions of the liquid. Therefore, for example, in a mixture of n-pentane and neopentane, the Hall mobility is expected to be independent of the mole fraction of n-pentane according to the percolation model. However, experimental measurement indicates that both Hall and drift mobilities fall with increase of n-pentane concentration.

The effect of pressure (0–2.5 kbar) on drift mobility has been summarized by Holroyd and Schmidt (1989) and by Munoz (1991). Pressure increases liquid density and decreases its compressibility; its effect on mobility is therefore complex. Experimental measurements (Munoz et al., 1985, 1987; Munoz and Holroyd, 1986) indicate that in TMS, a high-mobility liquid, pressure decreases mobility at all temperatures (18°–120°C), but the decrease is greater at higher temperatures; the isochoric mobility decreases roughly inveresely with absolute temperature. In low-mobility liquids such as n-hexane, 3-methylpentane, n-pentane, and cyclopentane, pressure also decreases the mobility, but the isochoric mobility increases with temperature in an activated fashion. Fitted to the Arrhenius equation, the activation energy is somewhat less than that for the liquid-vapor coexistence line. In low-mobility liquids, the decrease of mobility with pressure has a natural explanation in the trapping model (see Sect. 10.2.2) in terms of the change of the equilibrium constant between the quasi-free and trapped states with pressure. Thus, the volume change in trapping can be obtained from the equilibrium constant K of the reaction e_{qf} + trap $\longleftrightarrow$ e_t, as $\Delta v = -k_B T(\chi \ln K/\delta P)_T$. The experimental observation that the volume change is negative is interpreted as being due to electrostriction of the solvent by the trapped electron (Munoz et al., 1987). In intermediate mobility-liquids, such as 224-TMP and 22-DMB, pressure increases the mobility at lower temperatures (22–~60° C) when the mobility is relatively low. At higher temperatures, pressure decreases the mobility when the mobility is greater. The isochoric mobility generally increases with temperature, but the application of the Arrhenius equation has been questioned (Munoz, 1991). Obviously, a simple explanation of this comlex behavior is lacking.

10.2.1   HOPPING MODELS

In a hopping model, applied to low- and intermediate-mobility liquids, the electron is assumed to jump from one localized state to another either classically by thermal activation over a potential barrier, or by quantum-mechanical tunneling, sometimes with phonon assistance (Holstein, 1959; Bagley, 1970; Funabashi and Rao, 1976; Schmidt, 1977). Here, one can make a distiction between a localized and a trapped state. In the localized state, the electron does not perturb the liquid structure greatly, whereas in the trapped state it does (Cohen, 1977). Consider a hopping model where the rms jump length of the electron is Λ and the motion is activated over a barrier of energy ε_0 with an attempt frequency ν_0. In the presence of an electric field of strength E the frequencies of successful attempts along and opposite to the direction of the electric field are given respectively by

$v_0 \exp - [(-\varepsilon_0 - eE\Lambda/2)/k_B T]$ and $v_0 \exp - [(\varepsilon_0 + eE\Lambda/2)k_B T]$, noting that the activated complex is situated midway between the positions of jump. The drift velocity is then given by $\Lambda v_0 \exp(-\varepsilon_0/k_B T)[\exp(eE\Lambda/2k_B T) - \exp(-eE\Lambda/2k_B T)]$. The field–dependent mobility is now obtained as follows:

$$\mu(E) = \frac{e\Lambda^2}{k_B T} v_0 \exp\left(-\frac{\varepsilon_0}{k_B T}\right) \frac{\sinh \xi}{\xi} \quad \text{with } \xi = \frac{eE\Lambda}{2k_B T}. \quad (10.5)$$

Equation (10.5), derived in the Bagley model, can also be derived on a small polaron model (Holstein, 1959; Efros, 1967). At fields smaller than $2k_B T/e\Lambda$ these models predict a field-independent, but activated mobility given by $\mu(T) = (e\Lambda^2/k_B T)v_0 \exp(-\varepsilon_0/k_B T)$. Note that if Eq. (10.5) is applicable, then the jump length Λ can be evaluated from field dependence alone. Similarly, the activation energy ε_0 can be determined from temperature dependence only. However, these must be consistent to give the correct magnitude of the mobility. Such consistency has been attempted over the temperature interval ~100 K to ~300 K and over a wide span of field strength in ethane, propane, butane, and cyclopentane by Schmidt et al., (1974) and Sowada et al., (1976). The results have been summarized by Schmidt (1977), and they indicate that Eq. (10.5) is indeed applicable to these liquids at all temperatures. The activation energies for the low-field mobility are respectively 0.087, 0.155, 0.13, and 0.12 eV. The attempt frequency is computed to be 1.4×10^{13} s^{-1} in all cases except propane, for which it is 2.0×10^{14} s^{-1}. Propane is also special in that the value of Λ remains constant at 3.0 nm at all temperatures. In other liquids, Λ generally increases with temperature. The largest variation is seen for ethane, where Λ increases from 1.0 nm at 100 K to 4.5 nm at 200 K. Of course, these are fitted values and no particular physical significance has been attached to either the jump length, activation energy, or attempt frequency. The weakness of the model is not only the unexplained temperature variation of Λ, but also its extraordinarily large value.

It is interesting to note that although in principle ε_0 and Λ can have statistical variations, no experimental evidence has been found for dispersive transport in liquids (see Scher and Montroll, 1975). According to Schmidt (1977), all ionized electrons in liquid hydrocarbons can be collected at the anode with only minor distribution of arrival time due to diffusional broadening. Therefore, ε_0 should be taken as a constant. On the contrary, Funabashi and Rao (1976) consider electron hopping over fluctuating barrier heights. They justified the field-dependent mobility in liquid propane and liquid ethane (see Eq. 10.5) with a plausible distribution function for energy barriers and an entirely reasonable value of Λ = 0.75 nm. However it should be noted that the model of Funabashi and Rao is a one-dimensional one, and no three-dimensional generalization has been proposed. The point may be important, because in three-dimensional transport the electron may be able to avoid a high barrier by going around it.

10.2.2   Two-State Models

Just before Sect. 10.2.1, we introduced the percolation model as an example of inhomogeneous transport. Another popular example of inhomogeneous transport is the two-state model, first formulated by Frommhold (1968) for electron motion in a gas interrupted by random attachment-detachment processes, and then applied to liquid hydrocarbons by Minday *et al.*, (1971). According to this model (sometimes called the trapping model), the electron can exist either in a *quasi-free* state of high mobility or in a *localized* (or *trapped*) state of much lower mobility. At equilibrium, frequent transitions between these states are permitted; these determine the probabilities of finding the electron in these two states. If the transition rate from the quasi-free to the trapped state is denoted by k_{ft}, then the mean lifetime of the quasi-free state will be $\tau_f = k_{ft}^{-1}$. Similarly, the lifetime of the trapped state is given by $\tau_t = k_{tf}^{-1}$, where k_{tf} is the transition rate from the trapped to the quasi-free state. The probability that the electron will be found in the quasi-free state is therefore given by $P_f = \tau_f/(\tau_f + \tau_t)$ and that in the trapped state by $P_t = 1 - P_f$. The overall expected mobility is then given by

$$\mu = P_f\mu_{qf} + P_t\mu_t \approx P_f\mu_{qf}, \tag{10.6}$$

where μ_{qf} and μ_t are respectively the mobilities in the quasi-free and in the trapped state.

The extreme right-hand side of Eq. (10.6) applies in most low- and intermediate-mobility hydrocarbon liquids in which $\mu_t \ll P_f\mu_{qf}$, even though often $P_t \approx 1$. From the magnitude and temperature dependence of observed mobilities in various liquid hydrocarbons, the quasi-free mobility has often been taken to be ~100 cm^{2}v^{-1}s^{-1} (Minday *et al.*, 1971; Davis and Brown, 1975). There is no fundamental reason that the quasi-free mobility should be the same in different hydrocarbons. Indeed, Berlin *et al.*, (1978) obtained a quasi-free mobility ~30 cm^{2}v^{-1} s^{-1} in *n*-hexane and ~400 cm^{2}v^{-1}s^{-1} in neopentane using a density fluctuation model for electron scattering, whereas Davis *et al.*, (1972) suggested a value 150 cm^2 v^{-1}s^{-1} for both *n*-hexane and neopentane on the basis of a trapping model. Nevertheless, the often-used value $\mu_{qf} = 100$ cm^{2}v^{-1}s^{-1} has some justification as being the measured mobility in TMS, which may be considered to be trap-free on the basis of Hall mobility measurement (*vide supra*).

It is clear that in low- and intermediate-mobility liquids $\tau_t \gg \tau_f$ and $P_f \approx \tau_f/\tau_t$. If the trapped electron energy is lower than V_0, the smallest energy of quasi-free electrons, by an amount ε_0, the binding energy in the trap, then one gets approximately $\tau_t = k_{tf}^{-1} = \nu^{-1}\exp(\varepsilon_0/k_BT)$. In a classical activation process, ε_0 is an activation energy and ν would correspond to vibrational frequency in the trap. However, these associations are not precise, because of the stated approximations and because the condition $\varepsilon_0 \gg k_B T$ cannot always be fulfilled. Baird and Rehfeld (1987) have employed a thermodynamic model that interprets ε_0 in terms of trap concentration and the chemical potentials of the empty trap, the quasi-free electron, and the trapped electron. The actual evaluation of mobility in this model requires the knowledge of Hall mobility and a scattering factor by which it is related to the quasi-free mobility. Unfortunately, at present, the determination of the Hall mobility is limited to high-mobility liquids. In any case, with the knowledge of τ_t, the overall mobility can be written from Eq. (10.6) as

$$\mu = \mu_{qf} \frac{\nu}{k_{ft}} \exp\left(-\frac{\varepsilon_0}{k_B T}\right).$$

In the absence of a known trapping cross-section, k_{ft} can be determined by the use of detailed balancing (Ascarelli and Brown, 1960) as shown in the next subsection. It is easy to see that neither k_{ft} nor μ_{qf} would depend strongly on temperature. Therefore, according to this model, the activation energy obtained from experimental determination of mobility over a limited span of temperature should be approximately equal to the electron binding energy in the trap. Table 10.1 lists such activation energies ε_{ac}, which lie in the range 0.10 to 0.25 eV in most low- and intermediate-mobility liquids with the exception of 2,2,3,3-tetramethylpentane, for which an activation energy of 0.06 eV has been reported by Dodelet and Freeman (1972). Extrapolation of the Arrhenius form of temperature dependence of mobility to infinite temperature yields in most cases mobilities $\sim 100–1000 \ cm^2 v^{-1} s^{-1}$ which may be compared with the preexponential factor in the last equation. However, the activation energy itself may be temperature–dependent beyond the limited range of experiment, and the comparison may not be meaningful. In high-mobility liquids, the activation energy is low and comparable to $k_B T$ when the mobility actually increases with temperature (see Table 10.1). No simple explanation can be offered for such low activation energies.

The field dependence of mobility in the two-state trapping model was obtained by LeBlanc (1959) using a classical picture as $\mu(E)/\mu(0) = \exp(eEa/k_B T)$, where a is the trap diameter. According to Schmidt (1977), the experimental data in butane could not be fitted into this form with any value of a. LeBlanc's derivation, simply based on the reduction of activation energy along the field direction, might have been premature. If detrapping rate along a random direction to the field is considered, one would get the same form of field dependence (i.e., sinh-type) of mobility as in the hopping model (see the previous sub-section). However, just as in the hopping model, the value of a so–obtained would be extraordinarily large and hard to interpret as the trap diameter.

A variation of the two-state trapping model has been proposed by Schiller (1972) and his co-workers (Schiller et al., 1973; Berlin et al., 1978). The theoretical underpinning of this model is the concept of electron *localization due to energy fluctuation* in a certain region of the liquid. The localization probability is given by

$$P = (2\pi\sigma^2)^{-1/2} \int_{-\infty}^{V_0} \exp\left[-\frac{(\varepsilon - \varepsilon_t)^2}{2\sigma^2}\right] d\varepsilon$$

where ε_t is the energy of the localized state and σ is a parameter describing the gaussian distribution of energy fluctuation. The effective mobility is $\mu = \mu_{qf}$ $(1 - P)$, where again the mobility in the trapped state is negligible. Calculation of zero-field mobility in this model requires the knowledge of V_0 and ε_t as functions of density and temperature. The involvement of V_0 is explicit, and it is clear that the mobility will decrease with V_0. There actually exists an empirical relationship between V_0 and mobility given by Wada *et al.*, (1977) as follows:

$$\mu = \frac{125}{1 + 360 \exp(15V_0)}, \tag{10.7}$$

where V_0 is expressed in eV and μ in $cm^2v^{-1}s^{-1}$. Calculation of the quasi-free mobility in this model ranged from 30 $cm^2v^{-1}s^{-1}$ in *n*-hexane to 400 $cm^2v^{-1}s^{-1}$ in neopentane. This mobility is related to the mean geometric size of regions of fluctuation, for which the estimated value was obtained as 1–3 nm. Rather similar sizes were obtained in different hydrocarbons by Berlin and Schiller (1987), and the authors concluded that the quasi-free electrons interact with ~60–120 molecules at a time. When compared with experiments, the model is reasonably successful for nearly spherical molecules but not for straight-chain hydrocarbons (see Holroyd and Schmidt, 1989).

10.2.3   THE QUASI-BALLISTIC MODEL

Equation (10.6) for the mobility in the two-state model implicitly assumes that the electron lifetime in the quasi-free state is much greater than the velocity relaxation (or autocorrelation) time, so that a stationary drift velocity can occur in the quasi-free state in the presence of an external field. This point was first raised by Schmidt (1977), but no modification of the two-state model was proposed until recently. Mozumder (1993) introduced the quasi-ballistic model to correct for the competition between trapping and velocity randomization in the quasi-free state.

The new model is called quasi-ballistic because the electron motion in the quasi-free state is partly ballistic—that is, not fully diffusive, due to fast trapping. It is intended to be applied to low- and intermediate-mobility liquids, where the mobility in the trapped state is negligible. According to this, the mean velocity of the quasi-free electron in the presence of an external field E satifies the Langevin equation (Chandrasekhar, 1943): $d\langle v\rangle/dt = -\zeta\langle v\rangle + eE/m$, where ζ is the friction coefficient related to the quasi-free mobility by $\zeta = e/m\mu_{qf}$. Twice integrating the Langevin equation and averaging over random initial velocity gives the distance traveled along the field direction *in the absence of trapping* as $l(t) = v_d\{t - \zeta^{-1}[1 - \exp(-\zeta t)]\}$, with $v_d = \mu_{qf}E$. Trapping occurs with a probability density $\tau_f^{-1}\exp(-t/\tau_f)$ at time t, so that the average drift distance per trapping is given by

$$\langle \Delta x\rangle = \frac{\int_0^\infty l(t)\exp(-t/\tau_f)\,dt}{\tau_f} = \frac{(eE/m)\tau_f^2}{1 + \zeta\tau_f}$$

Since there are $(\tau_f + \tau_t)^{-1}$ cycles of trapping and detrapping per unit time, the drift velocity is $\langle\Delta x\rangle/(\tau_f + \tau_t)$, from which the effective mobility is derived to be

$$\mu_{eff} = \frac{(e/m)\tau_f^2}{(\tau_f + \tau_t)(1 + \zeta\tau_f)}, \quad \text{or} \quad \mu_{eff}^{-1} = <\mu>_T^{-1} + <\mu>_F^{-1}. \quad (10.8)$$

Here $\langle\mu\rangle_T = (e/m)\tau_f^2/(\tau_f + \tau_t)$ is called the *ballistic mobility* and $\langle\mu\rangle_F = \mu_{qf}\tau_f/(\tau_f + \tau_t)$ is the usual *trap-controlled mobility*. $\langle\mu\rangle_F$ is the applicable mobility when the velocity autocorrelation time (ζ^{-1}) is much less than the trapping time scale in the quasi-free state ($\zeta\tau_f\gg1$). In the converse limit, $\langle\mu\rangle_T$ applies, that is—trapping effectively controls the mobility and a finite mobility results due to random trapping and detrapping even if the quasi-free mobility is infinite (see Eq. 10.8).

For shallow traps of binding energy ε_0, Ascarelli and Brown (1960) give the ratio of trapping ($k_{ft} = \tau_f^{-1}$) and detrapping ($k_{tf} = \tau_t^{-1}$) rates, on the basis of detailed balancing, as $k_{ft}/k_{tf} = n_t h^3(2\pi m k_B T)^{-3/2}\exp(\varepsilon_0/k_B T)$, where n_t is the trap density. Using a random walk model (Chandrasekhar, 1943) and a harmonic potential for detrapping, one gets

$$k_{tf} = \frac{\varepsilon_0}{h}\exp\left(-\frac{\varepsilon_0}{k_B T}\right), \quad (10.9a)$$

from which the trapping rate is obtained upon application of detailed balancing as

$$k_{ft} = n_t h^2\varepsilon_0(2\pi m k_B T)^{-3/2}. \quad (10.9b)$$

Apart from fundamental constants and the liquid temperature, the variable parameters in the effective mobility equation are the quasi-free mobility, the trap density, and the binding energy in the trap. Figure 10.2, shows the variation of μ_{eff} with ε_0 at $T = 300$ K for $\mu_{qf} = 100$ cm^{3}v^{-1}s^{-1} and $n_t = 10^{19}$cm^{-3}. It is clear that the importance of the ballistic mobility $\langle\mu\rangle_T$ increases with the binding

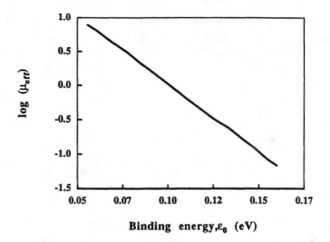

FIGURE 10.2   Variation of $\log \mu_{eff}$ at 300 K with binding energy ε_0 for $n_t = 10^{19} cm^{-3}$ and $\mu_{qf} = 100$ $cm^2 v^{-1} s^{-1}$. Notice the sensitivity of the movility to the binding energy. Reproduced from Mozumder (1993), with permission of Elsevier©.

energy, and it can never be ignored for low-mobilty liquids. Mozumder (1993) has shown that the dominant part of the variation of both ballistic and trap-controlled mobilities with temperature is of the Arrhenius type. The preexponential factors of these mobilities vary in opposite manner with temperature. It is then expected that the effective mobility will have an Arrhenius dependence on temperature with the activation energy close to the binding energy (*vide infra*).

In comparing the results of the quasi-ballistic model with experiment, generally $\mu_{qf} = 100$ cm^{2}v^{-1}s^{-1} has been used (Mozumder, 1995a) except in a case such as isooctane (Itoh *et al.*, 1989) where a lower Hall mobility has been determined when that value is used for the quasi-free mobility. There is no obvious reason that the quasi-free mobility should be the same in all liquids, and in fact values in the range 30–400 cm^{2}v^{-1}s^{-1} have been indicated (Berlin *et al.*, 1978). However, in the indicated range, the computed mobility depends sensitively on the trap density and the binding energy, and not so much on the quasi-free mobility if the effective mobility is less than 10 cm^{2}v^{-1}s^{-1}. A partial theoretical justification of 100 cm^2 v^{-1}s^{-1} for the quasi-free mobility has been advanced by Davis and Brown (1975). Experimentally, it is the measured mobility in TMS, which is considered to be trap-free (*vide supra*).

As for the trap density, a lower limit of ~10^{18} cm^{-3} has been taken, based on the fall of trapped electron yield in hydrocarbon glasses at a dose ~10^{20} eV/gm (Willard, 1975). An upper limit of trap density ~10^{20} may be argued on the basis of Berlin and Schiller's (1987) finding that a quasi-free electron interacts with

~10^2 molecules in a liquid hydrocarbon. Mozumder (1995a) finds that in most hydrocarbon liquids except benzene the trap density lies in the order of magnitude of 10^{19} cm^{-3}. In benzene, the room-temperature mobility is comparable to that in n-hexane, whereas the activation energy (0.32 eV) is much larger; this requires a much lower trap density ~10^{17} cm^{-3}.

Table 10.4 lists the values of trap density and binding energy obtained in the quasi-ballistic model for different hydrocarbon liquids by matching the calculated mobility with experimental determination at one temperature. The experimental data have been taken from Allen (1976) and Tabata et al., (1991). In all cases, the computed activation energy slightly exceeds the experimental value, and typically for n-hexane, $\varepsilon_0/E_{ac} = 0.89$. Some other details of calculation will be found in Mozumder (1995a). It is noteworthy that in low-mobility liquids ballistic motion predominates. Its effect on the mobility in n-hexane is 1.74 times greater than that of diffusive trap-controlled motion. As yet, there has been no calculation of the field dependence of electron mobility in the quasi-ballistic model.

In applying the quasi-ballistic model to electron scavenging, Mozumder (1995b) makes the plausible assumption that the electron reacts with the scavenger only in the quasi-free state with a specific rate k_s^f. Denoting the existence

TABLE 10.4   Electron Mobility, Trap Density, Binding Energy, and Activation Energy in the Quasi-ballistic Model

Liquid	T (K)	μ^a	n_t^b	ε_0^c	E_{ac}^d
Propane	273	1.46	0.14	0.137	0.175
Pentane	295	0.15	0.45	0.168	0.224
Cyclopentane	293	0.1	0.78	0.106	0.160
n-Hexane	295	0.1	1.0	0.150	0.170
Cyclohexane	295	0.45	1.2	0.111	0.170
Ethane	216	0.8	0.4	0.088	0.125
Butane	300	0.4	1.1	0.115	0.175
Benzene	300	0.114	0.012	0.284	0.324
Isooctane	296	4.5	0.675	0.044	0.055
3-Methylpentane	293	0.2	0.27	0.174	0.225
Isobutene	293	1.44	1.0	0.088	0.141

[a] Measured mobility in cm^2 v^{-1}s^{-1} from Allen (1976) and Tabata et al. (1991).

[b] Trap density in 10^{19} cm^{-3}.

[c] Binding energy in trap in eV.

[d] Calculated activation energy in eV, generally exceeding experimental value by ~0.03 eV.

probability of the electron in the quasi-free, trapped, and scavenged states respectively by π_f, π_t, and π_s, the kinetics of scavenging may be given as

$$\dot{\pi}_f = -(\lambda_s + k_{ft})\pi_f + k_{tf}\pi_t, \tag{10.10a}$$

$$\dot{\pi}_t = -k_{tf}\pi_t + k_{ft}\pi_f, \tag{10.10b}$$

where $\pi_f + \pi_t + \pi_s = 1$, $\lambda_s = k_s^f c_s$, c_s is the scavenger concentration, and the dots denote time derivatives. The long-time limit solution of Eqs. (10.10a, b) can be represented as $\pi_s(t) \approx 1 - \exp(-\beta t)$, $\beta = \lambda_s k_{tf}/(k_{ft} + k_{tf})$. Since the effective reaction rate may be defined through the relation $\pi_s(t) = 1 - \exp(-k_s^{ef} c_s t)$, one gets

$$k_s^{eff} = \frac{\beta}{c_s} = \frac{k_s^f k_{tf}}{k_{ft} + k_{tf}}. \tag{10.11}$$

Equation (10.11) has a simple interpretation in the stationary state of scavenging. Notice that $k_s^{eff} \propto k_s^f$, the constant of proportionality representing the fraction of time spent by the electron in the quasi-free state. In low- and intermediate-mobility liquids $k_{tf} \ll k_{ft}$, and the factor k_{ft}/k_{tf} needed to convert k_s^{eff} to k_s^f depends only on the liquid, being independent of the scavenger. It can be obtained from detailed balance if the trap density and binding energy are known. In practice, it is more convenient to use the measured mobility (μ_{eff}) as the independent variable. It has been shown (Mozumder, 1995a) that for a large class of hydrocarbons in which the effective mobility ranges from 0.06 to 8.0 cm²v⁻¹s⁻¹, $\mu_{qf} = 100$ cm²v⁻¹s⁻¹ and $n_t \sim 10^{19}$cm⁻³ describe the experimental mobility quite well. With these values and the data of Mozumder (1993), the dependence of k_{ft}/k_{tf} on μ_{eff} is shown in Figure 10.3 on a log–log plot. Within the specified range of mobility, the data can be fitted to a straight line, giving $k_{ft}/k_{tf} \sim \mu_{eff}^{-0.906}$. Experimental scavenging rates (k_s^{eff}) range from ~10^{10} to ~10^{14} M⁻¹s⁻¹ which then gives k_s^f in the range 10^{11} to ~5×10^{14} M⁻¹s⁻¹. The objective is to find the dependence of k_s^f on V_0 and to determine if a scavenging reaction is transport-controlled or not in a given case.

Following Noyes (1961), one may write $(k_s^f)^{-1} = k_{diff}^{-1} + k_{act}^{-1}$ and $k_s^f = \eta k_{diff}$, where k_{diff} is the diffusion-controlled rate, k_{act} is the rate of final the chemical step, and η is the reaction efficiency in that step. Denoting the *first* electron0-scavenger encounter probability from an initial separation r_0 by $P(r_0)$, the pair reaction probabilty in given by

$$\omega(r_0) = \rho P(r_0)[1 + (1 - \rho)P(L) + (1 - \rho)^2 P^2(L) + \cdots]$$

$$= \frac{\rho P(r_0)}{1 - (1 - \rho)P(L)} \equiv \eta P(r_0),$$

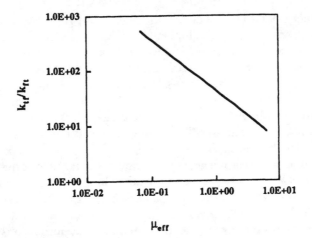

FIGURE 10.3   Log–log plot of k_{ft}/k_{tf} vs μ_{eff} in liquid hydrocarbons The slope of the fitted line is −0.906. See text for details. Reproduced from Mozumder (1995a), with the permission of Am. Chem. Soc. ©.

where ρ is the encounter reaction probability and $P(L)$ is the *first* arrival probability after the pair has attained an average separation L *following* an unsuccessful encounter. Both $P(L)$ and $P(r_0)$ are obtainable in principle from diffusion theory; however, a correction is necessary for the long mean free path $\langle l \rangle$ of the electron in the quasi-free state. Mozumder (1995b) derives $P(r_0) = (a/r_0)(1 + d/2r_0)/(1 + d/2a)$ and $P(L) = 3(a/d)^2/[1 = 2(a/d)]$, using a fractal model for this correction, where a is the encounter radius and $d \sim L \sim \langle l \rangle$ is the fractal length parameter. The probability of escaping scavenging reaction starting from an initial separation r_0 is then given by $\phi(r_0) = 1 - \eta P(r_0)$ with the foregoing expression for $P(r_0)$. The specific scavenging rate k_s^f in the quasi-free state may be given in terms of this escape probability using a well-known relationship (Sano, 1981) as

$$k_s^f = \left| 4\pi r^2 D(r) \frac{\partial \phi}{\partial r} \right|_{r=a} = \frac{\eta k_D}{(1 + d/2a)},\qquad (10.12)$$

where $D(r)$ is the distance-dependent diffusion coefficient (*fractal effect*), $k_D = 4\pi D_0 a$, and D_0 is the diffusion coefficient at infinite separation. Both D_0 and $\langle l \rangle$, and therefore L and d also, are obtainable from the quasi-free mobility using respectively the Nernst-Einstein and Lorentz equations. With known values of a, D_0, and d, η and k_{act} can be computed from k_s^f. Finally, the reaction can be judged as transport-controlled or not by comparing k_{act} with k_{diff}.

Mozumder (1995b) has analyzed the reaction of electrons with five "efficient" scavengers, CCl_4, C_2H_5Br, biphenyl, SF_6, and N_2O, in seven liquid hydrocarbons (in ascending order of mobility), n-hexane, n-pentane, cyclohexane, cyclopentane, isooctane, neopentane, and tetramethylsilane, using the procedure outlined in the previous discussion. In the last two liquids, the electron may be assumed to be mostly in the quasi-free state; the others are low- or intermediate-mobility liquids, to which the quasi-ballistic model is applicable. The details of the calculation will be found in the original reference; here, we will summarize the findings.

The reaction radius a is not accurately known. It is taken to correspond to the largest value of k_s^f for a given scavenger, usually in a low-mobility liquid such as n-hexane. This gives a = 1.00, 1.07, 1.00, 1.25, and 1.07 nm respectively for CCl_4, C_2H_5Br, biphenyl, SF_6, and N_2O. Since these values are $\ll d \sim \langle l \rangle$, which is computed to be ~6 nm from the quasi-free mobility, recollisions seem to be unimportant for electron scavenging reactions. Figure 10.4, shows the calculated reaction efficiency η for the scavengers CCl_4, C_2H_5Br, and biphenyl in various hydrcarbon liquids as a function of the effective electron mobility in these liquids. Values of V_0 according to Eq. (10.7) (Wada et al., 1977) are also displayed in the figure. *In general, the reaction efficiency falls with effective mobility in all cases.* The apparent peak for CCl_4 is probably due to computational uncertainty. The decrease of reaction efficiency is partly due to long mean free path of the electron in the quasi-free state, and partly due to reaction inefficiency, which

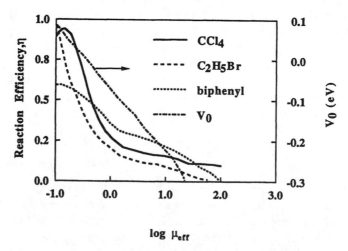

FIGURE 10.4   Scavenging reaction efficiency η vs. $\log \mu_{eff}$ ($cm^2v^{-1}s^{-1}$) for CCl_4, C_2H_5Br, and biphenyl in various liquid hydrocarbons. Corresponding V_0 values are shown on the right scale. Reproduced from Mozumder (1995b), with the permission of Am. Chem. Soc.©

increases with effective mobility. Few reactions are found to be diffusion-controlled, none in high-mobility liquids. In very low mobility liquids such as n-hexane most reactions are diffusion-controlled except for bipnenyl, with which the reaction does not seem to be diffusion-controlled in any liquid. The reaction efficiency is seen to increase with V_0, which may be an indirect effect due to its relationship with the effective mobility.

10.3 THERMODYNAMICS OF ELECTRON TRAPPING AND ELECTRON SOLVATION IN LIQUID HYDROCARBONS

Baird and Rehfeld (1987) have analyzed the thermodynamics of electron transport in the two-state trapping model. According to these authors, the effective mobility, ignoring the mobility in the trapped state, is given by

$$\mu_{eff} = \frac{r \cdot \mu_H}{1 + \exp(-\Delta\Phi^0/k_B T)}, \tag{10.13}$$

where μ_H is the Hall mobility, r is the Hall ratio (see Sect. 10.1.3)—that is, $\mu_{qf} = r \cdot_{\mu H}$—and $\Delta\Phi^0$ is the change in standard chemical potential of the electron on trapping. (The chemical potential of a solution is the rate of increase of its free energy at constant temperature and pressure per mole of added solute.) Since the denominator of the right-hand side of Eq. (10.13) is also given in this model by $(1 + k_{ft}/k_{tf})$ (see Sect. 10.2.2), one gets $k_{ft}/f_{tf} = \exp(-\Delta\Phi^0/k_B T)$.

Baird and Rehfeld express $\Delta\Phi^0$ in terms of the trap concentration and the chemical potentials of the empty trap and of the electron in the quasi-free and trapped states. Further, they indicate a statistical-mechanical procedure to calculate these chemical potentials. Although straightforward in principle, their actual evaluation is hampered by the paucity of experimental data. Nevertheless, Eq. (10.13) is of great importance in determining the relative stability of the quasi-free versus the trapped states of the electron if data on time-of-flight and Hall mobilities are available.

Such is the case of neopentane over the temperature interval 260–432 K (Munoz and Ascarelli, 1983). However r, which is not well known and which depends on the scattering mechanism (see Eq. 10.4), must be so chosen that the condition $\mu_{eff} \leq r\mu_H$ is satisfied over the entire temperature range (see Eq. 10.13). Baird and Rehfeld find that this condition is fulfilled in neopentane only for scatterings with constant mean free path ($r = 0.849$) or for scatterings with constant relaxation time ($r = 1.0$). In either case, the experimental data of Munoz and Ascarelli (1983) lead to the surprising conclusion that $\Delta\Phi^0 > 0$ for $T < 415$ K, indicating the quasi-free state is more stable in NP than the trapped state, but that it is less than 0 for $T > 420$ K indicating the relative stability of the trapped state. If this finding is contrary to expectation due to lower trap density at a higher temperature, there could be an explanation in terms of electron trapping by critical clusters whose density increases near the critical point. If a similar consideration is applied to isooctane, where the time-of-flight mobility is ~7 $cm^2v^{-1}s^{-1}$ (Schmidt and Allen, 1970) and the Hall mobility is a factor ~3.5 times greater over the substantial temperature range of measurement (Itoh *et al.*, 1989), both constant mean free path and constant mean free time of scattering models would predict the stability of the trapped state over the quasi-free state. It should be noted that in these liquids

$$|\Delta \Phi^0| \sim k_B T,$$

so that irrespective of which state is more stable, the population of the other state is *not* negligible even if the trapped state is essentially immobile.

Hamill (1981a) has questioned the usual thermodynamic meaning of V_0 as the *enthalpy* change upon transferring the electron from vacuum to the bottom of the conducting band in the liquid (Holroyd *et al.*, 1975; Holroyd, 1977). According to Hamill, V_0 should be the corresponding *free energy* change, since it is often *measured* as the change in the work function of a metal when it is immersed in the liquid and the work function is to be identified with free energy. (Note that there are other ways, such as spectroscopic methods, for determining V_0.) Equating V_0 with free energy change gives the entropy change for the process as $\Delta S_0 = -\delta V_0/\delta T$, which would be large and positive in several liquid hydrocarbons, using the data of Holroyd *et al.*, (1975). In a perfect crystalline lattice $V_0 = U_p + T_0$, where U_p is the potential energy of the electron and T_0 is its kinetic energy arising from coherent scattering by molecules with filled orbitals. Coherent scattering is ordered, and no entropy is generated by it. In the liquid, scattering is incoherent due to positional and other disorder. This generates an additional kinetic energy $T \Delta S_0$ to make up V_0. Hamill (1981a) asserts that $U_p + T_0$ should be common to all alkanes and the variation of V_0 is due entirely to the entropy term $T \Delta S_0$, which measures the departure from perfect crystalline order. A similar argument gives the free energy (V_{0t}) and the entropy of the trapped state, and Hamill takes the activation free energy for mobility as $\alpha(V_0 - V_{0t})$ with $0 < \alpha < 1$, invoking a hopping transport model (see Sect. 10.2.1). He considers mobilty expressions of the type $\mu = \mu_0 \exp(-E_{ac}/k_B T)$ incomplete because of the neglect of entropy effect. Electron scavenging reactions and attachment-detachment equilibria in solution have been treated by Hamill (1981a, b) using his thermodynamic model.

Reversible reactions of excess electrons with CO_2 and certain aromatic molecules

$$(e^- + A \xrightarrow[k_2]{k_1} A^-)$$

have been observed in several hydrocarbon liquids (Warman *et al.*, 1975; Holroyd *et al.*, 1975; Holroyd, 1977). Using these data and making some approximations and assumptions, Holroyd (1977) has determined the change in thermodynamic parameters for electron solvation in these liquids. Setting up the kinetic equations for attachment and detachment and neglecting drift time, Holroyd derives the electron concentration in a conductivity cell containing the solution as

$$\frac{n(t)}{n(0)} = \frac{w_1 + k_2}{w_1 - w_2} \exp(w_1 t) + \frac{w_2 + k_2}{w_2 - w_1} \exp(w_2 t),$$

where $2w_{1,2} = -(k_1' + k_2 + k_3') \pm [(k_1' + k_2 + k_3')^2 - 4k_2 k_3']^{1/2}$, $k_1' = k_1[A]$, $k_3' = k_3[S]$, and [S] is the concentration of impurities that reacts with the electron with specific rate k_3. A correction to $n(t)$ due to finite drift time, which is important at longer times, has also been indicated. The current $I(t)$ measured in the cell at time t is given by

$$I(t) = \frac{ev}{d} \int_0^d n(x, t) \, dx,$$

where d is the electrode spacing and v is the electron drift velocity.

The quantity k_3' may be considered as an instrumental constant to be determined in a blank experiment—that is, without added solute. In this case, the current is given by $I(t)/I(0) = (1 - vt/d) \exp(-k_3' t)$, from which k_3' can be determined. With the solute added, the current initially decays exponentially (fast decay) from which is determined $k_1' + k_2 + k_3'$, while the ratio of the initial plateau to the initial current gives $k_2/(k_1' + k_2 + k_3')$. The detachment rate k_2 is now obtained from the last two numbers, and then the attachment rate k_1 is also obtained since k_3' is already predetermined. In short, both attachment (k_1) and detachment (k_2) rates are obtainable from the time dependence of the cell current following a brief pulse of ionizing radiation.

From the equilibrium constant of this reversible reaction, $K = k_1/k_2$, the free energy and enthalpy of the reaction are given respectively by $\Delta G_r^0 = -k_B T \ln(K)$ and $\Delta H_r^0 = -k_B \partial(\ln K)/\partial(T^{-1})$. If desired the entropy change on reaction can be computed from the relation $\Delta G_r^0 = \Delta H_r^0 - T\Delta S_r^0$. By measuring the attachment and detachment rates to triphenylene, phenanthrene, naphthalene, styrene, and α-methylstyrene at different temperatures, Holroyd determined the changes in the thermodynamic potentials on reaction with these solutes in TMS, isooctane, and *n*-hexane. All these are found to be negative; in particular, the entropy changes are large. For styrene and α-methylstyrene, which are common in all the solutions, the entropy changes in TMS and isooctane are somewhat similar, ~−140 to −180 J/(mole∝K), while that in *n*-hexane is even greater: ~−200 J/(mole∝K).

Invoking a thermodynamic cycle for electron attachment in the gas and liquid phases, Holroyd (1977) gets $\Delta G_r^0(l) = \Delta G_r^0(g) + \Delta G_{sol}^0(A^-) - \Delta G_{sol}^0(A) - \Delta G_{sol}^0(e^-)$, where $\Delta G_{sol}^0(e^-)$ stands for the free energy change in transferring the electron from the gas to the lowest state in the liquid and $\Delta G_{sol}{}^0(A^-) - G_{sol}{}^0(A)$ is the difference between the free energies of solution of the anion and the neutral molecule. This difference is the stabilization energy of the anion by dielectric polarization, given approximately by Born's equation as $\Delta G_p(A^-) = -e^2(\varepsilon - 1)$ $/2a\varepsilon$, where the anion size parameter a may be obtained from the molar volume. One thus gets

$$\Delta G_r^0\left(1\right) = \Delta G_r^0\left(g\right) + \Delta G_P\left(A^-\right) - \Delta G_{sol}\left(e^-\right). \tag{10.14}$$

It is assumed that the free energy and the enthalpy of the attachment reaction in the gas phase are both numerically equal to the electron affinity, implying that the gas phase entropy of A and A^- are about the same. Equation (10.14) then allows the computation of the free energy of solution of the electron from the free energy of a reversible reaction in the solution and data on dielectric constant and gas phase electron affinity. The so-determined free energy of solution of the electron should be independent of the attaching solute. Actually, in TMS the thusly determined $\Delta G_{sol}{}^0(e^-)$ varies slightly with the solute, and a value -0.66 ± 0.08 eV has been established for it (Holroyd, 1977), which is roughly equal to V_0 for this liquid. The enthalpy of electron solvation can be determined from an equation analogous to Eq. (10.14); however, the entropic contribution to $\Delta H_p(A^-)$ must be included. Using the Born formula, this contribution is $T(e^2/2\varepsilon^2a)(\delta\varepsilon/\delta T)$. Electron solution enthalpy in TMS evaluated in this fashion is approximately equal to the change in free energy, implying that solution entropy is very small in this liquid. The free energies and enthalpies of electron solution in other liquids are determined relative to TMS. Since the dielectric constants at different temperatures are about the same in various nonpolar liquid hydrocarbons, the polarization terms are approximately equal. From Eq. (10.14) Holroyd then gets

$$\Delta X_{sol}^0\left(e^-\right)_2 - \Delta X_{sol}^0\left(e^-\right)_1 = \Delta X_r^0\left(1\right)_1 - \Delta X_r^0\left(1\right)_2; X = G, H. \tag{10.15}$$

Using Eq. (10.15), Holroyd finds $\Delta G_{sol}{}^0(e^-)\approx\Delta H_{sol}{}^0(e^-) = -0.47$ eV in isooctane; again, the entropy change in solution is small. In n-hexane $\Delta G_{sol}{}^0(e^-) = -0.36$ eV and $\Delta H_{sol}{}^0(e^-) = -0.18$ eV, giving the entropy change as $\Delta S_{sol}{}^0(e^-) = 5 \times 10^{-4}$eV/K, which indicates considerable disorder in the ground state of the electron in this liquid. It should be noted that the changes in the thermodynamic functions obtained in this manner depend somewhat on the solute used and on the inclusion or exclusion of polarization correction. Nevertheless, the relative change of any thermodynamic function from one liquid to another can be determined consistently and with greater reliability. Since the ground state electron energy in TMS is close to V_0, we may conclud that electron trapping is unimportant in this liquid. In isooctane, V_0 is measured in the range from -0.18 to -0.26 eV; since the ground state energy is at least 0.2 eV lower than V_0, this is clear evidence of the existence of traps. In n-hexane $0 < V_0 < 0.1$ eV; therefore, traps are even more important in this liquid.

Holroyd (1977) finds that generally the attachment reactions are very fast ($k_1 \sim 10^{12}-10^{13}$ M^{-1}s^{-1}), are relatively insensitive to temperature, and increase with electron mobility. The detachment reactions are sensitive to temperature and the nature of the liquid. Fitted to the Arrhenius equation, these reactions show very large preexponential factors, which allow the endothermic detachment reactions to occur despite high activation energy. Interpreted in terms of the transition state theory and taking the collision frequency as $\sim 10^{13}$ s^{-1}, these preexponential factors give activation entropies ~ 100 to ~ 200 J/(mole.K), depending on the solute and the solvent.

10.3.1 THERMODYNAMICS OF ELECTRON TRAPPING AND SOLVATION IN THE QUASI-BALLISTIC MODEL

Mozumder (1996) has discussed the thermodynamics of electron trapping and solvation, as well as that of reversible attachment-detachment reactions, within the context of the quasi-ballistic model of electron transport. In this model, as in the usual trapping model, the electron reacts with the solute mostly in the quasi-free state, in which it has an overwhelmingly high rate of reaction, even though it resides mostly in the trapped state (Allen and Holroyd, 1974; Allen et al., 1975; Mozumder, 1995b). Overall equilibrium for the reversible reaction with a solute A is then represented as

$$ e_{tr}^- \rightleftharpoons e_{qf}^- + A \rightleftharpoons A^- , \tag{I} $$

where e_{tr}^- and e_{qf}^- are respectively the trapped and quasi-free electron and A^- is the anion. The change in the standard thermodynamic potential of the reaction $\Delta X_r^0 (X = G, H,$ or $S)$ may then be given as

$$ \Delta X_r^0 = \Delta X_r^0 \left(qf \right) - \Delta X_{tr}^0 , \tag{10.16} $$

where ΔX_{tr}^0 refers to the trapping process associated with mobility, $\Delta X_r^0(qf)$ refers to the attachment-detachment reaction in the quasi-free state, and ΔX_r^0 refers to the overall reaction. Since ΔX_{tr}^0 is available from mobility data and ΔX_r^0 is given by experiment (Holroyd et al., 1975, 1979; Warman et al., 1975; Holroyd, 1977), Eq. (10.16) can be used to obtain the change in the thermodynamic potential for reaction in the quasi-free state, from which the reaction rate and efficiency of reaction in that state can be determined using a suitable diffusion theory. From Eqs. (10.9a, b), or directly from detailed balance, the equilibrium constant for the trapping process may be given as

$$K_{tr} = \frac{k_{ft}}{k_{tf}} = n_t h^3 (2\pi m k_B T)^{-3/2} \exp\left(\frac{\varepsilon_0}{k_B T}\right),$$ (10.17)

where all symbols have same meaning as before. The free energy change in trapping is given from (10.17) as $\Delta G_{tr} = -k_B T \ln K_{tr}$. For convenience of calculation, a standard state may be defined at a temperature $T_0 = 298$ K having a trap density of 10^{19} cm^{-3}, in terms of which the standard free energy change in trapping is given by (Mozumder, 1996)

$$\Delta G_{tr}^0(T, n_t) = -\varepsilon_0 - 2.3927 \frac{T}{T_0}\left[\ln n_t - \frac{3}{2}\ln\left(\frac{T}{T_0}\right) - 0.1728\right] kJ/mol,$$ (10.18)

where n_t is the trap density in unit of 10^{19} cm^{-3}. For the sake of comparison, the standard free energy of trapping in the usual trapping model is given by (see Baird and Rehfeld, 1987) $\Delta G_{tr}^0(T) = -k_B T \ln [\mu_{qf}/\mu(T) - 1]$, where $\mu(T)$ is the experimental mobility at temperature T and $\mu_{qf} = 100$ cm^{2}v^{-1}s^{-1} may be taken as the quasi-free mobility. In either model, if the free energy change varies linearly with temperature, then the standard enthalpy and entropy changes on trapping can be evaluated from the relationship $\Delta G_{tr}^0(T) = \Delta H_{tr}^0 - T\Delta S_{tr}^0$. In any case, these can also be evaluated from the thermodynamic equations $\Delta G_{tr}^0 = -T^2(\partial/\partial T)(\Delta G_{tr}^0/T)$ and $\Delta S_{tr}^0 = -(\partial/\partial T)(\Delta G_{tr}^0)$.

Fig.10.5 shows the standard free–energy change, as a function of temperature, upon electron trapping from the quasi-free state, in n-hexane, cyclohexane, n-pentane, propane, 3-methylpentane, and isooctane using the values of trap density and binding energy that describe electron mobility consistently in these liquids. A good linear dependence is obtained, from which changes in the standard enthalpy and entropy on trapping are computed and displayed in Table 10.5 along with the values of standard free energy change, trap density, and binding energy.

It is remarkable that the changes in all the thermodynamic potentials are negative upon trapping. The negative entropy change implies more disorder in the quasi-free state, which may be partially due to delocalization. An explanation may be found in the Anderson (1958) model of localization, but the point has not yet been established. The trapping process is enthalpy-dominated, although the contribution of entropy is not negligible. A similar calculation in the usual trapping model (*vide supra*) reveals somewhat more negative ΔG_{tr}^0 and ΔH_{tr}^0, but a little less negative ΔS_{tr}^0. A comparison between the quasi-ballistic and the usual trapping model is shown in Figure 10.6 with respect to the variation, of

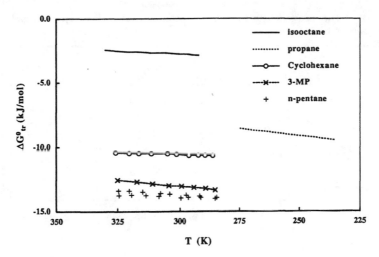

FIGURE 10.5 Standard free energy change, in various liquid hydrocarbons, versus temperature upon electron trapping from the quasi-free state according to the quasi-ballistic model. Reproduced from Mozumder, (1996), with the permission of Am. Chem. Soc.©

the free energy change upon trapping with the effective mobility. The two curves tend to converge at high mobility, due to the fact that in this limit trap-controlled mobility dominates transport and ballistic mobility makes a minor contribution. However, the choice between the quasi-ballistic and the usual trapping models is not so much in regard to thermodynamic parameters. It is that a fixed value of μ_{qf} for all hydrocarbon liquids and a set of n_t and ε_0 specific to a liquid explains the experimental variation of mobility with temperature in the quasi-ballistic model, which is very difficult to achieve in the usual trapping model.

Thermodynamic parameters of electron solvation would be calculable from those of trapping if an unequivocal meaning could be given to V_0. Earlier in this section, we alluded to two different meanings, one as a free energy change and another as an enthalpy change, in transferring an electron from vacuum to the lowest conducting state in the liquid. Precisely speaking, it is probably neither, as V_0 is one of the many states that are thermodynamically accessible. Its experimental determination as a work function differential is closer to enthalpy change than to the change in free energy. If V_0 is taken as a free energy change, then the calculated entropy of solvation would be unacceptably large and negative. Mozumder (1996) therefore takes $V_0 = \Delta H^0(qf)$, thereby getting the enthalpy change in solvation as $\Delta H_{sol}^0 = V_0 + \Delta H_{tr}^0$.

Taking n-hexane and isooctane as examples, the V_0 values are respectively +1.93 and −23.11 kJ/mol (Tabata $et\ al.$, 1991, ch. VII). Using the data of ΔH_{tr}^0 from Table 10.5A, one gets $\Delta H_{sol}^0 = -16.1$ and -31.0 kJ/mol respectively for these

TABLE 10.5   Thermodynamic Parameters of Electron in Hydrocarbon Liquids According to the Quasi-ballistic Model

			For Trapping		
Liquid	n_t^a	ε_0	ΔG_{tr}^0	ΔH_{tr}^0	ΔS_{tr}^0
n-Hexane	1.0	14.5	−14.1	−18.0	−13.5
Cyclohexane	1.2	10.7	−10.8	−14.4	−12.2
n-Pentane	0.45	16.2	−13.8	−19.8	−20.1
Propane[b]	0.14	13.2	−9.8	−16.2	−27.2
3-Methylpentane	0.27	16.8	−13.1	−20.3	−24.2
Isooctane	0.675	4.3	−2.9	−7.9	−16.8

All energies in kJ/mol and entropy in J/(mol· K). Uncertainties in ΔH_{tr}^0 and ΔS_{tr}^0 are respectively ~1 kJ/mol and 1 J/(mol· K).

[a]In units of 10^{19} cm^{-3}.

[b]At 234 K.

		For Solvation		
Liquid	V_0	ΔG_{sol}^0	ΔH_{sol}^0	ΔS_{sol}^0
n-Hexane	+1.9	−32	−16	53
Cyclohexane	+1.0	−27	−13	44
n-Pentane	0.0	−31	−20	36
Propane[a]	−6.7	−30	−23	29
3-Methylpentane	−13.5	−43	−34	32
Isooctane	−23.1	−40	−31	32
Neopentane[b]	−40.4	−56	−45	37
Tetramethylsilane	−61.7	−64	−62	6 (~0)

All energies in kJ/mol and entropy in J/(mol · K). Overall uncretainty in free energy and enthalpy is ~20% and that in entropy is ~30%.

[a]At 234 K.

[b]Thermodynamic parameters for electron trapping in NP from Baird and Rehfeld (1987) at $T = 296$ K.

liquids. Referring to Eq. (10.15), taking $X = H$ and designating isooctane as 1 and n-hexane as 2, one evaluates the left-hand side as 14.9 kJ/mol using the solution enthalpy data. Considering styrene and α-methylstyrene as solutes, the right-hand side of (10.15) calculates to between 13.5 and 15.4 kJ/mol (Holroyd, 1977), showing good agreement.

If experimental values (Holroyd, 1977), are taken, $\Delta G_{sol}^0 = -40.4$ and −31.8 kJ/mol respectively for isooctane and n-hexane, then from Table 10.5A and Eq. (10.16) with $X = G$, $\Delta G^0(qf) = -37.5$ and −17.7 kJ/mol respectively in isooctane and n-hexane. From these, the corresponding entropy changes are computed to

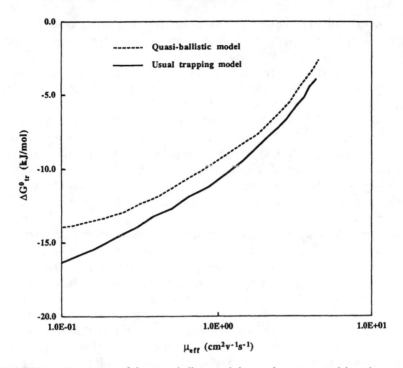

FIGURE 10.6 Comparison of the quasi-ballistic and the usual trapping models with respect to the variation of the free energy change upon trapping with the effective mobility. Reproduced from Mozumder (1996), with the permission of Am. Chem. Soc.©

be $\Delta S^0(qf) = 48.0$ and 66.0 J/(mole∝K) respectively. However, these values are relatively small and not very different in as diverse solvents as n-hexane and isooctane, considering uncertainties in measurement. Therefore, in all hydrocarbon liquids except TMS, $\Delta S^0(qf)$ is taken to be constant at 56 J/(mole∝K).

Note that the entropy change is positive in entering the disordered state of the liquid. In TMS, the measured mobility is almost the same as the quasi-free mobility, and the activation energy of mobility is nearly zero. This liquid may be considered trap-free. Therefore, ΔG_{sol}^0 in TMS is taken from solute reaction cycle (Holroyd, 1977) rather than using the fixed value of $\Delta S^0(qf)$. Since ΔG_{sol}^0 and ΔH_{sol}^0 are nearly equal in this liquid, there is very little entropy change in solution. For other liquids, the thermodynamic parameters of solvation are evaluated from the data of Table 10.5, taking $\Delta H^0(qf) = V_0$ and $\Delta S^0(qf) = 56$ J/(mole∝K). These values are shown in Table 10.5B along with V_0. In the three liquids, n-hexane, isooctane, and TMS, where comparison is possible, there is good agreement between thermodynamic parameters obtained from the solute reaction cycle and those obtained from the quasi-ballistic model. Comparing Tables 10.5A and B, it is apparent that although the entropy change in trapping from the quasi-free state is negative, the overall entropy change in electron solvation is positive. Thus, the solvation process is driven both by enthalpy and entropy.

For several reversible reactions, the thermodynamic parameters for reaction *in the quasi-free state* are given in Table 10.6 using Eq. (10.16) and the reaction scheme (I). Experimental data for $\Delta X_r^0 (X = G, H,$ or $S)$ are taken from Holroyd et al., (1975, 1979) and Holroyd (1977), while Table 10.5A provides data on ΔX_{tr}^0, except for TMS (*vide supra*). The chief uncertainty in these calculations is the experimental determination of V_0. It is remarkable that all thermodynamic parameters of reaction in the quasi-free state are negative in the same way as for the overall reaction. In particular, the entropy change is relatively large and probably for the same reason as for the overall reaction (Holroyd, 1977).

To obtain the attachment reaction efficiency in *the quasi-free state*, we denote the specific rates of attachment and detachment in the quasi-free state by k_1^f and k_2^f respectively and modify the scavenging equation (10.10a) by adding a term $k_2^f \pi_a$ on the right-hand side, where π_a is the existence probability of the electron in the attached state. From the stationary solution, one gets $k_1^f/k_2^f = (k_1/k_2)(k_{ft}/k_{tf})$, or in terms of equilibrium constants, $K_r(qf) = K_r.K_{tr}$, where k_1 and k_2 are the rates of overall attachment and detachment reactions, respectively. Furthermore, if one considers the attachment reaction as a scavenging process, then one gets (see Eq. 10.11) $k_1 = k_1^f k_{tf}/(k_{tf} + k_{ft}) = k_1^f/(1 + K_{tr})$ and consequently $k_2 = k_2^f K_{tr}/(1 + K_{tr})$. In low- and intermediate mobility liquids, $K_{tr} \gg 1$, giving $k_1^f \gg k_1$, but $k_2^f \sim k_2$. Thus, the rate-determining step in the detachment process is electron ejection in the quasi-free state, which is quickly followed by trapping.

TABLE 10.6   Thermodynamic Functions for Reversible Attachment–Detachment Reactions in Hydrocarbon Liquids

Solvent	Solute	$\Delta G_r^0(qf)$	$\Delta H_r^0(qf)$	$\Delta S_r^0(qf)$
n-Hexane	Styrene	−67	−127	−199
	α-Methylstyrene	−63	−129	−220
	p-$C_6H_4F_2$	−46	−88	−142
Cyclohexane	p-$C_6H_4F_2$	−43	−104	−202
Isooctane	Naphthalene	−57	−90	−111
	Styrene	−49	−103	−179
	α-Methylstyrene	−44	−104	−202
	CO_2	−56	−112	−188
Neopentane	CO_2	−35	−98	−213

All energies in kJ/mol with an uncertainty of 20%. Entropies in J/(mol·K) with an uncertainty of 30%.

After obtaining k_1^f from the measured value of k_1 by this procedure, one can determine the attachment efficiency in the quasi-free state, $\eta = k_1^f/k_{diff}$, by the same procedure as for scavenging reactions (see Eq. 10.11 et seq.). Mozumder (1996) classifies the attachment reactions somewhat arbitrarily as nearly diffusion-controlled, partially diffusion-controlled, and not diffusion-controlled depending on whether the efficiency $\eta > 0.5, 0.5 > \eta > 0.2$, or $\eta < 0.2$, respectively. By this criterion, the attachment reaction efficiency generally falls with electron mobility. Nearly diffusion-controlled reactions can only be seen in the liquids of the lowest mobility. Typical values of η are: (1) 0.65 and 0.72 respectively for styrene and $p\text{-}C_6H_4F_2$ in n-hexane; (2) 0.14 and 0.053 respectively for α-methylstyrene and naphthalene in isooctane; (3) 1.8×10^{-3} for CO_2 in neopentane; and (4) 0.043 and 0.024 respectively for triphenylene and naphthalene in TMS.

REFERENCES

Allen, A.O.(1976), *Drift Mobilities and Conduction Band Energies of Excess Electrons in Dielectric Liquids*, NSRDS–NBS 58, National Bureau of Standards, Washington, D.C.

Allen, A.O., and Holroyd, R.A. (1974), J. Phys. Chem. *78*, 796.

Allen, A.O., Gangwer, T.E., and Holroyd, R.A. (1975), J. Phys. Chem. *79*, 25.

Anderson, P.W. (1958), Phys. Rev. *109*, 1492.

Asaf, U., and Steinberger, I.T. (1974), Phys. Rev. *B10*, 4464.

Ascarelli, G. (1986), Phys. Rev. *B34*, 4278.

Ascarelli, G., and Brown, S. C. (1960), Phys. Rev. *120*, 1615.

Atrazev, V.M., and Dmitriev, E.G. (1986), J. Phys. *C19*, 4329.

Bagley, B.G. (1970), Solid State Commun. *8*, 345.

Baird, J.K., and Rehfeld, R H. (1987), J. Chem. Phys. *86*, 4090.

Bakale, G., and Schmidt, W.F. (1973a), Z. Naturforsch. Teil *A28*, 511.

Bakale, G., and Schmidt, W.F. (1973b), Chem. Phys. Lett. *22* 164.

Basak, S., and Cohen, M.H. (1979), Phys. Rev. *B20*, 3404.

Berlin, Y.A., and Schiller, R. (1987), Radiat. Phys. Chem. *30*, 71.

Berlin, Y.A., Nyikos, L., and Schiller, R. (1978), J. Chem. Phys. 69, 2401.

Chandrasekhar, S. (1943), Rev. Mod. Phys. *15*, 1.

Cohen, M.H. (1977), Can. J. Chem. *55*, 1906.

Cohen, M.H., and Lekner, J. (1967), Phys. Rev. *158, 305*.

Davis, H. T., and Brown, R. G. (1975), Adv. Chem. Phys. *31*, 329.

Davis, H. T., Schmidt, L. D., and Minday, R. M. (1972), Chem. Phys. Lett. *13*, 413.

Dekker, A.J. (1957), *Solid State Physics*, pp.326–329, Prentice Hall, New Jersey,.

Dodelet, J.–P., and Freeman, G.R. (1972), Can. J. Chem. *50*, 2667.

Dodelet, J.–P., Shinsaka, K., and Freeman, G.R. (1973), J. Chem. Phys. *59*, 1293.

Dodelet, J.–P., Shinsaka, K., and Freeman, G.R. (1976), Can. J. Chem. *54*, 744.

Doldissen, W., and Schmidt, W.F. (1979), Chem. Phys. Lett. *68*, 527.

Doldissen, W., Schmidt, W.F., and Bakale,G. (1980), J. Phys. Chem. *84*,1179.

Efros, A.L. (1967), Sov. Phys. Solid State *9*, 901.

Freeman, G.R. (1986), J. Chem. Phys. *84*, 4723.

Frommhold, L. (1968), Phys. Rev. *172*, 118.

Funabashi, K., and Rao, B.N. (1976), J. Chem. Phys. *64*, 1561.

Fuochi, P.G., and Freeman, G.R. (1972), J. Chem. Phys. 56, 2333.

Gee, N., and Freeman, G.R. (1983), J. Chem. Phys. *78*, 1951.

Gee, N., and Freeman, G.R. (1987), J. Chem. Phys. *86*, 5716.

Gyorgi, I., Gee, N., and Freeman, G.R. (1983), J. Chem. Phys. *79*, 2009.

Hamill, W.H. (1981a), J. Phys. Chem. *85*, 2074.

Hamill, W.H. (1981b), J. Phys. Chem. *85*, 3588.

Holroyd, R.A. (1977), Ber. Bunsenges. Phys. Chem. *81*, 298.

Holroyd, R.A., and Schmidt, W.F. (1989), Annu. Rev. Phys. Chem. *40*, 439.

Holroyd, R.A., Dietrich, B.K., and Schwarz, H.A. (1972), J. Phys. Chem. *76*, 3794.

Holroyd, R.A., Gangwer, T.E., and Allen, A.O. (1975), Chem. Phys. Lett. *31*, 520.

Holroyd, R.A., McCreary, R.D., and Bakale, G. (1979), J. Phys. Chem. *83*, 435.

Holstein, T. (1959), Ann. Phys. *8*, 343.

Huang, S.S.–S., and Freeman, G.R. (1978), J. Chem. Phys. *68*, 1355.

Huang, S.S.–S., and Freeman, G.R. (1981), Phys. Rev. *A24*, 714.

Ioffe, A.F., and Regel, A.R. (1960), in *Progress in Semiconductors*, v.4 (Gibson, A.F. ed.), Heywood, London.

Itoh, K., Munoz, R.C., and Holroyd, R.A. (1989), J. Chem. Phys. *90*, 1128.

Itoh, K., Holroyd, R.A., and Nishikawa, M. (1991), J. Chem. Phys. *94*, 2073.

Jacobsen, F.M., Gee, N., and Freeman, G.R. (1986), Phys. Rev. *A34*, 2329.

Jahnke, J.A., Meyer, L., and Rice, S.A. (1971), Phys. Rev. *A3*, 734.

Jortner, J. (1970), in *Actions Chimiques et Biologiques des Radiations*, v.14 (Haissinski, M., ed.), ch. I, Masson, Paris.

Jortner, J., and Kestner, N.R. (1973), J. Chem. Phys. *59*, 26.

Kalinowski, I., Rabe, J.G., and Schmidt, W.F. (1975), Z. Naturforsch. *30A*, 568.

Kaneko, K., Usami, Y., and Kitahara, K. (1988), J. Chem. Phys. *89*, 6420.

LeBlanc, O. (1959), J. Chem. Phys. *30*, 1443.

Meyer, L., Davis, H.T., Rice, S.A., and Donnelly, R.J. (1962), Phys. Rev. *126*, 1927.

Miller, L.S., Howe, S., and Spear, W.E. (1968), Phys. Rev. *166*, 871.

Minday, R.M., Schmidt, L.D., and Davis, H.T. (1971), J. Chem. Phys. *54*, 3112.

Minday, R.M., Schmidt, L.D., and Davis, H.T. (1972), J. Phys. Chem. *76*, 442.

Mott, N.F., and Davis, E.A. (1979), *Electronic Processes in Non-Crystalline Materials*, 2nd ed., ch. 1, Clarendon Press, Oxford, England.

Mozumder, A. (1993), Chem. Phys. Lett. *207*, 245.

Mozumder, A. (1995a), Chem. Phys. Lett. *233*, 167.

Mozumder, A. (1995b), J. Phys. Chem. *99*, 6557.

Mozumder, A. (1996), J. Phys. Chem. *100*, 5964.

Munoz, R.C. (1988), Radiat. Phys. Chem. *32*, 169.

Munoz, R.C. (1991), in *Excess Electrons in Dielectric Media* (Ferradind, C., and Jay–Gerin, J.–P., eds.) ch.6, CRC Press, Boca Raton, Florida.

Munoz, R.C., and Ascarelli,G. (1983), Phys. Rev. Lett. *51*, 215.

Munoz, R.C., and Ascarelli, G. (1984), J. Phys. Chem. *88*, 3712.

Munoz, R.C., and Holroyd, R.A. (1986), J. Chem. Phys. *84*, 5810.

Munoz, R.C. and Holroyd, R.A. (1987), Chem. Phys. Lett. *137*, 250.

Munoz, R.C., Holroyd, R.A., and Nishikawa, M. (1985), J. Phys. Chem. *89*, 2969.

Munoz, R.C., Holroyd, R.A., Itoh, K., Nishikawa, M., and Fueki, K. (1987), J. Phys. Chem. *91*, 4639.

Nakamura, S., Sakai, Y., and Tagashira, H. (1986), Chem. Phys. Lett. 130, 551.

Nishikawa, M. (1985), Chem. Phys. Lett. *114*, 271.

Noyes, R.M. (1961), Prog. React. Kinet. *1*, 129.

Nyikos, L., Zador, E., and Schiller, R. (1977), Proc. of *Fourth International Symp. Radiat. Chem.*, (P. Hedwig and R. Schiller, eds.) p. 179, Akad-Kiado, Budapest.

Redfield, A.G. (1954), Phys. Rev. *94*, 526.

Reininger, R., Asaf, U., Steinberger, I.T., and Basak, S. (1983), Phys. Rev. *B28*, 4426.

Robinson, M.G., and Freeman, G.R. (1974), Can J. Chem. <u>52</u>, 440.

Sakai, Y., Schmidt, W.F., and Khrapak, A. (1992), Chem. Phys. <u>164</u>, 139.

Sano, H. (1981), J. Chem. Phys. *74*, 1394.

Scher, H., and Montroll, E.W. (1975), Phys. Rev. *B12*, 2455.

Schiller, R. (1972), J. Chem. Phys. *52*, 2222.

Schiller, R., Vass, S., and Mandics, J. (1973), Int. J. Radiat. Phys. Chem. *5*, 491.

Schmidt, W.F. (1976), in *Electron-solvent and Anion-solvent Interactions*, (Kevan, L., and Webster, b.c., eds.) ch. 7, Elsevier, New York.

Schmidt, W.F. (1977), Can. J. Chem. *55*, 2197.

Schmidt, W.F., and Allen, A.O. (1969), J. Chem. Phys. *50*, 5037.

Schmidt, W.F., and Allen, A.O. (1970), J. Chem. Phys. *52*, 4788.

Schmidt, W.F., and Bakale, G. (1972), Chem. Phys. Lett. *17*, 617.

Schmidt, W.F., Bakale, G., and Sowada, U. (1974), J. Chem. Phys. *61*, 5275.

Shinsaka, K., and Freeman, G.R. (1974), Can. J. Chem. *52*, 3495.

Shockley, W. (1951), Bell Sys. Tech. Jour. *30*, 990.

Smith, R.A. (1978), *Semiconductors*, 2nd ed., ch5. and 8, Cambridge University Press, Cambridge, England.

Sowada, U. (1976), Thesis, Free University of Berlin, Germany.

Sowada, U., Bakale, G., and Schmidt, W.F. (1976), High Energy Chemistry *10*, 290.

Spear, W.E., and LeComber, P.G. (1969), Phys. Rev. 178, 1454.

Springett, B.E., Jortner, J., and Cohen, M.H. (1968), J. Chem. Phys. *48*, 2720.

Stephens, J.A. (1986), J. Chem. Phys. *84*, 4721.

Tabata, Y., Ito, Y., and Tagawa, S., eds. (1991), *Handbook of Radiation Chemistry*, pp. 398–407, CRC Press, Boca Raton, Florida.

Tewari, P.H. and Freeman, G.R., (1968), J. Chem. Phys. *49*, 4394.

Wada, T., Shinsaka, Y., Namba, H., and Hatano, Y. (1977), Can. J. Chem. *55*, 2144.

Warman, J.M., deHaas, M.P., Zador, E., and Hummel, A. (1975), Chem. Phys. Lett. *35*, 383.

Willard, J.E. (1975), J. Phys. Chem. *79*, 2966.

Yoshino, K., Sowada, U., and Schmidt, W.F. (1976), Phys. Rev. *A14*, 438.

Radiation Chemical Applications in Science and Industry

In the preceding ten chapters of this book, we have described various important chemical and physical changes brought about by the absorption of ionizing radiation in gaseous and condensed media. Wherever possible, we have tried to elucidate the underlying mechanism with a discussion of the properties and reactivities of the intermediate species. However, the book would remain incomplete without discussion of some of the various uses that have been found for radiation-induced reactions in science and industry.

Applied radiation chemistry has gained considerable momentum since the late sixties and early seventies, and it would be futile to describe all the progress made in a single chapter. Spinks and Woods (1990) have nicely summarized the synthetic and processing aspects of the field in the latest edition of their book. Proceedings of various international conferences on specific aspects of radiation applications have appeared sporadically in *Radiation Physics and Chemistry* starting from middle seventies, of which mention may be made of vol. 9 (1977), vol. 14 (1979), vol. 25 (1985), vol. 31 (1988), vol. 34 (1989), vols. 35–36 (1990), vol. 37 (1991), vol. 40 (1992), vol. 42 (1993), and vol. 46 (1995). A recent international symposium proceedings in Japan (1997), *JAERI-Conf 97–003*, deals with accelerator beam applications. Of the earlier books and other publications in the field that are still useful, mention may be made of Atomic Radiation and Polymers by A. Charlesby (1960), The Radiation Chemistry of Macromolecules v.1 and v.2 by M.Dole (1973), Academic Press, New York, Technical Report Ser.84 on Radiation Chemistry and its Applications, International Atomic Energy Agency (1968), and Silverman (1981). Other references will be cited in the sections on food irradiation and on the effect of low-energy ion irradiation.

Before proceeding further, we would like to make some general comments on radiation chemical applications.

1. All intermediate species produced by the absorption of radiation (electrons, ions, excited states, free radicals, etc.) may be potentially useful for synthesis. However, the most frequently used intermediates are the free radicals. Their yield is high and relatively insensitive to temperature or state of aggregation (Wagner, 1969).

2. While almost all the radiation-chemical changes can also be brought about by thermal or photochemical means, there are some advantages of using irradiation since it can be conducted at lower temperatures without contamination by catalysts or initiators (Vereshchinskii, 1972). The radiation is absorbed uniformly over the volume of the reactor, which can be made of metal or glass, and the medium can be transparent or opaque (Wilson, 1972). Further, the *G*-values of the products are more easily calculated than the quantum yields of corresponding photochemical reactions.

3. There is better control of the progress of a reaction obtained by interrupted, pulsed, or continuous irradiation.

4. The primary radical yields are often ~3. A much higher value (>10) indicates chain reaction. In fact, the chain reaction mechanism for the formation of HCl from a gaseous mixture of hydrogen and chlorine exposed to radium irradiation is one of the earliest example of this kind, although the detailed chemistry was later shown to involve dissociated atoms rather than electrons and ions, as was originally proposed (see Bansal and Freeman, 1971).

5. The chief disadvantages of using radiation for industrial processing seem to be cost (in some cases), safety, and public and governmental concern over the long-term effects of irradiation.

In this chapter we will consider the following as examples of radiation chemical applications: (1) dosimetry, (2) industrial synthesis and processing, (3) irradiation of food, waste, and medical equipment, and (4) low-energy ion interaction with matter. Dosimetry is of fundamental importance for yield calculations and also for personnel exposure. Industrial processing would include some polymerization. Irradiated food and radiation sterilization of medical equipment are gradually gaining acceptance. Finally, low-energy ion implantation has technological relevance in the semiconductor industry.

11.1 DOSIMETRY

Only the energy absorbed in a defined mass of a medium can bring about chemical changes. Therefore, dose usually means *absorbed dose*, defined as the amount of energy absorbed per unit mass of the irradiated material. The Système International (SI) unit for dose is joule per kilogram ($J.Kg^{-1}$), which is given the name gray (Gy). The earlier unit still in vogue is the rad or its derivatives such as the kilorad (Krad) or the megarad (Mrad). Originally defined as a dose of 100 ergs per gram, it is equivalent to 0.01 Gy or to $6.24.10^{13}$ eV g^{-1}, a unit that has also been used. The dose rate is expressed in unit of $Gy.s^{-1}$, $Gy.min^{-1}$, or in the corresponding units of convenience. Sometimes, as in biological applications, the distribution of absorbed energy within small volumes of cellular dimensions is neded, thus introducing the concept of *microdosimetry*.

The intensity of the radiation field is expressed by *exposure* for such electromagnetic radiations as X-rays or γ-rays and by *fluence* for particulate radiation. The exposure is defined by the absolute total charge of either sign produced by the ionizing radiation in unit mass of dry air at STP, under the assumption that all secondary electrons are stopped in air. Its SI unit is coulomb per kilogram ($C.Kg^{-1}$). This replaces the earlier unit called the roentgen (R), which was defined in terms of the release of 1 esu of charge per cm^3 of dry air at STP. Therefore, $1 R = 2.58{\sim}10^{-4}$ $C.Kg^{-1}$. The fluence is defined as the number of particles incident per unit area around a point. Its symbol is Φ, and it is measured in unit of m^{-2}. The rates of exposure and fluence are derived from the corresponding time derivatives. The fluence rate is given the name *flux density* and denoted ϕ; it is measured in units of $m^{-2}s^{-1}$.

Since 1925, The International Commission on Radiation Units and Measurements at Bethesda, Maryland has been publishing reports updating the definitions and units for measurements of various radiation-related quantities. Of these *ICRU Reports*, special mention may be made of reports no. 19 (1971) [radiation quantities and units], 33 (1980) [radiation quantities and units], 36 (1983) [microdosimetry], 47 (1992) [thermoluminiscent dosimetry], and 51 (1993) [radiation protection dosimetry]. A succinct description of various devices used in dosimetry, such as ionization chambers, chemical and solid-state dosimeters, and personnel (pocket) dosimeters, will be found in Spinks and Woods (1990). In this section, we will only consider some chemical dosimeters in a little detail. For a survey of the field the reader is referred to Kase *et al.*, (1985, 1987), McLaughlin (1982), and to the International Atomic Energy Agency (1977). Of the earlier publications, many useful information can still be gleaned from Hine and Brownell (1956), Holm and Berry (1970), and Shapiro (1972).

11.1.1   CHEMICAL DOSIMETERS

Chemical dosimeters measure the absorbed dose by the quantitative determination of chemical change—that is, the G value of a suitable product—in a known chemical system. These are secondary dosimeters in the sense that the corresponding G values must be established with reference to a primary, absolute dosimeter. The primary dosimeters are usually physical in nature: calorimeters, ionization chambers, or charge measuring devices with particles of known energy. However, the primary dosimeters are generally cumbersome, whereas the chemical dosimeters are convenient to handle. On the other hand, the chemical dosimeters are not suitable for low-dose measurements.

In principle, a large variety of radiation chemical changes of a semipermanent nature induced in a solution can be utilized for dosimetric purpose. The practical suitability is determined by satisfying as many of the following requirements as possible:

1. The response of the dosimeter should be linear to the dose—that is, the G value of the product should be independent of the dose or the dose rate. The useful dose range may be stipulated as ~1–10^6 Gy, and it may be difficult to design a chemical system that preserves linearity over the entire range. In any case a useful chemical dosimeter must have linearity over most of the intermediate dose range.

2. The product G value should be independent of the incident particle LET; otherwise, the LET dependence of the yield must have been well established.

3. The product yield should be independent of temperature and insensitive to variation of experimental conditions during the course of radiolysis such as accumulation of radiolytic products, change in pH and so forth.

4. The reagents used in the dosimeter should be standard and easily available without the necessity of stringent purification. The dosimeter should be easy to use and portable.

5. The performance of the dosimeter should be reproducible with a precision of a few percent. Dosimeter response should be insensitive to minor changes in its composition and it should be stable under normal conditions of exposure to air or light and the like.

A large variety of aqueous and a few nonaqueous solutions have been used or proposed as chemical dosimeters with respective dose ranges for use (Spinks and Woods, 1990; Draganić and Draganić, 1971). Of these, a special mention may be made of the hydrated electron dosimeter for pulse radiolytic use (10^{-2} to 10^{+2} Gy per pulse). It is composed of an aqueous solution of 10 mM ethanol (or 0.7 mM H_2) with 0.1 to 10 mM NaOH. Concentration of hydrated electrons formed in the solution by the absorption of radiation is monitored by fast spectrophotometry, which is then used for dosimetry with the known G value of the hydrated electron.

However, by far the most widely used chemical dosimeters are the Fricke (ferrous sulfate) and the ceric sulfate dosimeter, which will now be described.

11.1.1a   The Fricke Dosimeter

The chemical change in the Fricke dosimeter is the oxidation of ferrous ions in acidic aerated solutions. It is prepared from a ~1 mM solution of ferrous or ferroammonium sulfate with ~1 mM NaCl in air-saturated 0.4 M H_2SO_4. Addition of the chloride inhibits the oxidation of ferrous ions by organic impurities, so that elaborate reagent purification is not necessary. Nevertheless, the use of redistilled water is recommended for each extensive use. Absorption due to the ferric ion is monitored at its peak ~304–305 nm. The dose in the solution is calculated from the formula

$$D \ (Gy) = 9.65 \times 10^6 \left[\frac{OD_1 - OD_n}{\Delta\varepsilon \ \rho l G(Fe^{3+})} \right],$$

where OD_1 and OD_n are respectively the optical density of the irradiated and non-irradiated solution, $G(Fe^{3+})$ is the G value for the oxidation of the ferrous ion, l is the optical path length, ρ is the solution density, and $\Delta\varepsilon$ is the difference between the molar extinction coefficients of Fe^{3+} and Fe^{2+}. For ^{60}Co-γ, $G(Fe^{3+}) = 15.6$ and its LET variation has also been established (Spinks and Woods, 1990). Since the ferrous ion absorption at 304 nm is negligible, the molar extinction coefficient of the ferric ion at its peak, 2205 $M^{-1}cm^{-1}$, can be used safely. The standard Fricke dosimeter is useful in the dose range 20–400 Gy. The upper limit can be extended to ~2 KGy by saturating the solution with oxygen and using a higher Fe^{2+} concentration. Another modification of the theme is the ferrous-cupric dosimeter, which extends the upper limit to ~10 KGy.

11.1.1b   The Ceric Sulfate Dosimeter

The chemical change in the ceric sulfate dosimeter is the reduction of ceric ions in acidic aerated solution. It is prepared from ceric sulfate in 0.4 M H_2SO_4. However, it is necessary to choose a range of concentration of ceric sulfate (from ~0.2 to ~50 mM) for the anticipated range of dose measurement to minimize the error involved in the determination of the change of the molar extinction coefficient of the solution upon irradiation (Draganić and Draganić, 1971). Within a specified interval of concentration, the G value of the cerous ion should be independent of the concentration of the ceric ion. The G value of the cerous-on yield for $^{60}Co - \gamma$ irradiation is 2.3, and the molar extinction coefficient at the peak (320 nm) is 5600 $M^{-1}cm^{-1}$. However, the stock solution must be kept in the dark, since the solution is photosensitive. When carefully prepared, the response of the ceric sulfate dosimeter is linear to the absorbed dose in the interval 1–100 KGy, and it is independent of dose rate up to ~1 MGy.s^{-1}.

11.2   INDUSTRIAL SYNTHESIS
AND PROCESSING

Chemical synthesis is one important aspect of the application of radiation-chemical reactions in industry. Various kinds of radiation-induced syntheses are available, some of which will be described here. There are also nonsynthetic applications including, but not limited to, food irradiation, waste treatment, and sterilization by irradiation. Some of these will be taken up in the next section.

After a certain radiation-chemical reaction for product transformation or initiation has been tested in the laboratory, an important commercial consideration is the cost of processing. The quantity of radiation energy required to convert 1 kg of feed material completely to prouct may be stated as $10^6/MG$ (J kg^{-1}), where M is the molar mass of the material (kg.mol^{-1}) and G is the radiation yield expressed as μ mol.J^{-1}. The processing cost per unit mass ($.$kg^{-1}$) is then given by $10^6 s/MGf$, where s in the per-unit radiation energy cost ($.$J^{-1}$) and f is the fraction of radiation energy abosrbed by the irradiated material (Spinks and Woods, 1990). Because of scattering and transmission f is usually ~0.5, although the exact value will depend on the characteristics and geometry of radiation source and irradiation vessel.

It is clear from this discussion that the dose requirement and unit cost will be lower if the material has a higher molar mass M and the reaction has a high G value. Thus, the best candidates will be a polymeric material and a chain reaction. Quite often, a free-radical irradiation is used. The radiation source of choice is usually a $^{60}Co - \gamma$ facility, although electron beam irradiation is also used. Since most radiation-chemical reactions used in industry can also be brought about by other conventional means such as thermal, or photochemical processes, the processing cost must be below ~10¢ . kg^{-1} to be competitive, since it is unlikely that the cost of irradiation will come down in future. It should remembered that in figuring the irradiation cost one has to include the cost of operation, maintenance, and the like. (Danno, 1960).

Various industrial pilot plants and full-scale operations, using radiation-chemical processing have been reported, with production rates ~50 to ~1000 tons per year (Spinks and Woods, 1990; Chutny' and Kucera, 1974). Production rates less than ~50 tons per year are not considered viable. These operations are or have been conducted in countries such as the United States, the former U.S.S.R., Japan, and France. However, some operations have also been reported in the former Czechoslovakia and Romania, especially in connection with petroleum industry. In the United States, chlorination of benzene to gammexane (hexachlorocyclohexane) was hotly pursued at one time by radiation or photoinitiation. Since the early seventies the activity has dwindled, presumably due to lack of demand and environmental considerations.

The numerous radiation-chemical reactions used or proposed for industrial processing include, but are not limited to, the following:

1. Oxidation of hydrocarbons. Paraffin wax in the petroleum industry was oxidized in a gamma field at a higher temperature to produce higher alcohols and fatty acids. The yield envisaged was as a few thousand tons per year.

2. Oxidation of benzene to phenol. This was attempted in the former U.S.S.R. and Japan on a pilot-plant scale. High yields were reported, but full-scale operation apparently was discontinued because of destruction of product by irradiation and the possibility of explosion in the reaction vessel. The latter danger can be controlled in the oxidation of halogenated hydrocarbons such as trichloro- or tetrachloroethylenes, where a chain reaction leads to the formation of dichloro- or trichloro-acetic acid chlorides through the respective oxides.

3. Chlorination reactions. Chlorination of hydrocarbons has been carried out in Japan, chlorination of toluene in the United States, chlorination of tetrachloropentane in the former U.S.S.R. to give octachlorocyclopentane, and chlorination of propanoic acid in France to give chloropropanoic acid. Chlorination of methane by irradiation to give lower halomethanes was found to be cost-effective. Chlorination of various amorphous polymers such as polypropylene, polybutadiene, and PVC, has also been carried out.

4. Synthesis of bromoethane (ethane + HBr) and of dibutyltin dibromide (bromobutane + Sn). Other halogenation reactions such as bromination of p-xylene to tetrabromoxylenes is of importance as an efficient fire extinguishing agent for plastics.

5. Sulfoxidation and sulfochlorination of of hydrocarbons for the production of detergents. Starting from kerosene fractions, processing rates in the United States and the former U.S.S.R. reached several thousand tons per year.

6. Polymerization reactions. Polymerization of ethylene to polyethylene has been conducted at pilot-plant scales reaching a target of 1500 tons per year. Some reactions, including polymerization and copolymerization of polymers for grafting on textile fibers, have been successfully performed. Similarly, cross-linking of polyethylene to improve thermal properties has also been achieved.

Radiation synthesis has been reviewed by Wagner (1969), Vershchinskii (1972), Wilson (1972), and Chutny' and Kucera (1974). A good summary is also available in Spinks and Woods (1990). On the theoretical side, radiation-induced reactions of importance to industry can be classified as addition reactions, subsitution reactions, and other reactions including polymerization, cross-linking, and so on. In all of these, chain reactions play a very significant role. According to Wagner (1969), if the production cost is to be kept below a few cents per kilogram, one must get ~100 molecules of product per initiating free radical, on the assumption of a G value of radical production ~8 and a probable product molecular weight ~125. This indicates chain reactions with a chain length ~100.

11.2.1 ADDITION REACTIONS

For alkenes and alkynes, addition to the double or triple C–C bond is common. A typical example is

$$HBr + CH_2{=}CH_2 \rightarrow CH_3{-}CH_2Br.$$

This kind of reaction can proceed in solution by a carbocation mechanism, but in the radiation-induced case, it proceeds almost exclusively by a radical mechanism. In most cases, the radiation initiates reactions that are of chain character.

Addition to an unsaturated compound (U) can occur in an alternating chain as exemplified here:

$$A\cdot + U \rightarrow AU\cdot,$$

$$AU\cdot + AB \rightarrow AUB + A\cdot,$$

where A· is a radical with an unpaired electron on such atoms as C, Si, P, S, Cl, or Br, and B is either a hydrogen atom or a halogen. The radicals A· or AU· may disappear by recombination or disproportionation according to the scheme

$$AB \rightarrow A\cdot + B\cdot \left(\text{initiation}\right),$$

$$A\cdot + U \rightarrow AU\cdot \left(\text{addition}\right),$$

$$AU\cdot + AB \rightarrow AUB + A\cdot \left(\text{abstraction and propagation}\right),$$

$$2A\cdot \rightarrow A_2\left(\text{recombination and termination}\right).$$

The chain length is therefore adversely affected by the irradiation dose rate being inversely proportional to its square root. Wagner (1969) lists a large class of unsaturated compounds in which addition reactions can be induced by irradiation. Typical examples involving long chain lengths are for the addends HCl, Cl_2, and HBr in ethylene, benzene, toluene, and so on. where the products are telomers or hexachlorides.

11.2.1a Alkylation

Alkane free radical addition to aklenes and alkynes generates products with more carbon atoms than the reactants—that is, the result is alkylation. Generally, alkylation gives mixture of products. However, under controlled conditions, it is possible to get a desired product in predominance (Chutney' and Kucera, 1974). Thus, the addition of methane to ethene, under irradiation, gives a product mixture in which the yield of methylpropane predominates ($\sim$100 μmol.J^{-1}). Similarly, the reaction between propane and ethene gives butane and methylbutane, although the yield is smaller. In some cases and under suitable conditions, the radiation-induced alkylation can proceed with rates much greater than thermal reactions.

11.2.1b Halogenation

Chlorination of aromatic compounds under irradiation has been studied extensively (Wagner, 1969). With benzene, the product is a mixture of stereoisomeric hexachlorocyclohexanes with yields $\sim$10^4μmol.J^{-1}. This certainly points to chain reaction with the initiation either from a dissociation, $Cl_2 \rightarrow 2Cl$, or from the participation of the first excited singlet state of benzene ($^1B_{2u}$) giving

$$C_6H_6^* + Cl_2 \rightarrow C_6H_6 + 2Cl.$$

One of the setereoisomers formed, commercially knowm as gammexane, is an insectiside, which was once quite popular. These addition reactions are initiated by radicals. In the presence of ionic catalysts such as $FeCl_3$, substitution reactions occur instead of addition.

Hydrogen halides will easily add to unsaturated compounds under radiolysis or photolysis. The free-radical chain reaction process is initiated by the dissociation of the halide or by the radiolytic production of radicals from the halide or the organic compound. Thus, for the radiolysis of a mixture of HBr and ethene the postulated initiation is

$$HBr \rightarrow H + Br,$$

$$C_2H_4 \rightarrow C_2H_2 + 2HHBr,$$

$$H + HBr \rightarrow H_2 + Br,$$

and so on. The propagation is the sequence

$$Br + CH_2{=}CH_2 \rightarrow \cdot CH_2 - CH_2Br,$$

$$\cdot CH_2 - CH_2Br + HBr \rightarrow CH_3CH_2Br + Br,$$

and the termination is by the biradical recombination $R + R \rightarrow R_2$, where R is Br or $\cdot CH_2 - CH_2Br$. The reaction is of great commercial importance and has been carried out in the gaseous, liquid, and solid phases as well as in solutions of halogenated organic solvents with yields $\sim$10^3 to $\sim$10^4 μmol.J^{-1} (Armstrong and Spinks, 1959). Polyhalogenated compounds can also add to alkenes by chain reactions initiated by radicals. Chutney' and Kucera (1974) report a reaction between CCl_4 and ethene as an industrial process in the former U.S.S.R. with annual production rate of several hundred tonnes.

Liquid phase chlorination work in the former U.S.S.R. has been summarized by Vereshchinskii (1972). With tetradecane, the reaction is nearly or partially diffusion-controlled at a dose rate of 0.1–0.4 rad s^{-1}. However, during the chlorination process, the liquid phase properties change continuously because of chlorine absorption accompanying the chemical reactions. Due to long chain reactions the chlorination G value is high and can reach ~10^5 per 100 eV of energy absorption. At around 10–30°C the reaction rate is found to vary as the square root of the dose rate. A set of consecutive reactions has been reported in the liquid phase chlorination of 1,1,1,5-tetrachloropentane (Vereshchinskii, 1972).

11.2.1c   Oxidation, Sulfoxidation, and Sulfochlorination

Hydroperoxides can form in fairly long chain reactions with the reaction sequence indicated as follows:

$$R\cdot + O_2 \rightarrow RO_2\cdot,$$

$$RO_2\cdot + RH \rightarrow ROOH + R\cdot.$$

Thus, the overall reaction may be written as RH + $O_2 \rightarrow$ ROOH. The G values for hydroperoxide formation at 50°C range from ~16 for 2,2,4-trimethylpentene-1 to ~400 for cyclohexene (Wagner, 1969). Although this temperature is somewhat lower than the temperature of decomposition of the hydroperoxide, in practice the reactions are conducted at elevated temperatures. In such cases, the radition-induced initiation either eliminates the induction period or allows the recations to proceed at somethat lower temperatures than would be otherwise required.

Sulfoxidation has been carried out in paraffins (Black and Baxter, 1958) and in cyclohexane (Hummel et al., 1964). In sulfoxidation, SO_2 first adds to a free radical, but the reaction is not complete with the unstable $\tilde{R}SO_4H$ species. The individual steps of the chain are indicated as follows:

$$R\cdot + SO_2 \rightarrow RSO_2\cdot,$$

$$RSO_2\cdot + O_2 \rightarrow RSO_4\cdot,$$

$$RSO_4\cdot + RH \rightarrow RSO_4H + R\cdot,$$

$$\cdot OH + RH \rightarrow H_2O + R,$$

$$RSO_3\cdot + RH \rightarrow RSO_3H + R\cdot,$$

with the overall reaction being represented as $RH + SO_2 + O_2 \rightarrow RSO_3H$.

The sulfochlorination chain reaction is not a branching process, but the chain can be very long as the reactions of the indivudual steps indicated in the following are all presumably exothermic:

$$R \cdot + SO_2 \rightarrow RSO_2 \cdot,$$

$$RSO_2 \cdot + Cl_2 \rightarrow RSO_2Cl + Cl \cdot,$$

$$Cl \cdot + RH \rightarrow HCl + R \cdot.$$

The overall reaction is then represented by $RH + SO_2 + Cl_2 \rightarrow RSO_2Cl + HCl$.

11.2.1d Other Addition Reactions

Reactions, under irradiation, involving the addition to unsaturated hydrocarbons of nitrogen and phosphorus compounds and of silanes have been summarized by Spinks and Woods (1990). Some cycloaddition reactions have also been reviewed. These will not be detailed here.

11.2.2 SUBSTITUTION REACTIONS

Substitution reactions usually occur with saturated molecules. A typical case is the reaction of chlorine and methane in which the hydrogen atoms of methane are replaced by chlorine in sequence—for example,

$$CH_4 + Cl_2 \rightarrow CH_3Cl + HCl,$$

$$CH_3Cl + Cl_2 \rightarrow CH_2Cl_2 + HCl,$$

and so on. Under favorable energetic conditions, substitutive (double abstraction) reactions involving saturated molecules can occur. Most such chain reactions involving saturated molecules proceed via the scheme

$$A \cdot + RH \rightarrow R \cdot + AH,$$

$$R \cdot + AX \rightarrow A \cdot + RX.$$

For long chains of these reactions, each step must be exothermic. Typical examples include Cl_2 in carboxylic acids, and $BrCCl_3$ in alkyl aromatics. The products are variable substitution and side chain substitution compounds, respectively. Radiation-induced substitution reactions have been reviewed by Wilson (1972) with examples of nitration, nitrosation, sulfochlorination, and others. These generally proceed by a free-radical mechanism. The free radicals are generated by the action of radiation on the reagent, which is present in large excess—for example, as a solvent. A special class of substitution reactions, between chlorine and saturated hydrocarbons, may be initiated by thermal, photochemical, or radiation-induced processes. The free-radical mechanism is common among these, but there can be other initiating mechanisms in the thermal case. The initiation, propagation, and termination in the case of the reaction between chlorine and methane, proceeding via the free-radical mechanism, may be indicated as follows:

Initiation

$$CH_4 \rightarrow \cdot CH_3 + H,$$

$$Cl_2 \rightarrow 2Cl,$$

$$H + Cl_2 \rightarrow HCl + Cl.$$

Propagation

$$\cdot CH_3 + Cl_2 \rightarrow CH_3Cl + Cl,$$

$$Cl + CH_4 \rightarrow HCl + CH_3.$$

Termination

$$2Cl \rightarrow Cl_2,$$

$$2 \cdot CH_3 \rightarrow C_2H_6,$$

$$\cdot CH_3 + Cl \rightarrow CH_3Cl.$$

Molecules containing two or more chlorine atoms may be produced by the reaction of chlorine atoms or molecules with products generated in the earlier stage of the process. Product yields depend on irradiation conditions and can reach as high as 10^5 μmol.J^{-1}. With bromine and iodine, not all of the individual steps of the reaction are exothermic. Therefore, a sustained chain reaction is not expected, and the yields are low.

11.2.3   OTHER REACTIONS

Various other radiation-induced reactions have been studied for potential use in the industry on a pilot-plant scale. Among these may be mentioned hydrocarbon cracking (i.e., production of lower-molecular-weight hydrocarbons from higher-molecular-weight material), isomerization of organic molecules, and synthesis of labeled compounds with radioactive nuclei. When organic compounds are irradiated in the pure state or in aqueous solution, dimeric compounds are commonly formed. Although this suggests a possible route for the synthesis of such compounds, the yield is usually low.

In this brief section, we have not touched the vast field of radiation-induced polymerization and radiation effects on polymers. Fortunately, the field has been surveyed very well in international conference proceedings published in *Radiation Chemistry and Physics* referred in the beginning of this section. The earlier books by Charlesby (1960) and by Dole (1973) provide adequate background information.

11.3 STERILIZATION OF MEDICAL EQUIPMENT AND DISPOSABLES

One of the more successful applications of applied radiation technology has been the sterilization of sutures and disposable medical supplies. Sterilization is chemically equivalent to the disruptive action on large biological molecules, and as such the irradiation process is economically feasible because of the high molecular weight of the material. At present, ~30% of all medical, single-use, disposable products manufactured in North America are sterilized by irradiation (Spinks and Woods, 1990). Of the different kinds of disposables used in the medical industry, a large variety (including surgical gowns, gloves, syringes, bandages, sutures, dressings, catheters, and transfusion sets) can be sterilized by irradiation. Since the cost of cleaning and resterilizing nondisposables has long since increased to the point of favoring single-use disposables, the use of irradiation to achieve sterilization has become particularly meaningful.

The term *sterile* has now been given a statistical meaning by international convention (Handlos, 1981). An item may be called sterile if the probability of finding a microorganism in it is $<10^{-6}$. Generally, sterilization requires an absorbed dose ~25 Gy. However, because of legal requirements in different countries and because of the occasional presence of radiation-resistant microorganisms, sometimes higher doses, ~50 Gy, may be needed. For this, both $^{60}Co - \gamma$ and electron beam (EB) irradiation have been used (Handlos, 1981). Under relatively high dose conditions some polymers such as polyethylene and polystyrene may be stable, whereas others such as PVC or polypropylene may not. In the latter case, some additives must be included to improve the radiation stability.

Sterilization by irradiation was introduced by mid-fifties. In about 20 years, it was fully operational. When compared with the traditional methods of sterilization such as using formaldehyde, ethylene oxide (a toxic gas), or heating in an autoclave, several advantages of irradiation may be noted (Artandi, 1977):

1. Design simplicity and outstanding reliability. Control of radiation output with proven mechanical conveyors is easy and more reliable than conventional process equipment. For most radiation sources using isotopes, the conveyor speed is the controlling factor.

2. The process is continuous, allowing smooth product flow and requiring minimum human handling. Therefore, it is operable with low maintenance costs.
3. It is easily applicable to single-use medical products.
4. It provides a high degree of sterility assurance without the intervention of specially qualified personnel.
5. A great choice is available in product design and packaging material because sterilization depends on radiation penetration, which is relatively independent of the chemical composition of the packaging and product material.
6. It is possible to sterilize after packaging, which gives some flexibility of operation.
7. Irradiation avoids the use of toxic chemicals (e.g., ethylene oxide) or high temperatures (as in the autoclave). Throughout the world, there are ~100 plants for sterilization of medical products by irradiation.

As for the relative suitability of an electron beam (EB) facility vis-à-vis a cobalt-60 gamma facility, a key point is that although the ultimate chemistry is nearly identical in both cases, there is a notable difference in the penetration of the radiations. Another point is that the large capacity and consequent cost of EB machines require a relatively large production rate to justify their use. On the other hand, the EB machine, not being a radioactive source, is completely safe when switched off. Overall, since the sixties, sterilization by irradiation has steadily increased. However, most of this is by cobalt-60 gamma irradiation, the EB machines accounting for about a fifth or sixth of the total number of facilities.

11.4   WASTE TREATMENT BY IRRADIATION

With increasing population growth and urbanization, solid municipal waste disposal has become an important aspect of civilized life. It is now generally recognized that solid waste is simultaneously a major problem and an underutilized resource of material and energy. Traditional methods of treating sewage sludge—that is, land application by burying directly or following incineration—have some environmental and economic disadvantages. There are problems of odor and other sanitary hazards to the neighborhood. On the other hand, excessive dependency on chemical fertilizers in agriculture leads to humus deficiency in the soil, and some kind of organic fertilization of the farm land, at least in part, is considered desirable. With urbanization of rural areas, direct application of raw sludge, even after disinfection, may not be acceptable in many countries. Some stabilization treatment—for example, composting—together with disinfection is considered essential for land application. It is at this stage that radiation treatment offers a viable solution either by itself or, more likely, in combination with some traditional method. Considerable efforts in this direction have been made in developed countries as well as in some developing countries (Brenner et al., 1977; Kawakami et al., 1981; Kawakami and Hasimoto, 1984; Iya and Krishnamurthy, 1984).

To make sewage sludge compatible with plant life in an environmentally friendly way and also to make it usable as a safe soil conditioner, it is desirable to convert it into living humus by aerobic composting. Irradiation of sludge cake and subsequent composting have revealed the following (see Kawakami et al., 1981):

1. The optimum processing conditions are at temperatures ~50 °C , pH ~7–8 and an irradiation dose around 5 KGy. Irradiation of dewatered sludge can reduce irradiation costs.
2. Since irradiated sludge does not compost by itself, it is necessary to introduce a seed. The composted product can act as a seed. There is no significant difference between the composting characteristics of unirradiated or irradiated sludge when seeded similarly. Repeated use of composted sludge as seeds has no detrimental effects
3. CO_2 and NH_3 evolutions occur almost simultaneously. Their rate of evolution reaches a maximum at about ten hours after irradiation and evolution ceases after about three days under optimal conditions.
4. Irradiation shortens the composting period, because it is not necessary to keep an elevated fermentation temperature to reduce the pathogens in the sludge.

Irradiation of solid waste offers possibilties for improved handling of such organic waste components as putrescible matter and cellulose (paper products, chaff, straw, sawdust, etc.). In the petrochemical industry, the abundant supply of cellulose material can be utilized as alternative feedstock for generating fuels and for intermediate chemicals as well as for synthesis of single cell proteins. The basic step is the conversion of cellulose to its monomeric glucose, for which a convenient pretreatment consists of irradiation of hydropulped cellulosic waste. The reaction is completed by acid or enzymatic hydrolysis. The putrescible matter can be cold sterilized by irradiation for the generation of animal feed. Such feeds usually have a high water content and, if untreated or inadequately treated, these can harbor rapidly multiplying microorganisms some of which are pathogenic. Such feeds include animal waste, which had been used for soil conditioning and as an animal feed supplement. There is always some hazard in these applications because of pathogenic organisms and harmful chemical residues. The traditional method of counteracting these hazards (e.g., using steam autoclaves and processing some several thousand tons of waste per day) is considered too expensive. Therefore, there is a real need for an effective technical solution at a reasonable cost, and irradiation provides an alternative. In this connection, it should be remarked that irradiation is useful in breaking down organic pollutants that are not easily biodegradable, such as dyes, detergents, and phenols. These organic materials can react with radicals generated in water radiolysis, and the reaction products are more easily removable by ordinary biological or chemical treatment.

11.4.1   TREATMENT OF FLUE GAS

Electron beam (EB) irradiation of exhaust gases from coal-fired power stations for the reduction of oxides of sulfur and nitrogen was first started in Japan at the Japanese Atomic Energy Research Institute (JAERI)'s research laboratory in Takasaki (Tokunaga and Suzuki, 1984). (Some preliminary work was also performed in Japan to remove harmful oxides in the iron-ore sintering process.) The process has now arrived at the pilot-plant level in Japan and Germany, with some work also done in the United States (Helfritch and Feldman, 1984). The primary purpose is of course environmental—namely, intercepting harmful oxides before they are released into the atmosphere. Since it appears that the use of coal for the generation of electrical energy will not diminish in the near future, the concept is fundamentally sound, the only complicating factor being economics. To the cooled exhaust gas, water and ammonia or calcium carbonate are added prior to irradiation. The addition of water increases humidity. After irradiation, solid products—ammonium sulfate or ammonium nitrate—are formed; these are easily separated from the gas stream and bagged as fertilizers. This is a secondary by-product use; the cost of the fertilizers has been estimated in Japan to be at least twice that produced by a standard procedure. Yet it can be justified by environmental considerations.

Flue gas treatment requires an absorbed dose ~10 kGy. EB machines of somewhat less than 1 MeV energy with a current ~250 mA can process ~10^4–10^5 m^3 of flue gas per hour. This is comparable to the exhaust gas output of a fairly large coal-fired station generating ~500 MW of electrical power. Irradiation is done in a channel kept underground to provide radiation shielding, and it is achieved by multiple machines. The irradiation energy used in such a process is estimated at ~1% of the equivalent electrical output of such stations. Work in the United States suggests that ~90% or more of the oxides can be removed by irradiation. The equipment can be built from commercially available components, and the relatively low temperature and relatively high humidity favor economic operation.

The reaction mechanism in the irradiated flue gas is probably quite complex, but basically the EB excites the gas molecules and promotes reactions that convert the oxides to acids. These then react with ammonia or calcium compounds to give solid products that are removed by the filter. The initiation reaction is believed to be brought about by radical formation, such as OH,

O, or HO_2, in the irradiated flue gas. The reactions with the oxides may be given as follows:

$$NO + HO_2 \rightarrow NO_2 + OH,$$

$$NO_2 + OH + M(\text{third body}) \rightarrow HNO_3 + M,$$

$$SO_2 + OH + M \rightarrow HSO_3 + M,$$

$$HSO_3 + O_2 \rightarrow SO_3 + HO_2,$$

$$SO_3 + H_2O \rightarrow H_2SO_4.$$

Certain other reactions are also possible, some of which do not give solid products. For example:

$$NO + OH + M \rightarrow HNO_2 + M,$$

$$2HNO_2 \xrightarrow{\ H_2SO_4\ } NO + NO_2 + H_2O$$

$$NH_3 + OH \rightarrow NH_2 + H_2O,$$

$$NH_2 + NO \rightarrow N_2 + H_2O,$$

$$SO_2 + OH + H_2O \rightarrow \text{various products.}$$

Of the major components of flue gas, oxygen and water vapor influence the reactions of the oxides considerably, but carbon dioxide does not. Under irradiation, NO is oxidized by reactions with O, OH, and HO_2 radicals. The resultant NO_2 is oxidized to HNO_3 by reaction with OH radicals. SO_2 is similarly oxidized to H_2SO_4 by reactions with O and OH. The products can also be converted to aerosols and collected in electrostatic precipitators.

11.4.2 WASTE WATER TREATMENT

Considerable research has been done in many industrial countries, especially in Japan and in the former U.S.S.R., on the radiation treatment of waste water and other liquid wastes (see Pikaev and Shubin, 1984; Sakumoto and Miyata, 1984). Apart from disinfection or sterilization, the processes involve the radiation treatment of polluted water, the radiation-induced decomposition of dyes, phenols, cyanides, and so forth. (*vide supra*). At the basis of purification of aqueous waste by irradiation lies the strongly oxidizing and reducing actions of water radicals—that is, e_h, H atoms, and OH. These species induce various radiolytic transformations of solutes, which can be used either directly or in combination with traditional procedures. The transformations include, but are not limited to, decompositions of organic compounds, discoloration of dyes, precipitate formation, and redox reactions. Radiation discoloration of aqueous solutions of humus compounds needs a dose ~5 kGy or more, but with oxygenation (air bubbling) the required dose can be brought down to ~1 kGy.

The radiation treatment of waste water has some advantages: It makes use of highly reactive species, which require no chemical catalysts, and nearly complete decomposition into CO_2 and H_2O can be achieved. However, by itself, radiation treatment seems to be prohibitively expensive, because of relatively the high dose requirement. Synergistic studies made in Japan, the former Soviet Union, and other countries indicate that a combination of irradiation with biological oxidation, coagulation, flotation, ozonation, and other procedures. can reduce the dose requirement by as much as an order of magnitude. Under suitable conditions, broad-range and complete oxidation of organic pollutants in waste water to carbon dioxide and water can be achieved by the combined use of irradiation and ozone. This process is therefore generally applicable to the treatment of all dissolved organic pollutants, making it one of the best tertiary treatments.

11.5   FOOD IRRADIATION

The purpose of irradiation of foods or, for that matter, of any food treatment can be manifold, such as (1) delaying or preventing sprouting, or inactivating molds and bacteria to prolong shelf life; (2) elimination of insects and parasites in foods and spices; (3) delaying ripening of fruits; and (4) production of sterile products, which can be stored without refrigeration. According to the *CRC Handbook of Radiation Chemistry* (Tabata, *et al.,* 1991), the absorbed dosage requirement may be classified as low (<1 kGy), intermediate (1–10 kGy) or high (10–50 kGy) depending on the specific purpose. Doses above 100 kGy are generally not recommended. Inhibition of sprouting in onions, potatoes, and the like requires the lowest dose, ~0.02–0.15 kGy. Understandably, many countries have first approved this kind of food irradiation. Next is the elimination of insects and parasites in cereals, fruits, dried fish, some fresh meats, requiring ~0.1–1.0 kGy. Delay of physiological processes in fresh fruits and vegetables needs ~1 kGy. In the intermediate-dose category, a dose ~1.0–5.0 kGy will be needed to prolong the shelf life of fresh fish, meat, and meat products. Control of pathogenic microorganisms and spoilage decontamination in seafoods (fresh or frozen), poultry (raw or frozen), spices and animal feed require ~3–10 kGy. A similar dosage is also needed for the improvement of food properties such as restoration of dried vegetables, agar production, wine and whiskey maturation. A high-level dose

(10–50 kGy) may be needed for sterilization in poultry, meat, seafoods, prepared foods, hospital diets, and of course in animal feed to allow germ-free research.

The advantages of food irradiation parallel somewhat the radiation treatment of waste products. (1) The process can be carried out without significant rise in temperature due to radiation absorption or subsequent chemical reactions. (2) Large volumes of food can be handled continuously with commercially available equipment. (3) Certain types of irradiation from radioisotopes, X-rays, and γ-rays can penetrate wide packages. Therefore, irradiation may often be done after packaging. (4) No toxic chemicals such ethylene oxide or carcinogens or mutagens need be used. Irradiation probably produces fewer harmful substances than standard chemical treatment to achieve the same end result. The quantity of toxic products formed on irradiation is often expressed as URP (unique radiolytic product). The relative amount of URP—that is the ratio of the mass of the product (kg) per unit mass of food (kg)—can be expressed as $DGM \sim 10^{-6}$, where D is the absorbed dose in Gy, G is the radiation-chemical yield (μ mol.J^{-1}) and M is the molecular weight (kg.mol^{-1}) of the product. For an absorbed dose of 10 kGy and taking $G = 0.5$ and $M = 0.1$ as plausible values, one gets the relative product yield as 5×10^{-4} kg per kg of starting material, which is very small indeed. The combined radiolytic load from all sources would then be rather insignificant, unless some chain reaction were initiated. However, well-established radiation-induced chain reactions have shown that if there are chain reactions in irradiated food, these would be of conventional type, as encountered in oxidative changes during storage or as initiated by UV. It is unlikely that the total load of URP will exceed 0.1% of the food mass up to the highest dose permitted in food irradiation. The only possible detrimental effect of such food irradiation may be on the vitamin components—for example, vitamin C and vitamins A and E, which act as radical scavengers. It should be noted, however, that such vitamins are also damaged by any other method of food processing.

Public concern about the safety and wholesomeness of irradiated food has always played an important role in the licensing, production, and consumption of irradiated food. Generally speaking, most governments have been over-cautious in approving irradiated food for comsumption. Different countries have adopted different safe dose levels for the varied uses of food irradiation. However, some consensus has also emerged. After exhaustive chemical, biological, and physiological investigations, an international standard was adopted in 1983 by the Codex Alimentarius Commission, which is a joint body formed out of World Health Organization (WHO) and the Food and Agriculture Organization (FAO) of the United Nations. A joint expert committee of irradiated food (JECFI) representing the International Atomic Energy Commission (IAEA), WHO, and FAO concluded in 1980 that irradiated foods of all kinds are safe up to a dose of 10 kGy and recommended that no testing on undesirable effects of irradiation is necessary at this or lower dose levels.

At this stage, it should be noted that almost invariably the desired effect of irradiation of foods can also be attained by other standard methods, such as heat treatment, use of chemicals such as ethylene oxide, and so on. Apart from toxicity and carcinogenic or mutagenic considerations to which we have earlier alluded to, there is also the consideration of energy requirement. In large scale operations, there may be advantages in processes that combine irradiation with some standard form of food preservation, just as in the case of waste treatment. According to Josephson (1981), the energy required for sterilization by irradiation is about 1/5 of that needed by heat sterilization and about 1/50 of that required for blast freezing of chicken of equal mass. In addition, there is the cost of storing the frozen foods. Although the cost of radiation energy per unit is higher than that of alternative sources of energy needed for heating, freezing, and the like, the overall economics is certainly in favor of irradiation. Brynjolfsson (1979) estimated the cost of irradiated sterilization of bacon to be around 1–3 cents per pound. For irradiation, use can be made of radioisotopes (gamma rays from ^{60}Co or ^{137}Cs), X-rays or electron beam (EB) machines. While the end result may be the same the penetrations are different for electrons on one hand and X-rays and γ-rays on the other. Electrons will only irradiate the surface, whereas X-rays and γ-rays will penetrate the whole sample. Therefore, by choosing the radiation type, it is possible to irradiate the food either on the surface or throughout its volume.

The low dose requirement for inhibition of sprouting in onions, potatoes, and the like is partially predicated by increased rotting and interference with the healing process at higher doses. Onions are better preserved by irradiation as early as possible during the dormant stage of the tuber, in order to prevent the discoloration of the interior. The optimum dose level is around 0.1 kGy. Conditions are somewhat similar in other root vegetables, such as carrots, and beets, but the optimum dose varies from species to species and may also vary seasonally. Ripening control—that is, delay of ripening—has been established in various fruits such as mangoes, bannanas, and papayas. With some species and some varieties, skin damage has been observed upon irradiation at 1–2 kGy, which has been attributed to enzymatic reactions induced by the radiation. Control of ripening extends the shelf life of these fruits and certain vegetables such as tomatoes and mushrooms. Irradiation of mushrooms at ~1 kGy can delay their cap opening by a week, with obvious commercial implications. Disinfection of parasites and control of insects in grains, fruits, dried fish, animal feed and so on, requires different dose levels depending on the objective. For outright killing, a dose ~5 kGy may be needed. However, the dose requirement would be an order of magnitude less if the desired result is lethality within a few weeks or sterility of the surviving organisms.

11.5.1   RADIATION CHEMISTRY
OF FOOD COMPONENTS

Unique radiolytical products (URP) in irradiated food are usually formed by the secondary reactions of water radicals, e_h , H, and OH, and to a lesser extent by the direct action of radiation, especially for foods with considerable water content. Due to concerns about the toxicological hazards to humans, the earlier procedures consisted almost entirely of animal experiments, often after the sacrifice of the animal or other tests performed when the animal has produced a certain number of litters. It was later realized that a more suitable procedure would be to analyze the effects of radiation on individual food components. A brief summary of these effects will now be presented.

11.5.1a   Carbohydrates

Irradiation lowers the viscosity of polysaccharides and in general changes other physical and chemical properties, some of which (like browning) are similar to the effects of burning. Upon radiolysis, glucose yields a variety of products including gluconic acid, glucuronic acid, saccharic acid, D-xylose, deoxycarbonyls, glucosome, and so on. In aqueous solution, the main effect is the reaction of OH radicals with C-H bonds. When irradiated in the solid state, water has a protective effect, either by the reaction of water with the radicals formed from sugars, resulting in the re-formation of the initial substance, or by energy transfer and consequent degradation via H bonds.

11.5.1b   Proteins

Denaturation of proteins by irradiation can occur in small amounts at moderate or high doses. Some reduction in the availability of certain amino acids (lysine, methionine, etc.) has been reported in the proteins of irradiated foods. However, these effects are rather small, even at high radiation doses.

Generally speaking, the peptide linkage is relatively radioresistant. The main products of radiolysis are generated from side group reactions involving amino acids in the peptide chain. A number of radiolytic products may be expected upon the radiolysis of proteins, but none of these in a large or alarming quantity. On the other hand, irradiation of poultry and meat generates considerable volatile products attributed to the indirect action of radiation. Of these, the larger fractions are carbon dioxide and ammonia. Some fatty acids, keto acids, and mercaptans are formed and also to a lesser degree benzene, hydrogen sulfide, dimethyl sulfide, and others.

11.5.1c   Fats

The main products of the irradiation of beef fat at high doses are such hydrocarbons as alkanes, alkenes, alkynes, and alkadienes. Of these, the alkanes and alkenes consitute the most significant part of the volatile products.

There has not been found any substantial difference between the effects of irradiation on the lipid fraction of complex foods and that on model systems analogous to to fats. The autooxidation products of either natural fats or of model systems on irradiation are the same as those present in oxidized fats that have not been irradiated. The extent of oxidation of fats under irradiation depends on the degree of unsaturation. Peroxides and other comlex oxidation products can be detected in the irradiation of fats in foods at a dose of few kGy.

11.5.1d   Vitamins

Some effects of irradiation on vitamins have been mentioned earlier. It appears that irradiation and heat treatment affect vitamins differently. Apparently, vitamins B_1, B_6, B_{12} and folic acid decompose less under as high a radiation dose as 60 kGy than under autoclaving at 120°C for 20 minutes. On the other hand, vitamin C is much more sensitive to irradiation. Generally, the radiosensitive vitamins are also sensitive to light, heat, and oxygen. In fresh foods, the vitamins that are most susceptible to irradiation are A and E. There is also some decomposition of vitamins B_1 and C. Other vitamins are fairly stable under irradiation. However, for the most part, the vitamins are more susecptible to heat treatment than to irradiation.

11.5.2   IDENTIFICATION OF IRRADIATED FOOD

From the viewpoint of public concern, it is desirable to have a relatively simple way to identify irradiated food products. However, this is not an easy problem, partly due to the low yields of the radiolytic products and partly due to cumbersome physical apparatus needed for the detection of, for example, radicals by electron spin resonance (ESR) techniques. Two methods of detection in common use are: (1) thermoluminiscence (Sattar et al., 1987) and (2) ESR (Dodd et al., 1985). By ESR techniques one can detect the radicals trapped in the bones of irradiated meat, fish, and poultry and also in the shells of shellfish and in the seeds of fruit. These radicals are usually long-lived and can survive normal handling and storage. However, this method is not suitable for onions and the like, because in this case the signal, due to radicals generated in the outer skins, decays with time. Many other methods of detection have been proposed. Indeed, almost any physical or chemical change brought about by irradiation has been tried or suggested, including chromatographic analysis of volatile products, conductivity measurements on potatoes, viscosity measurements of spices, and the determination of levels of vitamins and enzymatic activities.

11.5.3   CLEARANCE AND WHOLESOMENESS OF IRRADIATED FOOD

By wholesomeness is generally implied safe consumption. In this respect, irradiated foods have made great progress (vide supra). On the other hand in most countries there are additional demands on the color, texture and the like, of the food to be consumed. This is a somewhat more difficult problem to handle, and to some extent it is a subjective cultural problem as well. That foods irradiated under controlled conditions are safe to eat has been amply demonstrated by analysis and actual usage. Irradiated foods have been used as diets for germ-free animals for 25 years without any observed ill effects.

Some 40 countries have cleared irradiated foods of certain types for human consumption, or have given provisional clearance. Large scale ($\sim 10^4$ tons per year) irradiation of potatoes has been approved in Japan, and very large scale ($\sim 10^5$ tons per year) irradiation of grains has been reported from the former Soviet Union for insect control. However, it must be admitted that clearances with associated legal complications have come slowly in most countries, and even today there are ongoing debates regarding the ethics and economics of food irradiation.

11.6 OTHER USE OF LOW-LET RADIATION

There are miscellaneous uses of low-LET irradiation other than what we have discussed so far. However, for lack of space, we have not treated the vast field of radiation-induced polymerization and the effect of radiation on polymers. Some useful references have been cited in the beginning of this chapter. Here we will only describe briefly two topics—grafting and curing.

11.6.1 GRAFTING

Under suitable conditions, it is possible to graft a second polymer onto the original polymer by the reaction of free radicals generated by the irradiation of the polymer. The monomer could be included in the original sample, or the irradiated polymer can be subsequently immersed. Some permanent-press materials have been produced in this manner by grafting substituted acrylamides to cotton-polyester fabrics. Different kinds of monomeric substances can be grafted to various fibers, natural or synthetic, to attain desired properties. These grafted polymers are useful in ion-exchange processes, as adsorbents, and in the textile industry. In addition, there is also use of grafting in immobilizing enzymes and antigens by binding these onto polymeric surfaces.

11.6.2 CURING

Curing processes are essentially surface processes, often of the nature of hardening or polishing. Radiation curing has the advantage that it does not use a solvent or heat. The polymerization rate is high and can be controlled. For such surface applications, it is preferable to use an electron beam (EB) machine. Because of the low penetration of electron beams only the surface receives a considerable dose. Curing of paints, varnishes, inks, and so forth, requires a dose around a few tens of kGy to achieve hardening. In related applications, irradiation can be used to bond coatings to a number of surfaces, including metals, paper, fibers, and films.

11.7   LOW ENERGY ION IMPLANTATION

Ion implantation using low-energy ion beams is now in the advanced stage of research, and some applications are forthcoming. The lower energy of the beam is associated with relatively high LET. The ions can be implanted in shallow regions to achieve desired properties. Ion implantations have been carried out successfully in metals, semiconductors, and insulators. In microelectronics, there are various uses of the ion beam method for advanced integrated circuit fabrication, including but not limited to low-energy implantation for shallow junctions, germanium implantation in silicon for bandgap engineering in high-performance integrated circuits, implantation in silicon for well engineering, and so forth.

In an international conference held at the Japan Atomic Energy Research Institute (JAERI) in 1996, several authors have summarized their experiences on ion beam implantation (Namba, 1997; Nashiyama et al., 1997; White et al., 1997; Angert and Trautmann, 1997). These researches include applications to advanced microelectronics materials ion, beam processing of optical material, and the design of special apparatus for single event effects (SEE) for use in simulated space flight. In ion implantation, positive ions are usually used. However, a problem arises during implantation due to charge-up. The conventional solution is the use of an electron shower. Another novel method is the use of negative ion beam generated by a radiofrequency-plasma sputter-type source. With such a device, the surface voltage can be controlled within manageable limits. In some cases, cluster-ion beams have also been used.

Ion beams used for implantation can have wide variations in energy, spot size, and current depending on the intended use (Namba, 1997). Beam energy ranges from the order of eV for deposition/epitaxy to the order of MeV for well engineering. In between, the more commonly used energies are the KeV regime for etching and ~100 KeV for doping. Beam spot size can be a few nm for a focused ion beam, and in the other limit, ~0.5 m for a sheet ion beam. Beam current can vary from that due to a single ion to several amperes.

For the purpose of ultrashallow sub-tenth-micron boron doping, energies of a few KeV may be used. The distribution of boron concentration decreases very rapidly with depth, typical values being ~10^{20} to 10^{22} atoms.cm^{-3} at the surface falling almost exponentially to ~10^{13} to 10^{14} atoms.cm^{-3} at a depth of 20 nm. If very low energy (~100 eV or less) ions are used, then these can be deposited on a substrate by soft landing to give depositions of high isotopic purity. However, this technique cannot be used by itself to give binary compound films; for this, a dual beam technique has benn suggested and developed.

Ge-ion implant in Si narrows the bandgap in the source region, which enhances hole flow in that region. The procedure improves performance by lowering the drain breakdown voltage. In a low-gate bias, this voltage improvement ~1 eV has been achieved by an ion implantation method.

One of the more important applications of ion implantation is well engineering at relatively high ion energies. For the protection of dynamic random access memories from soft errors, different structures have been proposed and employed. The conventional procedure is to use epitaxial wafers; well engineering beneath the active p- and n-channels provides a less expensive alternative.

Using ion beams to dope semiconductors has the obvious advantages of dopant purity due to mass analysis of the ion beam and profile reproducibility due to control of energy and dose. While such procedures have advanced a great deal in the semiconductor industry, the use of ion beams to modify the properties of metals, alloys, and the like, is relatively new. Still newer is the ion beam processing of optical materials, but progress is being made in that direction as well (White et al., 1997). There are various physical ways by which ion beams modify the properties of optical materials. One way is the generation of stable defects in the solid, which will absorb or emit light at characteristic wavelengths. Another way derives from the fact that ion irradiation often results in decreased local density. In the material, this low-density region represents a region of lower refractive index surrounded by a region of higher refractive index. At appropriate wavelenths, light will be confined in the higher refractive index region, creating a light waveguide effect. Such an effect has been demonstrated in many optical materials. An inverse effect can also occur, if the irradiation changes the chemical composition in such a way that creates regions of higher local density. This also can be used as a light waveguide. An example is the implantation of Ti ions at a high enough concentration into the near surface of $LiNbO_3$. On annealing Ti goes into the lattice in substitutional sites, which locally increases the refractive index. Doping, by ion implantation, of optically active ions is naturally of great interest in laser technology. Doping by such rare earth ions as Er, for which optical transition occurs at 1540 nm, can be readily utilized in optical telecommunications. Another use of ion implantation at high dose with thermal processing is the generation of nanocrystals and quantum dots (2–20 nm) near the surfaces of optical materials. Because these nanocrystals and quantum dots have an enormous surface-to-volume ratio, they possess unique properties that can be utilized. Due to the quantum confinement effect, their electronic energy levels are size-dependent and are shifted drastically from bulk values. Control of size distribution to achieve a specially desirable property is crucial. Ion implantation results in a supersaturated solution at the near surface of the optical material. Thermal annealing leads to precipitation, forming nanocrystals at high density.

One of the uses of ion beam irradiation is the testing of electronic circuits for malfunctioning under single event effects (SEE) of high-energy heavy ions, as may be expected in space flights or in artificial satellites (Nashiyama *et al.*, 1997). Of the various types of such effects, the most important appears to be the single event upset (SEU), which is triggered when the charge released by the energetic ion in the electronic device exceeds the critical value for the malfunctioning of its memory circuits. The release time is also important; quicker charge release is more detrimental. Cyclotron ion irradiation has been used for testing purpose using one of the three methods: (1) direct beam irradiation; (2) scattered ion irradiation; and (3) recoil atom irradiation. These terms are self-explanatory and for heavy ions above 10 MeV/amu energy, only the direct method is applicable due to the possibility of inelastic nuclear interaction. In the scattered ion method, the heavy ions are scattered from a Au target, and then these are incident on the electronic device placed at different scattering angles for SEE testing. In the recoil ion method, the primary ion beam is more massive than the target. Target atoms recoiled in the forward direction is used for testing the electronic device. If the device is placed at angles greater than the maximum scattering angle, then complications due to scattered ion irradiation can be avoided. An obvious advantage of this method is that irradiation by different ions can be achieved simply by changing the target material. Experiments with different electronic devices have established a threshold LET and a saturated upset cross-section for each device, but between different devices these do not differ by orders of magnitude. For example, the threshold LET lies in the interval 5–17 MeV/(mg.cm^{-2}) and the saturated upset cross section in the range of 3×10^{-8} to 6×10^{-7} cm^2/bit for most devices (Nashiyama *et al.*, 1997).

Certain other uses of heavy ion irradiation that relate to track structure have been discussed by Angert and Trautmann (1997). These depend on selective etching of heavy-ion tracks in solids or films. The damaged material along the ion track dissolves at a faster rate than the bulk with a suitable chemical etchant. Etched tracks in polymers can be used as high-precision membranes. Further, the technique of ion track membrane can be combined with hydrogels. A responsive gel is grafted onto the surface of a single-pore membrane. The flow through the pore can be controlled by varying the pH or temperature. Single-ion irradiation and etching techniques have been used to form selective surface gels in extremely small pores considered as model systems for biological membranes. Etched pores in polymers can be used as microstructures. The pores are filled with a metallic material in a galvanic process. The plastic material may then be removed by an organic solvent. Microstructures with desired properties may thus be produced. For example, oriented needles of 100 μm length with a diameter of a few micrometers can be made. Under suitable conditions, it is possible to obtain different shapes, such as parallel or funnel columns, tips, and tubes, using different metals. The process can be controlled during the growth process, and in combination with an ion microprobe, either ordered or statistically distributed structures can be made.

REFERENCES

Angert, N., and Truatmann, C. (1997), in Proc. 7th International Symposium on Advanced Nuclear Energy Research, p. 47, Takasaki, Japan.

Armstrong, D.A., and Spinks, J.W.T.(1959), Can. J. Chem. 37, 1002, 1210.

Artandi, C. (1977), Radiat. Phys. Chem. 9, 183.

Bansal, K.M., and Freeman, G.R. (1971), Radiat. Res. Revs. 3, 209.

Black, J.F., and Baxter, E.F., Jr. (1958), in Second International Conference on Peaceful Uses of Atomic Energy, V. XXIX, p. 797, United Nations.

Brenner, W., Rugg, B., and Rogers, C. (1977), Radiat. Phys. Chem. 9, 389.

Brynjolfsson, A. (1979), Tech. Report FEL–89, U.S. Army Natick Research and Development Command, Natick, Massachusetts.

Charlesby, A. (1960) Atomic Radiation and Polymers, Pergamon, Oxford, England.

Chutny, B., and Kucera, J. (1974), Radiat. Res. Revs. 5, 1.

Danno, A. (1960), in Actions Chimique et Biologique des Radiations, v. 13, (Haissinsky, M., ed.), ch. 4, Masson, Paris.

Dodd, N.J.F., Swallow, A.J., and Ley, F.J. (1985), Radiat. Phys. Chem. 26, 451.

Dole, M. (1973) The Radiation Chemistry of Macromolecules, v. 1 and 2, Academic Press, New York.

Dragnic, I., and Dragnic, Z. D. (1971), The Radiation Chemistry of Water, Academic, New York.

Handlos, V. (1981), Radiat. Phys. Chem. 18, 175.

Helfritch, D.J., and Feldman, P.L. (1984), Radiat. Phys. Chem. 24, 129.

Hine, G.J., and Brownell, G.L. (1956), Radiation Dosimetry, Academic Press, New York.

Holm, N.W., and Berry, R.J., eds. (1970), Manual of Radiation Dosimetry, Marcel Dekker, New York.

Hummel, D.O., Mentzel, W., and Schneider, C. (1964), Ann. Chem. 673, 15.

International Atomic Energy Agency (1968), Radiation Chemistry and Its Applications, Technical Ser. 84, Vienna.

International Atomic Energy Agency (1968), Manual of Food Irradiation Dosimetry, Technical Report 178, Vienna.

Iya, V.K., and Krishnamurthy, K. (1984), Radiat. Phys. Chem. 24, 67.

Josephson, E.S. (1981), Radiat. Phys. Chem. 18, 223.

Kase, K.R., Bjarngard, B.E., and Attix, F.H. eds. (1985), "The Dosimetry of Ionizing Radiation" v.1, Academic Press, New York.

Kase, K.R., Bjarngard, B.E. and Attix, F.H., eds. (1987), The Dosimetry of Ionizing Radiation, v.2, Academic Press, New York.

Kawakami, W., Hasimoto, S., Watanabe, H., Nishimura, K., Watanabe, H., Ito, H., and Takehisa, M. (1981), Radiat. Phys. Chem. 18, 771.

Kawakami, W., and Hasimoto, S. (1984), Radiat. Phys. Chem. 24, 29.

McLaughlin, W.L. ed. (1982) Trends of Radiation Dosimetry, in Int. J. Appl. Radiat. and Isotopes 33, 953.

Namba, S. (1997), in Proc. 7th International Symposium on Advanced Nuclear Energy Research, p. 13, Takasaki, Japan.

Nashiyama, I., Hirao, T., Itoh, H., and Ohshima, T. (1997), in Proc. 7th International Symposium on Advanced Nuclear Energy Research, p. 22, Takasaki, Japan.

Pikaev, A.K., and Shubin, V.N. (1984), Radiat. Phys. Chem. 24, 77.

Chapter 11 Radiation Chemical Applications in Science and Industry

Sakumoto, A., and Miyata, T. (1984), Radiat. Phys. Chem. *24*, 99.

Sattar, A., Delinceé, H., and Diehl, J.F. (1987), Radiat. Phys. Chem. *29*, 215.

Shapiro, J. (1972), *Radiation Protection,* Harvard University Press, Cambridge, Mass.

Silverman, J. (1981), J. Chem. Educ. *58*, 168

Spinks, J.W.T., and Woods, R.J. (1990), *An Introduction to Radiation Chemistry,* 3rd ed.,
    Wiley–Interscience, New York.

Tabata, Y., Ito, Y., and Tagawa, S., eds. (1991), *CRC Handbook of Radiation Chemistry,* ch.
    XVI.C, CRC Press, Boca Raton, Florida

Tokunaga, O., and Suzuki, N. (1984), Radiat. Phys. Chem. *24*, 145.

Vereshchinskii, I.V. (1972), in *Advances in Radiation Chemistry* v.3 (Burton, M., and Magee, J.L.,
    eds.), pp.75–123, Wiley–Interscience, New York.

Wagner, C.D. (1969), in *Advances in Radiation Chemistry* v.1 (Burton, M., and Magee, J.L eds.),
    pp.199–244, Wiley–Interscience, New York.

White, C.W., Budai, J.D., Zhu, J.G., and Withrow, S.P. (1997), in *Proc. 7th International
    Symposium on Advanced Nuclear Energy Research,* p.33 Takasaki, Japan.

Wilson, J.G. (1972), Radiat. Res. Revs. *4*, 71.

INDEX